U0171225

R 语言机器学习 第 3 版（影印版）
Machine Learning with R, Third Edition

Brett Lantz 著

南京　东南大学出版社

图书在版编目（CIP）数据

R 语言机器学习：第 3 版= Machine Learning with
R, 3rd Edition：英文/（美）布雷特·兰茨（Brett Lantz）
著. —影印本. —南京：东南大学出版社，2020.8
　　ISBN　978－7－5641－8954－9

　　Ⅰ.①R…　Ⅱ.①布…　Ⅲ.①程序语言-程序设
计-英文　Ⅳ.①TP312

　　中国版本图书馆 CIP 数据核字（2020）第 109694 号
　　图字：10－2020－162 号

© 2019 by PACKT Publishing Ltd.

Reprint of the English Edition, jointly published by O'Reilly Media, Inc. and Southeast University
Press, 2020. Authorized reprint of the original English edition, 2020 PACKT Publishing Ltd, the owner
of all rights to publish and sell the same.

All rights reserved including the rights of reproduction in whole or in part in any form.

英文原版由 PACKT Publishing Ltd 出版 2019。

英文影印版由东南大学出版社出版 2020。此影印版的出版和销售得到出版权和销售权的所
有者 —— PACKT Publishing Ltd 的许可。

版权所有，未得书面许可，本书的任何部分和全部不得以任何形式重制。

R 语言机器学习 第 3 版（影印版）

出版发行：东南大学出版社
地　　址：南京四牌楼 2 号　　邮编：210096
出 版 人：江建中
网　　址：http://www.seupress.com
电子邮件：press@seupress.com
印　　刷：常州市武进第三印刷有限公司
开　　本：787 毫米×980 毫米　　16 开本
印　　张：28.75
字　　数：563 千字
版　　次：2020 年 8 月第 1 版
印　　次：2020 年 8 月第 1 次印刷
书　　号：ISBN 978－7－5641－8954－9
定　　价：118.00 元

本社图书若有印装质量问题，请直接与营销部联系。电话（传真）：025 - 83791830

`mapt.io`

Mapt is an online digital library that gives you full access to over 5,000 books and videos, as well as industry leading tools to help you plan your personal development and advance your career. For more information, please visit our website.

Why subscribe?

- Spend less time learning and more time coding with practical eBooks and Videos from over 4,000 industry professionals

- Learn better with Skill Plans built especially for you

- Get a free eBook or video every month

- Mapt is fully searchable

- Copy and paste, print, and bookmark content

Packt.com

Did you know that Packt offers eBook versions of every book published, with PDF and ePub files available? You can upgrade to the eBook version at `www.Packt.com` and as a print book customer, you are entitled to a discount on the eBook copy. Get in touch with us at `customercare@packtpub.com` for more details.

At `www.Packt.com`, you can also read a collection of free technical articles, sign up for a range of free newsletters, and receive exclusive discounts and offers on Packt books and eBooks.

Contributors

About the authors

Brett Lantz (@DataSpelunking) has spent more than 10 years using innovative data methods to understand human behavior. A sociologist by training, Brett was first captivated by machine learning during research on a large database of teenagers' social network profiles. Brett is a DataCamp instructor and a frequent speaker at machine learning conferences and workshops around the world. He is known to geek out about data science applications for sports, autonomous vehicles, foreign language learning, and fashion, among many other subjects, and hopes to one day blog about these subjects at dataspelunking.com, a website dedicated to sharing knowledge about the search for insight in data.

This book could not have been written without the support of my family. In particular, my wife Jessica deserves many thanks for her endless patience and encouragement. My sons Will and Cal were born in the midst of the first and second editions, respectively, and supplied much-needed diversions while writing this edition. I dedicate this book to them in the hope that one day they are inspired to tackle big challenges and follow their curiosity wherever it may lead.

I am also indebted to many others who supported this book indirectly. My interactions with educators, peers, and collaborators at the University of Michigan, the University of Notre Dame, and the University of Central Florida seeded many of the ideas I attempted to express in the text; any lack of clarity in their expression is purely mine. Additionally, without the work of the broader community of researchers who shared their expertise in publications, lectures, and source code, this book might not exist at all. Finally, I appreciate the efforts of the R and RStudio teams and all those who have contributed to R packages, whose work have helped bring machine learning to the masses. I sincerely hope that my work is likewise a valuable piece in this mosaic.

About the reviewer

Raghav Bali is a Senior Data Scientist at one of the world's largest healthcare organization. His work involves research and development of enterprise level solutions based on machine learning, deep learning and natural language processing for healthcare and insurance related use cases. In his previous role at Intel, he was involved in enabling proactive data driven IT initiatives using natural language processing, deep learning and traditional statistical methods. He has also worked in finance domain with American Express, solving digital engagement and customer retention use cases.

Raghav has also authored multiple books with leading publishers, the recent one on latest advancements in transfer learning research.

Raghav has a master's degree (gold medalist) in Information Technology from International Institute of Information Technology, Bangalore. Raghav loves reading and is a shutterbug capturing moments when he isn't busy solving problems.

Table of Contents

Preface

Machine learning, at its core, is concerned with algorithms that transform information into actionable intelligence. This fact makes machine learning well-suited to the present-day era of big data. Without machine learning, it would be nearly impossible to keep up with the massive stream of information.

Given the growing prominence of R — a cross-platform, zero-cost statistical programming environment — there has never been a better time to start using machine learning. R offers a powerful but easy-to-learn set of tools that can assist you with finding the insights in your own data.

By combining hands-on case studies with the essential theory that you need to understand how things work under the hood, this book provides all the knowledge needed to start getting to work with machine learning.

Who this book is for

This book is intended for anybody hoping to use data for action. Perhaps you already know a bit about machine learning, but have never used R; or, perhaps you know a little about R, but are new to machine learning. In any case, this book will get you up and running quickly. It would be helpful to have a bit of familiarity with basic math and programming concepts, but no prior experience is required. All you need is curiosity.

What this book covers

Chapter 1, Introducing Machine Learning, presents the terminology and concepts that define and distinguish machine learners, as well as a method for matching a learning task with the appropriate algorithm.

Chapter 2, Managing and Understanding Data, provides an opportunity to get your hands dirty working with data in R. Essential data structures and procedures used for loading, exploring, and understanding data are discussed.

Chapter 3, Lazy Learning – Classification Using Nearest Neighbors, teaches you how to understand and apply a simple yet powerful machine learning algorithm to your first real-world task: identifying malignant samples of cancer.

Chapter 4, Probabilistic Learning – Classification Using Naive Bayes, reveals the essential concepts of probability that are used in cutting-edge spam filtering systems. You'll learn the basics of text mining in the process of building your own spam filter.

Chapter 5, Divide and Conquer – Classification Using Decision Trees and Rules, explores a couple of learning algorithms whose predictions are not only accurate, but also easily explained. We'll apply these methods to tasks where transparency is important.

Chapter 6, Forecasting Numeric Data – Regression Methods, introduces machine learning algorithms used for making numeric predictions. As these techniques are heavily embedded in the field of statistics, you will also learn the essential metrics needed to make sense of numeric relationships.

Chapter 7, Black Box Methods – Neural Networks and Support Vector Machines, covers two complex but powerful machine learning algorithms. Though the math may appear intimidating, we will work through examples that illustrate their inner workings in simple terms.

Chapter 8, Finding Patterns – Market Basket Analysis Using Association Rules, exposes the algorithm used in the recommendation systems employed by many retailers. If you've ever wondered how retailers seem to know your purchasing habits better than you know yourself, this chapter will reveal their secrets.

Chapter 9, Finding Groups of Data – Clustering with k-means, is devoted to a procedure that locates clusters of related items. We'll utilize this algorithm to identify profiles within an online community.

Chapter 10, Evaluating Model Performance, provides information on measuring the success of a machine learning project and obtaining a reliable estimate of the learner's performance on future data.

Chapter 11, Improving Model Performance, reveals the methods employed by the teams at the top of machine learning competition leaderboards. If you have a competitive streak, or simply want to get the most out of your data, you'll need to add these techniques to your repertoire.

Chapter 12, Specialized Machine Learning Topics, explores the frontiers of machine learning. From working with big data to making R work faster, the topics covered will help you push the boundaries of what is possible with R.

What you need for this book

The examples in this book were written for and tested with R version 3.5.2 on Microsoft Windows and Mac OS X, though they are likely to work with any recent version of R.

Download the example code files

You can download the example code files for this book from your account at http://www.packtpub.com. If you purchased this book elsewhere, you can visit http://www.packtpub.com/support and register to have the files emailed directly to you.

You can download the code files by following these steps:

1. Log in or register at http://www.packtpub.com.
2. Select the **SUPPORT** tab.
3. Click on **Code Downloads & Errata**.
4. Enter the name of the book in the **Search** box and follow the on-screen instructions.

Once the file is downloaded, please make sure that you unzip or extract the folder using the latest version of:

- WinRAR/7-Zip for Windows
- Zipeg/iZip/UnRarX for Mac
- 7-Zip/PeaZip for Linux

The code bundle for the book is also hosted on GitHub at https://github.com/PacktPublishing/Machine-Learning-with-R-Third-Edition, and at https://github.com/dataspelunking/MLwR/. We also have other code bundles from our rich catalog of books and videos available at https://github.com/PacktPublishing/. Check them out!

Download the color images

We also provide a PDF file that has color images of the screenshots/diagrams used in this book. You can download it here: https://www.packtpub.com/sites/default/files/downloads/9781788295864_ColorImages.pdf.

Conventions used

In this book, you will find a number of text styles that distinguish between different kinds of information. Here are some examples of these styles and an explanation of their meaning.

Code in text, function names, filenames, file extensions, user input, and R package names are shown as follows: "The knn() function in the class package provides a standard, classic implementation of the k-NN algorithm."

R user input and output is written as follows:

```
> table(mushrooms$type)

   edible poisonous
     4208      3916
```

New terms and **important words** are shown in bold. Words that you see on the screen, for example, in menus or dialog boxes, appear in the text like this: "The **Task Views** link on the left side of the CRAN page provides a curated list of packages."

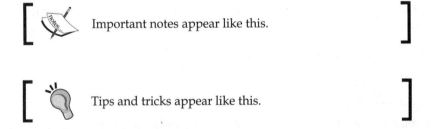

Important notes appear like this.

Tips and tricks appear like this.

Get in touch

Feedback from our readers is always welcome.

General feedback: If you have questions about any aspect of this book, mention the book title in the subject of your message and email us at customercare@packtpub.com.

Errata: Although we have taken every care to ensure the accuracy of our content, mistakes do happen. If you have found a mistake in this book we would be grateful if you would report this to us. Please visit, http://www.packt.com/submit-errata, selecting your book, clicking on the Errata Submission Form link, and entering the details.

Piracy: If you come across any illegal copies of our works in any form on the Internet, we would be grateful if you would provide us with the location address or website name. Please contact us at copyright@packt.com with a link to the material.

If you are interested in becoming an author: If there is a topic that you have expertise in and you are interested in either writing or contributing to a book, please visit http://authors.packtpub.com.

Reviews

Please leave a review. Once you have read and used this book, why not leave a review on the site that you purchased it from? Potential readers can then see and use your unbiased opinion to make purchase decisions, we at Packt can understand what you think about our products, and our authors can see your feedback on their book. Thank you!

For more information about Packt, please visit packt.com.

1

Introducing Machine Learning

If science fiction stories are to be believed, the invention of artificial intelligence inevitably leads to apocalyptic wars between machines and their makers. The stories begin with today's reality: computers being taught to play simple games like tic-tac-toe and to automate routine tasks. As the stories go, machines are later given control of traffic lights and communications, followed by military drones and missiles. The machines' evolution takes an ominous turn once the computers become sentient and learn how to teach themselves. Having no more need for human programmers, humankind is then "deleted."

Thankfully, at the time of this writing, machines still require user input.

Though your impressions of machine learning may be colored by these mass-media depictions, today's algorithms are too application-specific to pose any danger of becoming self-aware. The goal of today's machine learning is not to create an artificial brain, but rather to assist us with making sense of the world's massive data stores.

Putting popular misconceptions aside, by the end of this chapter, you will gain a more nuanced understanding of machine learning. You will also be introduced to the fundamental concepts that define and differentiate the most commonly used machine learning approaches. You will learn:

- The origins, applications, and pitfalls of machine learning
- How computers transform data into knowledge and action
- Steps to match a machine learning algorithm to your data

The field of machine learning provides a set of algorithms that transform data into actionable knowledge. Keep reading to see how easy it is to use R to start applying machine learning to real-world problems.

The origins of machine learning

Beginning at birth, we are inundated with data. Our body's sensors — the eyes, ears, nose, tongue, and nerves — are continually assailed with raw data that our brain translates into sights, sounds, smells, tastes, and textures. Using language, we are able to share these experiences with others.

From the advent of written language, human observations have been recorded. Hunters monitored the movement of animal herds; early astronomers recorded the alignment of planets and stars; and cities recorded tax payments, births, and deaths. Today, such observations, and many more, are increasingly automated and recorded systematically in ever-growing computerized databases.

The invention of electronic sensors has additionally contributed to an explosion in the volume and richness of recorded data. Specialized sensors, such as cameras, microphones, chemical noses, electronic tongues, and pressure sensors mimic the human ability to see, hear, smell, taste, and feel. These sensors process the data far differently than a human being would. Unlike a human's limited and subjective attention, an electronic sensor never takes a break and has no emotions to skew its perception.

Although sensors are not clouded by subjectivity, they do not necessarily report a single, definitive depiction of reality. Some have an inherent measurement error due to hardware limitations. Others are limited by their scope. A black-and-white photograph provides a different depiction of its subject than one shot in color. Similarly, a microscope provides a far different depiction of reality than a telescope.

Between databases and sensors, many aspects of our lives are recorded. Governments, businesses, and individuals are recording and reporting information, from the monumental to the mundane. Weather sensors record temperature and pressure data; surveillance cameras watch sidewalks and subway tunnels; and all manner of electronic behaviors are monitored: transactions, communications, social media relationships, and many others.

This deluge of data has led some to state that we have entered an era of **big data**, but this may be a bit of a misnomer. Human beings have always been surrounded by large amounts of data. What makes the current era unique is that we have vast amounts of *recorded* data, much of which can be directly accessed by computers. Larger and more interesting datasets are increasingly accessible at the tips of our fingers, only a web search away. This wealth of information has the potential to inform action, given a systematic way of making sense of it all.

The field of study interested in the development of computer algorithms for transforming data into intelligent action is known as **machine learning**. This field originated in an environment where the available data, statistical methods, and computing power rapidly and simultaneously evolved. Growth in the volume of data necessitated additional computing power, which in turn spurred the development of statistical methods for analyzing large datasets. This created a cycle of advancement allowing even larger and more interesting data to be collected, and enabling today's environment in which endless streams of data are available on virtually any topic.

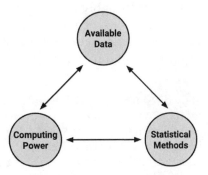

Figure 1.1: The cycle of advancement that enabled machine learning

A closely related sibling of machine learning, **data mining**, is concerned with the generation of novel insight from large databases. As the term implies, data mining involves a systematic hunt for nuggets of actionable intelligence. Although there is some disagreement over how widely machine learning and data mining overlap, a potential point of distinction is that machine learning focuses on teaching computers how to use data to solve a problem, while data mining focuses on teaching computers to identify patterns that humans then use to solve a problem.

Virtually all data mining involves the use of machine learning, but not all machine learning requires data mining. For example, you might apply machine learning to data mine automobile traffic data for patterns related to accident rates. On the other hand, if the computer is learning how to drive the car itself, this is purely machine learning without data mining.

 The phrase "data mining" is also sometimes used as a pejorative to describe the deceptive practice of cherry-picking data to support a theory.

Uses and abuses of machine learning

Most people have heard of Deep Blue, the chess-playing computer that in 1997 was the first to win a game against a world champion. Another famous computer, Watson, defeated two human opponents on the television trivia game show Jeopardy in 2011. Based on these stunning accomplishments, some have speculated that computer intelligence will replace workers in information technology occupations, just as machines replaced workers in fields and assembly lines.

The truth is that even as machines reach such impressive milestones, they are still relatively limited in their ability to thoroughly understand a problem. They are pure intellectual horsepower without direction. A computer may be more capable than a human of finding subtle patterns in large databases, but it still needs a human to motivate the analysis and turn the result into meaningful action.

Without completely discounting the achievements of Deep Blue and Watson, it is important to note that neither is even as intelligent as a typical five-year-old. For more on why "comparing smarts is a slippery business," see the *Popular Science* article *FYI: Which Computer Is Smarter, Watson Or Deep Blue?*, by *Will Grunewald, 2012*: https://www.popsci. com/science/article/2012-12/fyi-which-computer- smarter-watson-or-deep-blue.

Machines are not good at asking questions, or even knowing what questions to ask. They are much better at answering them, provided the question is stated in a way that the computer can comprehend. Present-day machine learning algorithms partner with people much like a bloodhound partners with its trainer: the dog's sense of smell may be many times stronger than its master's, but without being carefully directed, the hound may end up chasing its tail.

Figure 1.2: Machine learning algorithms are powerful tools that require careful direction

To better understand the real-world applications of machine learning, we'll now consider some cases where it has been used successfully, some places where it still has room for improvement, and some situations where it may do more harm than good.

Machine learning successes

Machine learning is most successful when it augments, rather than replaces, the specialized knowledge of a subject-matter expert. It works with medical doctors at the forefront of the fight to eradicate cancer; assists engineers and programmers with efforts to create smarter homes and automobiles; and helps social scientists to build knowledge of how societies function. Toward these ends, it is employed in countless businesses, scientific laboratories, hospitals, and governmental organizations. Any effort that generates or aggregates data likely employs at least one machine learning algorithm to help make sense of it.

Though it is impossible to list every use case for machine learning, a look at recent success stories identifies several prominent examples:

- Identification of unwanted spam messages in email
- Segmentation of customer behavior for targeted advertising
- Forecasts of weather behavior and long-term climate changes
- Reduction of fraudulent credit card transactions
- Actuarial estimates of financial damage of storms and natural disasters
- Prediction of popular election outcomes
- Development of algorithms for auto-piloting drones and self-driving cars
- Optimization of energy use in homes and office buildings
- Projection of areas where criminal activity is most likely
- Discovery of genetic sequences linked to diseases

By the end of this book, you will understand the basic machine learning algorithms that are employed to teach computers to perform these tasks. For now, it suffices to say that no matter what the context is, the machine learning process is the same. Regardless of the task, an algorithm takes data and identifies patterns that form the basis for further action.

The limits of machine learning

Although machine learning is used widely and has tremendous potential, it is important to understand its limits. Machine learning, at this time, emulates a relatively limited subset of the capabilities of the human brain. It offers little flexibility to extrapolate outside of strict parameters and knows no common sense. With this in mind, one should be extremely careful to recognize exactly what an algorithm has learned before setting it loose in the real world.

Without a lifetime of past experiences to build upon, computers are also limited in their ability to make simple inferences about logical next steps. Take, for instance, the banner advertisements seen on many websites. These are served according to patterns learned by data mining the browsing history of millions of users. Based on this data, someone who views websites selling shoes is interested in buying shoes and should therefore see advertisements for shoes. The problem is that this becomes a never-ending cycle in which, even after shoes have been purchased, additional shoe advertisements are served, rather than advertisements for shoelaces and shoe polish.

Many people are familiar with the deficiencies of machine learning's ability to understand or translate language, or to recognize speech and handwriting. Perhaps the earliest example of this type of failure is in a 1994 episode of the television show *The Simpsons*, which showed a parody of the Apple Newton tablet. For its time, the Newton was known for its state-of-the-art handwriting recognition. Unfortunately for Apple, it would occasionally fail to great effect. The television episode illustrated this through a sequence in which a bully's note to "Beat up Martin" was misinterpreted by the Newton as "Eat up Martha."

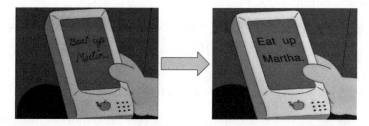

Figure 1.3: Screen captures from *Lisa on Ice, The Simpsons, 20th Century Fox (1994)*

Machine language processing has improved enough in the time since the Apple Newton that Google, Apple, and Microsoft are all confident in their ability to offer voice-activated virtual concierge services such as Google Assistant, Siri, and Cortana. Still, these services routinely struggle to answer relatively simple questions. Furthermore, online translation services sometimes misinterpret sentences that a toddler would readily understand, and the predictive text feature on many devices has led to a number of humorous "autocorrect fail" sites that illustrate computers' ability to understand basic language but completely misunderstand context.

Some of these mistakes are surely to be expected. Language is complicated, with multiple layers of text and subtext, and even human beings sometimes misunderstand context. In spite of the fact that machine learning is rapidly improving at language processing, the consistent shortcomings illustrate the important fact that machine learning is only as good as the data it has learned from. If context is not explicit in the input data, then just like a human, the computer will have to make its best guess from its limited set of past experiences.

Machine learning ethics

At its core, machine learning is simply a tool that assists us with making sense of the world's complex data. Like any tool, it can be used for good or for evil. Where machine learning goes most wrong is when it is applied so broadly, or so callously, that humans are treated as lab rats, automata, or mindless consumers. A process that may seem harmless can lead to unintended consequences when automated by an emotionless computer. For this reason, those using machine learning or data mining would be remiss not to at least briefly consider the ethical implications of the art.

Due to the relative youth of machine learning as a discipline and the speed at which it is progressing, the associated legal issues and social norms are often quite uncertain, and constantly in flux. Caution should be exercised when obtaining or analyzing data in order to avoid breaking laws; violating terms of service or data use agreements; or abusing the trust or violating the privacy of customers or the public.

 The informal corporate motto of Google, an organization that collects perhaps more data on individuals than any other, was at one time, "don't be evil." While this seems clear enough, it may not be sufficient. A better approach may be to follow the *Hippocratic Oath*, a medical principle that states, "above all, do no harm."

Retailers routinely use machine learning for advertising, targeted promotions, inventory management, or the layout of the items in a store. Many have equipped checkout lanes with devices that print coupons for promotions based on a customer's buying history. In exchange for a bit of personal data, the customer receives discounts on the specific products he or she wants to buy. At first, this appears relatively harmless, but consider what happens when this practice is taken a bit further.

One possibly apocryphal tale concerns a large retailer in the United States that employed machine learning to identify expectant mothers for coupon mailings. The retailer hoped that if these mothers-to-be received substantial discounts, they would become loyal customers who would later purchase profitable items such as diapers, baby formula, and toys.

Equipped with machine learning methods, the retailer identified items in the customer purchase history that could be used to predict with a high degree of certainty not only whether a woman was pregnant, but also the approximate timing for when the baby was due.

After the retailer used this data for a promotional mailing, an angry man contacted the chain and demanded to know why his daughter received coupons for maternity items. He was furious that the retailer seemed to be encouraging teenage pregnancy! As the story goes, when the retail chain called to offer an apology, it was the father who ultimately apologized after confronting his daughter and discovering that she was indeed pregnant!

Whether completely true or not, the lesson learned from the preceding tale is that common sense should be applied before blindly applying the results of a machine learning analysis. This is particularly true in cases where sensitive information, such as health data, is concerned. With a bit more care, the retailer could have foreseen this scenario and used greater discretion when choosing how to reveal the pattern its machine learning analysis had discovered.

For more detail on how retailers use machine learning to identify pregnancies, see the *New York Times Magazine* article, titled *How Companies Learn Your Secrets*, by *Charles Duhigg, 2012*: `https://www.nytimes.com/2012/02/19/magazine/shopping-habits.html`.

As machine learning algorithms are more widely applied, we find that computers may learn some unfortunate behaviors of human societies. Sadly, this includes perpetuating race or gender discrimination and reinforcing negative stereotypes. For example, researchers have found that Google's online advertising service is more likely to show ads for high-paying jobs to men than women, and is more likely to display ads for criminal background checks to black people than white people.

Proving that these types of missteps are not limited to Silicon Valley, a Twitter chatbot service developed by Microsoft was quickly taken offline after it began spreading Nazi and anti-feminist propaganda. Often, algorithms that at first seem "content neutral" quickly start to reflect majority beliefs or dominant ideologies. An algorithm created by Beauty.AI to reflect an objective conception of human beauty sparked controversy when it favored almost exclusively white people. Imagine the consequences if this had been applied to facial recognition software for criminal activity!

For more information about the real-world consequences of machine learning and discrimination see the New York Times article *When Algorithms Discriminate*, by *Claire Cain Miller, 2015*: https://www.nytimes.com/2015/07/10/upshot/when-algorithms-discriminate.html.

To limit the ability of algorithms to discriminate illegally, certain jurisdictions have well-intentioned laws that prevent the use of racial, ethnic, religious, or other protected class data for business reasons. However, excluding this data from a project may not be enough because machine learning algorithms can still inadvertently learn to discriminate. If a certain segment of people tends to live in a certain region, buys a certain product, or otherwise behaves in a way that uniquely identifies them as a group, machine learning algorithms can infer the protected information from other factors. In such cases, you may need to *completely* de-identify these people by excluding any *potentially* identifying data in addition to the already-protected statuses.

Apart from the legal consequences, inappropriate use of data may hurt the bottom line. Customers may feel uncomfortable or become spooked if aspects of their lives they consider private are made public. In recent years, a number of high-profile web applications have experienced a mass exodus of users who felt exploited when the applications' terms of service agreements changed or their data was used for purposes beyond what the users had originally intended. The fact that privacy expectations differ by context, by age cohort, and by locale adds complexity to deciding the appropriate use of personal data. It would be wise to consider the cultural implications of your work before you begin on your project, in addition to being aware of ever-more-restrictive regulations such as the European Union's newly-implemented **General Data Protection Regulation (GDPR)** and the inevitable policies that will follow in its footsteps.

The fact that you *can* use data for a particular end does not always mean that you *should*.

Finally, it is important to note that as machine learning algorithms become progressively more important to our everyday lives, there are greater incentives for nefarious actors to work to exploit them. Sometimes, attackers simply want to disrupt algorithms for laughs or notoriety—such as "Google bombing," the crowd-sourced method of tricking Google's algorithms to highly rank a desired page.

Other times, the effects are more dramatic. A timely example of this is the recent wave of so-called fake news and election meddling, propagated via the manipulation of advertising and recommendation algorithms that target people according to their personality. To avoid giving such control to outsiders, when building machine learning systems, it is crucial to consider how they may be influenced by a determined individual or crowd.

Social media scholar danah boyd (styled lowercase) presented a keynote at the *Strata Data Conference 2017* in New York City that discussed the importance of hardening machine learning algorithms to attackers. For a recap, refer to: `https://points.datasociety.net/your-data-is-being-manipulated-a7e31a83577b`.

The consequences of malicious attacks on machine learning algorithms can also be deadly. Researchers have shown that by creating an "adversarial attack" that subtly distorts a street sign with carefully chosen graffiti, an attacker might cause an autonomous vehicle to misinterpret a stop sign, potentially resulting in a fatal crash. Even in the absence of ill intent, software bugs and human errors have already led to fatal accidents in autonomous vehicle technology from Uber and Tesla. With such examples in mind, it is of the utmost importance and ethical concern that machine learning practitioners should worry about how their algorithms will be used and abused in the real world.

How machines learn

A formal definition of machine learning attributed to computer scientist Tom M. Mitchell states that a machine learns whenever it is able to utilize its experience such that its performance improves on similar experiences in the future. Although this definition is intuitive, it completely ignores the process of exactly how experience can be translated into future action—and, of course, learning is always easier said than done!

Where human brains are naturally capable of learning from birth, the conditions necessary for computers to learn must be made explicit. For this reason, although it is not strictly necessary to understand the theoretical basis of learning, this foundation helps us to understand, distinguish, and implement machine learning algorithms.

As you compare machine learning to human learning, you may find yourself examining your own mind in a different light.

Regardless of whether the learner is a human or a machine, the basic learning process is similar. It can be divided into four interrelated components:

- **Data storage** utilizes observation, memory, and recall to provide a factual basis for further reasoning.

- **Abstraction** involves the translation of stored data into broader representations and concepts.

- **Generalization** uses abstracted data to create knowledge and inferences that drive action in new contexts.

- **Evaluation** provides a feedback mechanism to measure the utility of learned knowledge and inform potential improvements.

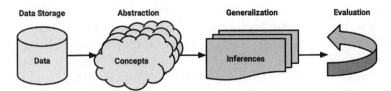

Figure 1.4: The learning process

Although the learning process has been conceptualized here as four distinct components, they are merely organized this way for illustrative purposes. In reality, the entire learning process is inextricably linked. In human beings, the process occurs subconsciously. We recollect, deduce, induct, and intuit within the confines of our mind's eye, and because this process is hidden, any differences from person to person are attributed to a vague notion of subjectivity. In contrast, computers make these processes explicit, and because the entire process is transparent, the learned knowledge can be examined, transferred, utilized for future action, and treated as a data "science."

The **data science** buzzword suggests a relationship among the data, the machine, and the people who guide the learning process. The term's growing use in job descriptions and academic degree programs reflects its operationalization as a field of study concerned with both statistical and computational theory, as well as the technological infrastructure enabling machine learning and its applications. The field often asks its practitioners to be compelling storytellers, balancing an audacity in the use of data with the limitations of what one may infer and forecast from the data. To be a strong data scientist, therefore, requires a strong understanding of how the learning algorithms work.

Data storage

All learning begins with data. Humans and computers alike utilize **data storage** as a foundation for more advanced reasoning. In a human being, this consists of a brain that uses electrochemical signals in a network of biological cells to store and process observations for short- and long-term future recall. Computers have similar capabilities of short- and long-term recall using hard disk drives, flash memory, and random-access memory (RAM) in combination with a central processing unit (CPU).

It may seem obvious, but the ability to store and retrieve data alone is insufficient for learning. Stored data is merely ones and zeros on a disk. It is a collection of memories, meaningless without a broader context. Without a higher level of understanding, knowledge is purely recall, limited to what has been seen before and nothing else.

To better understand the nuances of this idea, it may help to think about the last time you studied for a difficult test, perhaps for a university final exam or a career certification. Did you wish for an eidetic (photographic) memory? If so, you may be disappointed to know that perfect recall would unlikely be of much assistance. Even if you could memorize material perfectly, this rote learning would provide no benefit without knowing the exact questions and answers that would appear on the exam. Otherwise, you would need to memorize answers to every question that could *conceivably* be asked, on a subject in which there is likely to be an infinite number of questions. Obviously, this is an unsustainable strategy.

Instead, a better approach is to spend time selectively, and memorize a relatively small set of representative ideas, while developing an understanding of how the ideas relate and apply to unforeseen circumstances. In this way, important broader patterns are identified, rather than memorizing each and every detail, nuance, and potential application.

Abstraction

This work of assigning a broader meaning to stored data occurs during the **abstraction** process, in which raw data comes to represent a wider, more abstract concept or idea. This type of connection, say between an object and its representation, is exemplified by the famous René Magritte painting *The Treachery of Images*:

Figure 1.5: "This is not a pipe." Source: http://collections.lacma.org/node/239578

The painting depicts a tobacco pipe with the caption *Ceci n'est pas une pipe* ("This is not a pipe"). The point Magritte was illustrating is that a representation of a pipe is not truly a pipe. Yet, in spite of the fact that the pipe is not real, anybody viewing the painting easily recognizes it as a pipe. This suggests that observers are able to connect the *picture* of a pipe to the *idea* of a pipe, to a memory of a *physical* pipe that can be held in the hand. Abstracted connections like this are the basis of **knowledge representation**, the formation of logical structures that assist with turning raw sensory information into meaningful insight.

During a machine's process of knowledge representation, the computer summarizes stored raw data using a **model**, an explicit description of the patterns within the data. Just like Magritte's pipe, the model representation takes on a life beyond the raw data. It represents an idea greater than the sum of its parts.

There are many different types of models. You may already be familiar with some. Examples include:

- Mathematical equations
- Relational diagrams, such as trees and graphs
- Logical if/else rules
- Groupings of data known as clusters

The choice of model is typically not left up to the machine. Instead, the learning task and the type of data on hand inform model selection. Later in this chapter, we will discuss in more detail the methods for choosing the appropriate model type.

The process of fitting a model to a dataset is known as **training**. When the model has been trained, the data has been transformed into an abstract form that summarizes the original information.

You might wonder why this step is called "training" rather than "learning." First, note that the process of learning does not end with data abstraction—the learner must still generalize and evaluate its training. Second, the word "training" better connotes the fact that the human teacher trains the machine student to understand the data in a specific way.

It is important to note that a learned model does not itself provide new data, yet it does result in new knowledge. How can this be? The answer is that imposing an assumed structure on the underlying data gives insight into the unseen. It supposes a new concept that describes a manner in which data elements may be related.

Take, for instance, the discovery of gravity. By fitting equations to observational data, Sir Isaac Newton inferred the concept of gravity, but the force we now know as gravity was always present. It simply wasn't recognized until Newton expressed it as an abstract concept that relates some data to other data—specifically, by becoming the g term in a model that explains observations of falling objects.

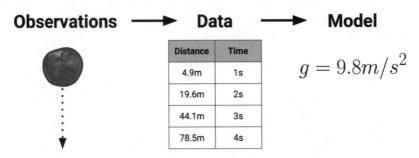

Distance	Time
4.9m	1s
19.6m	2s
44.1m	3s
78.5m	4s

$$g = 9.8m/s^2$$

Figure 1.6: Models are abstractions that explain observed data

Most models will not result in the development of theories that shake up scientific thought for centuries. Still, your abstraction might result in the discovery of important, but previously unseen, patterns and relationships among data. A model trained on genomic data might find several genes that when combined are responsible for the onset of diabetes, banks might discover a seemingly innocuous type of transaction that systematically appears prior to fraudulent activity, or psychologists might identify a combination of personality characteristics indicating a new disorder. These underlying patterns were always present, but by presenting information in a different format, a new idea is conceptualized.

Generalization

The next step in the learning process is to use the abstracted knowledge for future action. However, among the countless underlying patterns that may be identified during the abstraction process and the myriad ways to model those patterns, some patterns will be more useful than others. Unless the production of abstractions is limited to the useful set, the learner will be stuck where it started, with a large pool of information but no actionable insight.

Formally, the term **generalization** is defined as the process of turning abstracted knowledge into a form that can be utilized for future action, on tasks that are similar, but not identical, to those the learner has seen before. It acts as a search through the entire set of models (that is, theories or inferences) that *could* be established from the data during training.

If you can imagine a hypothetical set containing every possible way the data might be abstracted, generalization involves the reduction of this set into a smaller and more manageable set of important findings.

In generalization, the learner is tasked with limiting the patterns it discovers to only those that will be most relevant to its future tasks. Normally, it is not feasible to reduce the number of patterns by examining them one-by-one and ranking them by future utility. Instead, machine learning algorithms generally employ shortcuts that reduce the search space more quickly. To this end, the algorithm will employ **heuristics**, which are educated guesses about where to find the most useful inferences.

 Heuristics utilize approximations and other rules of thumb, which means they are not guaranteed to find the best model of the data. However, without taking these shortcuts, finding useful information in a large dataset would be infeasible.

Heuristics are routinely used by human beings to quickly generalize experience to new scenarios. If you have ever utilized your gut instinct to make a snap decision prior to fully evaluating your circumstances, you were intuitively using mental heuristics.

The incredible human ability to make quick decisions often relies not on computer-like logic, but rather on emotion-guided heuristics. Sometimes, this can result in illogical conclusions. For example, more people express fear of airline travel than automobile travel, despite automobiles being statistically more dangerous. This can be explained by the availability heuristic, which is the tendency for people to estimate the likelihood of an event by how easily examples can be recalled. Accidents involving air travel are highly publicized. Being traumatic events, they are likely to be recalled very easily, whereas car accidents barely warrant a mention in the newspaper.

The folly of misapplied heuristics is not limited to human beings. The heuristics employed by machine learning algorithms also sometimes result in erroneous conclusions. The algorithm is said to have a **bias** if the conclusions are *systematically* erroneous, which implies that they are wrong in a consistent or predictable manner.

For example, suppose that a machine learning algorithm learned to identify faces by finding two dark circles representing eyes, positioned above a straight line indicating a mouth. The algorithm might then have trouble with, or be *biased against*, faces that do not conform to its model. Faces with glasses, turned at an angle, looking sideways, or with certain skin tones might not be detected by the algorithm. Similarly, it could be *biased toward* faces with other skin tones, face shapes, or characteristics that conform to its understanding of the world.

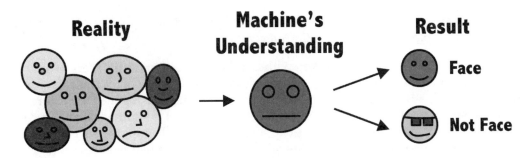

Figure 1.7: The process of generalizing a learner's experience results in a bias

In modern usage, the word "bias" has come to carry quite negative connotations. Various forms of media frequently claim to be free from bias, and claim to report the facts objectively, untainted by emotion. Still, consider for a moment the possibility that a little bias might be useful. Without a bit of arbitrariness, might it be a little difficult to decide among several competing choices, each with distinct strengths and weaknesses? Indeed, studies in the field of psychology have suggested that individuals born with damage to the portions of the brain responsible for emotion may be ineffectual at decision-making and might spend hours debating simple decisions, such as what color shirt to wear or where to eat lunch. Paradoxically, bias is what blinds us from some information, while also allowing us to utilize other information for action. It is how machine learning algorithms choose among the countless ways to understand a set of data.

Evaluation

Bias is a necessary evil associated with the abstraction and generalization processes inherent in any learning task. In order to drive action in the face of limitless possibility, all learning must have a bias. Consequently, each learning strategy has weaknesses; there is no single learning algorithm to rule them all. Therefore, the final step in the learning process is to **evaluate** its success, and to measure the learner's performance in spite of its biases. The information gained in the evaluation phase can then be used to inform additional training if needed.

Once you've had success with one machine learning technique, you might be tempted to apply it to every task. It is important to resist this temptation because no machine learning approach is best for every circumstance. This fact is described by the **No Free Lunch** theorem, introduced by *David Wolpert* in 1996. For more information, visit: http://www.no-free-lunch.org.

Generally, evaluation occurs after a model has been trained on an initial **training dataset**. Then, the model is evaluated on a separate **test dataset** in order to judge how well its characterization of the training data generalizes to new, unseen cases. It's worth noting that it is exceedingly rare for a model to perfectly generalize to every unforeseen case—mistakes are almost always inevitable.

In part, models fail to generalize perfectly due to the problem of **noise**, a term that describes unexplained or unexplainable variations in data. Noisy data is caused by seemingly random events, such as:

- Measurement error due to imprecise sensors that sometimes add or subtract a small amount from the readings
- Issues with human subjects, such as survey respondents reporting random answers to questions in order to finish more quickly
- Data quality problems, including missing, null, truncated, incorrectly coded, or corrupted values
- Phenomena that are so complex or so little understood that they impact the data in ways that appear to be random

Trying to model noise is the basis of a problem called **overfitting**; because most noisy data is unexplainable by definition, attempting to explain the noise will result in models that do not generalize well to new cases. Efforts to explain the noise also typically result in more complex models that miss the true pattern the learner is trying to identify.

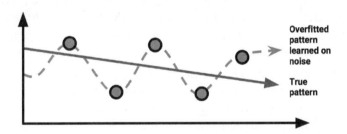

Figure 1.8: Modeling noise generally results in more complex models and misses underlying patterns

A model that performs relatively well during training but relatively poorly during evaluation is said to be **overfitted** to the training dataset because it does not generalize well to the test dataset. In practical terms, this means that it has identified a pattern in the data that is not useful for future action; the generalization process has failed. Solutions to the problem of overfitting are specific to particular machine learning approaches. For now, the important point is to be aware of the issue. How well the methods are able to handle noisy data and avoid overfitting is an important point of distinction among them.

Machine learning in practice

So far, we've focused on how machine learning works in theory. To apply the learning process to real-world tasks, we'll use a five-step process. Regardless of the task, any machine learning algorithm can be deployed by following these steps:

1. **Data collection**: The data collection step involves gathering the learning material an algorithm will use to generate actionable knowledge. In most cases, the data will need to be combined into a single source, such as a text file, spreadsheet, or database.

2. **Data exploration and preparation**: The quality of any machine learning project is based largely on the quality of its input data. Thus, it is important to learn more about the data and its nuances during a practice called data exploration. Additional work is required to prepare the data for the learning process. This involves fixing or cleaning so-called "messy" data, eliminating unnecessary data, and recoding the data to conform to the learner's expected inputs.

3. **Model training**: By the time the data has been prepared for analysis, you are likely to have a sense of what you are capable of learning from the data. The specific machine learning task chosen will inform the selection of an appropriate algorithm, and the algorithm will represent the data in the form of a model.

4. **Model evaluation**: Each machine learning model results in a biased solution to the learning problem, which means that it is important to evaluate how well the algorithm learned from its experience. Depending on the type of model used, you might be able to evaluate the accuracy of the model using a test dataset, or you may need to develop measures of performance specific to the intended application.

5. **Model improvement**: If better performance is needed, it becomes necessary to utilize more advanced strategies to augment the model's performance. Sometimes it may be necessary to switch to a different type of model altogether. You may need to supplement your data with additional data or perform additional preparatory work, as in step two of this process.

After these steps have been completed, if the model appears to be performing well, it can be deployed for its intended task. As the case may be, you might utilize your model to provide score data for predictions (possibly in real time); for projections of financial data; to generate useful insight for marketing or research; or to automate tasks, such as mail delivery or flying aircraft. The successes and failures of the deployed model might even provide additional data to train your next-generation learner.

Types of input data

The practice of machine learning involves matching the characteristics of the input data to the biases of the available learning algorithms. Thus, before applying machine learning to real-world problems, it is important to understand the terminology that distinguishes between input datasets.

The phrase **unit of observation** is used to describe the smallest entity with measured properties of interest for a study. Commonly, the unit of observation is in the form of persons, objects or things, transactions, time points, geographic regions, or measurements. Sometimes, units of observation are combined to form units, such as person-years, which denote cases where the same person is tracked over multiple years, and each person-year comprises a person's data for one year.

> The unit of observation is related, but not identical, to the **unit of analysis**, which is the smallest unit from which inference is made. Although it is often the case, the observed and analyzed units are not always the same. For example, data observed from people (the unit of observation) might be used to analyze trends across countries (the unit of analysis).

Datasets that store the units of observation and their properties can be described as collections of:

- **Examples**: Instances of the unit of observation for which properties have been recorded
- **Features**: Recorded properties or attributes of examples that may be useful for learning

It is easiest to understand features and examples through real-world scenarios. For instance, to build a learning algorithm to identify spam emails, the unit of observation could be email messages, examples would be specific individual messages, and the features might consist of the words used in the messages.

For a cancer detection algorithm, the unit of observation could be patients, the examples might include a random sample of cancer patients, and the features may be genomic markers from biopsied cells in addition to patient characteristics, such as weight, height, or blood pressure.

People and machines differ in the types of complexity they are suited to handle in the input data. Humans are comfortable consuming **unstructured data**, such as free-form text, pictures, or sound. They are also flexible handling cases in which some observations have a wealth of features, while others have very little.

On the other hand, computers generally require data to be **structured**, which means that each example of the phenomenon has the same features, and these features are organized in a form that a computer may understand. To use the brute force of the machine on large, unstructured datasets usually requires a transformation of the input data to a structured form.

The following spreadsheet shows data that has been gathered in **matrix format**. In matrix data, each row in the spreadsheet is an example and each column is a feature. Here, the rows indicate examples of automobiles for sale, while the columns record each automobile's features, such as the price, mileage, color, and transmission type. Matrix format data is by far the most common form used in machine learning. As you will see in later chapters, when forms of data are encountered in specialized applications, they are ultimately transformed into a matrix prior to machine learning.

features

year	model	price	mileage	color	transmission
2011	SEL	21992	7413	Yellow	AUTO
2011	SEL	20995	10926	Gray	AUTO
2011	SEL	19995	7351	Silver	AUTO
2011	SEL	17809	11613	Gray	AUTO
2012	SE	17500	8367	White	MANUAL
2010	SEL	17495	25125	Silver	AUTO
2011	SEL	17000	27393	Blue	AUTO
2010	SEL	16995	21026	Silver	AUTO
2011	SES	16995	32655	Silver	AUTO

examples

Figure 1.9: A simple dataset in matrix format describing automobiles for sale

A dataset's features may come in various forms. If a feature represents a characteristic measured in numbers, it is unsurprisingly called **numeric**. Alternatively, if a feature comprises a set of categories, the feature is called **categorical** or **nominal**. A special type of categorical variable is called **ordinal**, which designates a nominal variable with categories falling in an ordered list. One example of an ordinal variable is clothing sizes, such as small, medium, and large; another is a measurement of customer satisfaction on a scale from "not at all happy" to "somewhat happy" to "very happy." For any given dataset, thinking about what the features represent, their types, and their units, will assist with determining an appropriate machine learning algorithm for the learning task.

Types of machine learning algorithms

Machine learning algorithms are divided into categories according to their purpose. Understanding the categories of learning algorithms is an essential first step toward using data to drive the desired action.

A **predictive model** is used for tasks that involve, as the name implies, the prediction of one value using other values in the dataset. The learning algorithm attempts to discover and model the relationship between the **target** feature (the feature being predicted) and the other features.

Despite the common use of the word "prediction" to imply forecasting, predictive models need not necessarily foresee events in the future. For instance, a predictive model could be used to predict past events, such as the date of a baby's conception using the mother's present-day hormone levels. Predictive models can also be used in real time to control traffic lights during rush hour.

Now, because predictive models are given clear instructions on what they need to learn and how they are intended to learn it, the process of training a predictive model is known as **supervised learning**. The supervision does not refer to human involvement, but rather to the fact that the target values provide a way for the learner to know how well it has learned the desired task. Stated more formally, given a set of data, a supervised learning algorithm attempts to optimize a function (the model) to find the combination of feature values that result in the target output.

The often-used supervised machine learning task of predicting which category an example belongs to is known as **classification**. It is easy to think of potential uses for a classifier. For instance, you could predict whether:

- An email message is spam
- A person has cancer
- A football team will win or lose
- An applicant will default on a loan

In classification, the target feature to be predicted is a categorical feature known as the **class**, which is divided into categories called **levels**. A class can have two or more levels, and the levels may or may not be ordinal. Classification is so widely used in machine learning that there are many types of classification algorithms, with strengths and weaknesses suited for different types of input data. We will see examples of these later in this chapter and throughout this book.

Supervised learners can also be used to predict numeric data, such as income, laboratory values, test scores, or counts of items. To predict such numeric values, a common form of **numeric prediction** fits linear regression models to the input data. Although regression is not the only method for numeric prediction, it is by far the most widely used. Regression methods are widely used for forecasting, as they quantify in exact terms the association between the inputs and the target, including both the magnitude and uncertainty of the relationship.

Since it is easy to convert numbers to categories (for example, ages 13 to 19 are teenagers) and categories to numbers (for example, assign 1 to all males and 0 to all females), the boundary between classification models and numeric prediction models is not necessarily firm.

A **descriptive model** is used for tasks that would benefit from the insight gained from summarizing data in new and interesting ways. As opposed to predictive models that predict a target of interest, in a descriptive model, no single feature is more important than any other. In fact, because there is no target to learn, the process of training a descriptive model is called **unsupervised learning**. Although it can be more difficult to think of applications for descriptive models — after all, what good is a learner that isn't learning anything in particular — they are used quite regularly for data mining.

For example, the descriptive modeling task called **pattern discovery** is used to identify useful associations within data. Pattern discovery is the goal of **market basket analysis**, which is applied to retailers' transactional purchase data. Here, retailers hope to identify items that are frequently purchased together, such that the learned information can be used to refine marketing tactics. For instance, if a retailer learns that swimming trunks are commonly purchased at the same time as sunscreen, the retailer might reposition the items more closely in the store or run a promotion to "up-sell" customers on associated items.

Originally used only in retail contexts, pattern discovery is now starting to be used in quite innovative ways. For instance, it can be used to detect patterns of fraudulent behavior, screen for genetic defects, or identify hotspots for criminal activity.

The descriptive modeling task of dividing a dataset into homogeneous groups is called **clustering**. This is sometimes used for **segmentation analysis**, which identifies groups of individuals with similar behavior or demographic information in order to target them with advertising campaigns based on their shared characteristics. With this approach, the machine identifies the clusters, but human intervention is required to interpret them. For example, given a grocery store's five customer clusters, the marketing team will need to understand the differences among the groups in order to create a promotion that best suits each group. Despite this human effort, this is still less work than creating a unique appeal for each customer.

Lastly, a class of machine learning algorithms known as **meta-learners** is not tied to a specific learning task, but rather is focused on learning how to learn more effectively. A meta-learning algorithm uses the result of past learning to inform additional learning.

This encompasses learning algorithms that learn to work together in teams called **ensembles**, as well as algorithms that seem to evolve over time in a process called **reinforcement learning**. Meta-learning can be beneficial for very challenging problems or when a predictive algorithm's performance needs to be as accurate as possible.

Some of the most exciting work being done in the field of machine learning today is in the domain of meta-learning. For instance, **adversarial learning** involves learning about a model's weaknesses in order to strengthen its future performance or harden it against malicious attack. There is also heavy investment in research and development efforts to make bigger and faster ensembles, which can model massive datasets using high-performance computers or cloud-computing environments.

Matching input data to algorithms

The following table lists the general types of machine learning algorithms covered in this book. Although this covers only a fraction of the entire set of machine learning algorithms, learning these methods will provide a sufficient foundation for making sense of any other methods you may encounter in the future.

Model	Learning task	Chapter
Supervised learning algorithms		
k-nearest neighbors	Classification	*Chapter 3*
Naive Bayes	Classification	*Chapter 4*
Decision trees	Classification	*Chapter 5*
Classification rule learners	Classification	*Chapter 5*
Linear regression	Numeric prediction	*Chapter 6*
Regression trees	Numeric prediction	*Chapter 6*
Model trees	Numeric prediction	*Chapter 6*
Neural networks	Dual use	*Chapter 7*
Support vector machines	Dual use	*Chapter 7*
Unsupervised learning algorithms		
Association rules	Pattern detection	*Chapter 8*
k-means clustering	Clustering	*Chapter 9*
Meta-learning algorithms		
Bagging	Dual use	*Chapter 11*
Boosting	Dual use	*Chapter 11*
Random forests	Dual use	*Chapter 11*

To begin applying machine learning to a real-world project, you will need to determine which of the four learning tasks your project represents: classification, numeric prediction, pattern detection, or clustering. The task will drive the choice of algorithm. For instance, if you are undertaking pattern detection, you are likely to employ association rules. Similarly, a clustering problem will likely utilize the k-means algorithm, and numeric prediction will utilize regression analysis or regression trees.

For classification, more thought is needed to match a learning problem to an appropriate classifier. In these cases, it is helpful to consider the various distinctions among the algorithms—distinctions that will only be apparent by studying each of the classifiers in depth. For instance, within classification problems, decision trees result in models that are readily understood, while the models of neural networks are notoriously difficult to interpret. If you were designing a credit scoring model, this could be an important distinction because the law often requires that the applicant must be notified about the reasons he or she was rejected for the loan. Even if the neural network is better at predicting loan defaults, if its predictions cannot be explained, then it is useless for this application.

To assist with algorithm selection, in every chapter the key strengths and weaknesses of each learning algorithm are listed. Although you will sometimes find that these characteristics exclude certain models from consideration, in many cases the choice of algorithm is arbitrary. When this is true, feel free to use whichever algorithm you are most comfortable with. Other times, when predictive accuracy is the primary goal, you may need to test several models and choose the one that fits best, or use a meta-learning algorithm that combines several different learners to utilize the strengths of each.

Machine learning with R

Many of the algorithms needed for machine learning are not included as part of the base R installation. Instead, the algorithms are available via a large community of experts who have shared their work freely. These must be installed on top of base R manually. Thanks to R's status as free open-source software, there is no additional charge for this functionality.

A collection of R functions that can be shared among users is called a **package**. Free packages exist for each of the machine learning algorithms covered in this book. In fact, this book only covers a small portion of all of R's machine learning packages.

If you are interested in the breadth of R packages, you can view a list at **Comprehensive R Archive Network (CRAN)**, a collection of web and FTP sites located around the world to provide the most up-to-date versions of R software and packages. If you obtained the R software via download, it was most likely from CRAN. The CRAN website is available at http://cran.r-project.org/index.html.

If you do not already have R, the CRAN website also provides installation instructions and information on where to find help if you have trouble.

The **Packages** link on the left side of the CRAN page will take you to a page where you can browse the packages in alphabetical order or sorted by publication date. At the time of this writing, a total of 13,904 packages were available—over two times the number since the second edition of this book was written, and over three times since the first edition! Clearly, the R community is thriving, and this trend shows no sign of slowing!

The **Task Views** link on the left side of the CRAN page provides a curated list of packages by subject area. The task view for machine learning, which lists the packages covered in this book (and many more), is available at `https://CRAN.R-project.org/view=MachineLearning`.

Installing R packages

Despite the vast set of available R add-ons, the package format makes installation and use a virtually effortless process. To demonstrate the use of packages, we will install and load the RWeka package developed by Kurt Hornik, Christian Buchta, and Achim Zeileis (see *Open-Source Machine Learning: R Meets Weka, Computational Statistics Vol. 24, pp 225-232* for more information). The RWeka package provides a collection of functions that give R access to the machine learning algorithms in the Java-based Weka software package by Ian H. Witten and Eibe Frank. For more information on Weka, see `http://www.cs.waikato.ac.nz/~ml/weka/`.

 To use the RWeka package, you will need to have Java installed, if it isn't already (many computers come with Java preinstalled). Java is a set of programming tools, available for free, that allow for the use of cross-platform applications such as Weka. For more information and to download Java for your system, visit `http://www.java.com`.

The most direct way to install a package is via the `install.packages()` function. To install the RWeka package, at the R command prompt simply type:

```
> install.packages("RWeka")
```

R will then connect to CRAN and download the package in the correct format for your operating system. Some packages, such as RWeka, require additional packages to be installed before they can be used. These are called **dependencies**. By default, the installer will automatically download and install any dependencies.

The first time you install a package, R may ask you to choose a CRAN mirror. If this happens, choose the mirror residing at a location close to you. This will generally provide the fastest download speed.

The default installation options are appropriate for most systems. However, in some cases, you may want to install a package to another location. For example, if you do not have root or administrator privileges on your system, you may need to specify an alternative installation path. This can be accomplished using the `lib` option as follows:

```
> install.packages("RWeka", lib = "/path/to/library")
```

The installation function also provides additional options for installing from a local file, installing from source, or using experimental versions. You can read about these options in the help file by using the following command:

```
> ?install.packages
```

More generally, the question mark operator can be used to obtain help on any R function. Simply type `?` before the name of the function.

Loading and unloading R packages

In order to conserve memory, R does not load every installed package by default. Instead, packages are loaded by users with the `library()` function as they are needed.

The name of this function leads some people to incorrectly use the terms "library" and "package" interchangeably. However, to be precise, a library refers to the location where packages are installed and never to a package itself.

To load the RWeka package installed previously, you can type the following:

```
> library(RWeka)
```

Aside from RWeka, there are several other R packages that will be used in later chapters. Installation instructions will be provided as these additional packages are needed.

To unload an R package, use the `detach()` function. For example, to unload the RWeka package shown previously, use the following command:

```
> detach("package:RWeka", unload = TRUE)
```

This will free up any resources used by the package.

Installing RStudio

Before you begin working with R, it is highly recommended to also install the open-source **RStudio** desktop application. RStudio is an additional interface to R that includes functionalities that make it far easier, more convenient, and more interactive to work with R code. It is available free of charge at https://www.rstudio.com/.

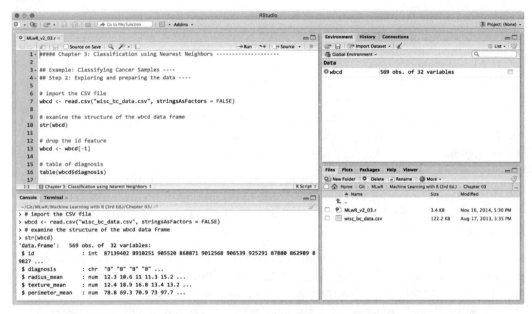

Figure 1.10: The RStudio desktop environment makes R easier and more convenient to use

The RStudio interface includes an integrated code editor, an R command-line console, a file browser, and an R object browser. R code syntax is automatically colorized, and the code's output, plots, and graphics are displayed directly within the environment, which makes it much easier to follow long or complex statements and programs. More advanced features allow R project and package management; integration with source control or version control tools, such as Git and Subversion; database connection management; and the compilation of R output to HTML, PDF, or Microsoft Word formats.

RStudio is a key reason why R is a top choice for data scientists today. It wraps the power of R programming and its tremendous library of machine learning and statistical packages in an easy-to-use and easy-to-install development interface. It is not only ideal for learning R, but can also grow with you as you learn R's more advanced functionality.

Summary

Machine learning originated at the intersection of statistics, database science, and computer science. It is a powerful tool, capable of finding actionable insight in large quantities of data. Still, as we have seen in this chapter, caution must be used in order to avoid common abuses of machine learning in the real world.

Conceptually, the learning process involves the abstraction of data into a structured representation, and the generalization of the structure into action that can be evaluated for utility. In practical terms, a machine learner uses data containing examples and features of the concept to be learned, then summarizes this data in the form of a model, which is used for predictive or descriptive purposes. These purposes can be grouped into tasks including classification, numeric prediction, pattern detection, and clustering. Among the many possible methods, machine learning algorithms are chosen on the basis of the input data and the learning task.

R provides support for machine learning in the form of community-authored packages. These powerful tools are available to download at no cost, but need to be installed before they can be used. Each chapter in this book will introduce such packages as they are needed.

In the next chapter, we will further introduce the basic R commands that are used to manage and prepare data for machine learning. Though you might be tempted to skip this step and jump directly into applications, a common rule of thumb suggests that 80 percent or more of the time spent on typical machine learning projects is devoted to the step of data preparation, also known as "data wrangling." As a result, investing some effort into learning how to do this effectively will pay dividends for you later on.

2

Managing and Understanding Data

A key early component of any machine learning project involves managing and understanding data. Although this may not be as gratifying as building and deploying models — the stages in which you begin to see the fruits of your labor — it is unwise to ignore this important preparatory work.

Any learning algorithm is only as good as its input data, and in many cases, the input data is complex, messy, and spread across multiple sources and formats. Due to this complexity, often the largest portion of effort invested in machine learning projects is spent on data preparation and exploration.

This chapter approaches data preparation in three ways. The first section discusses the basic data structures R uses to store data. You will become very familiar with these structures as you create and manipulate datasets. The second section is practical, as it covers several functions that are used for getting data in and out of R. In the third section, methods for understanding data are illustrated while exploring a real-world dataset.

By the end of this chapter, you will understand:

- How to use R's basic data structures to store and manipulate values
- Simple functions to get data into R from common source formats
- Typical methods to understand and visualize complex data

The ways R manages data will dictate the ways you must work with data, so it is helpful to know R's data structures before jumping directly into data preparation. However, if you are already familiar with R programming, feel free to skip ahead to the section on data preprocessing.

R data structures

There are numerous types of data structures across programming languages, each with strengths and weaknesses suited to specific tasks. Since R is a programming language used widely for statistical data analysis, the data structures it utilizes were designed with this type of work in mind.

The R data structures used most frequently in machine learning are vectors, factors, lists, arrays, matrices, and data frames. Each is tailored to a specific data management task, which makes it important to understand how they will interact in your R project. In the sections that follow, we will review their similarities and differences.

Vectors

The fundamental R data structure is the **vector**, which stores an ordered set of values called **elements**. A vector can contain any number of elements. However, all of its elements must be of the same type; for instance, a vector cannot contain both numbers and text. To determine the type of vector v, use the typeof(v) command.

Several vector types are commonly used in machine learning: integer (numbers without decimals), double (numbers with decimals), character (text data), and logical (TRUE or FALSE values). There are also two special values: NA, which indicates a missing value, and NULL, which is used to indicate the absence of any value. Although these two may seem to be synonymous, they are indeed slightly different. The NA value is a placeholder for something else and therefore has a length of one, while the NULL value is truly empty and has a length of zero.

Some R functions will report both integer and double vectors as numeric, while others will distinguish between the two. As a result, although all double vectors are numeric, not all numeric vectors are double type.

It is tedious to enter large amounts of data by hand, but simple vectors can be created by using the c() combine function. The vector can also be given a name using the arrow <- operator. This is R's assignment operator, used much like the = assignment operator in many other programming languages.

For example, let's construct a set of vectors containing data on three medical patients. We'll create a character vector named `subject_name` to store the three patient names, a numeric vector named `temperature` to store each patient's body temperature in degrees Fahrenheit, and a logical vector named `flu_status` to store each patient's diagnosis (TRUE if he or she has influenza, FALSE otherwise). As shown in the following code, the three vectors are:

```
> subject_name <- c("John Doe", "Jane Doe", "Steve Graves")
> temperature <- c(98.1, 98.6, 101.4)
> flu_status <- c(FALSE, FALSE, TRUE)
```

Values stored in R vectors retain their order. Therefore, data for each patient can be accessed using his or her position in the set, beginning at 1, then supplying this number inside square brackets (that is, [and]) following the name of the vector. For instance, to obtain the temperature value for patient Jane Doe, the second patient, simply type:

```
> temperature[2]
[1] 98.6
```

R offers a variety of methods to extract data from vectors. A range of values can be obtained using the colon operator. For instance, to obtain the body temperature of the second and third patients, type:

```
> temperature[2:3]
[1] 98.6 101.4
```

Items can be excluded by specifying a negative item number. To exclude the second patient's temperature data, type:

```
> temperature[-2]
[1]   98.1 101.4
```

It is also sometimes useful to specify a logical vector indicating whether or not each item should be included. For example, to include the first two temperature readings but exclude the third, type:

```
> temperature[c(TRUE, TRUE, FALSE)]
[1] 98.1 98.6
```

As you will see shortly, the vector provides the foundation for many other R data structures. Therefore, knowing the various vector operations is crucial for working with data in R.

Downloading the example code

You can download the example code files for all Packt books you have purchased from your account at `http://www.packtpub.com`. If you purchased this book elsewhere, you can visit `http://www.packtpub.com/support` and register to have the files emailed directly to you.

The example code is also available via GitHub at `https://github.com/dataspelunking/MLwR/`. Check here for the most up-to-date R code, as well as issue tracking and a public wiki. Please join the community!

Factors

If you recall from *Chapter 1, Introducing Machine Learning*, nominal features represent a characteristic with categories of values. Although it is possible to use a character vector to store nominal data, R provides a data structure specifically for this purpose. A **factor** is a special case of vector that is solely used for representing categorical or ordinal variables. In the medical dataset we are building, we might use a factor to represent gender because it uses two categories: male and female.

Why use factors rather than character vectors? One advantage of factors is that the category labels are stored only once. Rather than storing MALE, MALE, FEMALE, the computer may store 1, 1, 2, which can reduce the memory needed to store the values. Additionally, many machine learning algorithms treat nominal and numeric features differently. Coding categorical variables as factors ensures that R will handle categorical data appropriately.

A factor should not be used for character vectors that are not truly categorical. In particular, if a vector stores mostly unique values such as names or identification codes, keep it as a character vector.

To create a factor from a character vector, simply apply the `factor()` function. For example:

```
> gender <- factor(c("MALE", "FEMALE", "MALE"))
> gender
[1] MALE    FEMALE MALE
Levels: FEMALE MALE
```

Notice that when the gender factor was displayed, R printed additional information about its levels. The levels comprise the set of possible categories the factor could take, in this case, MALE or FEMALE.

When we create factors, we can add additional levels that may not appear in the original data. Suppose we created another factor for blood type, as shown in the following example:

```
> blood <- factor(c("O", "AB", "A"),
          levels = c("A", "B", "AB", "O"))
> blood
[1] O  AB A
Levels: A B AB O
```

When we defined the `blood` factor, we specified an additional vector of four possible blood types using the `levels` parameter. As a result, even though our data includes only blood types O, AB, and A, all four types are retained with the `blood` factor, as the output shows. Storing the additional level allows for the possibility of adding patients with the other blood type in the future. It also ensures that if we were to create a table of blood types, we would know that type B exists, despite it not being found in our initial data.

The factor data structure also allows us to include information about the order of a nominal variable's categories, which provides a method for creating ordinal features. For example, suppose we have data on the severity of patient symptoms, coded in increasing order of severity from mild, to moderate, to severe. We indicate the presence of ordinal data by providing the factor's levels in the desired order, listed ascending from lowest to highest, and setting the `ordered` parameter to TRUE as shown:

```
> symptoms <- factor(c("SEVERE", "MILD", "MODERATE"),
            levels = c("MILD", "MODERATE", "SEVERE"),
            ordered = TRUE)
```

The resulting `symptoms` factor now includes information about the requested order. Unlike our prior factors, the levels of this factor are separated by < symbols to indicate the presence of a sequential order from MILD to SEVERE:

```
> symptoms
[1] SEVERE   MILD      MODERATE
Levels: MILD < MODERATE < SEVERE
```

A helpful feature of ordered factors is that logical tests work as you would expect. For instance, we can test whether each patient's symptoms are more severe than moderate:

```
> symptoms > "MODERATE"
[1]  TRUE FALSE FALSE
```

Machine learning algorithms capable of modeling ordinal data will expect ordered factors, so be sure to code your data accordingly.

Lists

A **list** is a data structure, much like a vector, in that it is used for storing an ordered set of elements. However, where a vector requires all its elements to be the same type, a list allows different R data types to be collected. Due to this flexibility, lists are often used to store various types of input and output data and sets of configuration parameters for machine learning models.

To illustrate lists, consider the medical patient dataset we have been constructing, with data for three patients stored in six vectors. If we wanted to display all the data for the first patient, we would need to enter five R commands:

```
> subject_name[1]
[1] "John Doe"
> temperature[1]
[1] 98.1
> flu_status[1]
[1] FALSE
> gender[1]
[1] MALE
Levels: FEMALE MALE
> blood[1]
[1] O
Levels: A B AB O
> symptoms[1]
[1] SEVERE
Levels: MILD < MODERATE < SEVERE
```

If we expect to examine the patient's data again in the future, rather than retyping these commands, a list allows us to group all of the values into one object we can use repeatedly.

Similar to creating a vector with `c()`, a list is created using the `list()` function as shown in the following example. One notable difference is that when a list is constructed, each component in the sequence should be given a name. The names are not strictly required, but allow the values to be accessed later on by name rather than by numbered position. To create a list with named components for all of the first patient's data, type the following:

```
> subject1 <- list(fullname = subject_name[1],
                    temperature = temperature[1],
                    flu_status = flu_status[1],
                    gender = gender[1],
                    blood = blood[1],
                    symptoms = symptoms[1])
```

This patient's data is now collected in the `subject1` list:

```
> subject1
$fullname
[1] "John Doe"

$temperature
[1] 98.1

$flu_status
[1] FALSE

$gender
[1] MALE
Levels: FEMALE MALE

$blood
[1] O
Levels: A B AB O

$symptoms
[1] SEVERE
Levels: MILD < MODERATE < SEVERE
```

Note that the values are labeled with the names we specified in the preceding command. As a list retains order like a vector, its components can be accessed using numeric positions, as shown here for the `temperature` value:

```
> subject1[2]
$temperature
[1] 98.1
```

The result of using vector-style operators on a list object is another list object, which is a subset of the original list. For example, the preceding code returned a list with a single `temperature` component. To instead return a single list item in its *native* data type, use double brackets (`[[` and `]]`) when selecting the list component. For example, the following command returns a numeric vector of length one:

```
> subject1[[2]]
[1] 98.1
```

For clarity, it is often better to access list components by name, by appending a $ and the component name to the list name as follows:

```
> subject1$temperature
[1] 98.1
```

Like the double-bracket notation, this returns the list component in its native data type (in this case, a numeric vector of length one).

 Accessing the value by name also ensures that the correct item is retrieved even if the order of the list elements is changed later on.

It is possible to obtain several list items by specifying a vector of names. The following returns a subset of the subject1 list, which contains only the temperature and flu_status components:

```
> subject1[c("temperature", "flu_status")]
$temperature
[1] 98.1

$flu_status
[1] FALSE
```

Entire datasets could be constructed using lists, and lists of lists. For example, you might consider creating a subject2 and subject3 list, and grouping these into a list object named pt_data. However, constructing a dataset in this way is common enough that R provides a specialized data structure specifically for this task.

Data frames

By far the most important R data structure utilized in machine learning is the **data frame**, a structure analogous to a spreadsheet or database in that it has both rows and columns of data. In R terms, a data frame can be understood as a list of vectors or factors, each having exactly the same number of values. Now, because the data frame is literally a list of vector-type objects, it combines aspects of both vectors and lists.

Let's create a data frame for our patient dataset. Using the patient data vectors we created previously, the data.frame() function combines them into a data frame:

```
> pt_data <- data.frame(subject_name, temperature,
                        flu_status, gender, blood, symptoms,
                        stringsAsFactors = FALSE)
```

You might notice something new in the preceding code. We included an additional parameter: `stringsAsFactors = FALSE`. If we do not specify this option, R will automatically convert every character vector to a factor.

This feature is occasionally useful, but also sometimes unwarranted. Here, for example, the `subject_name` field is definitely not categorical data, as names are not categories of values. Therefore, setting the `stringsAsFactors` option to `FALSE` allows us to convert character vectors to factors only where it makes sense for the project.

When we display the `pt_data` data frame, we see that the structure is quite different from the data structures we worked with previously:

```
> pt_data
  subject_name temperature flu_status gender blood symptoms
1      John Doe        98.1      FALSE   MALE     O   SEVERE
2      Jane Doe        98.6      FALSE FEMALE    AB     MILD
3 Steve Graves        101.4       TRUE   MALE     A MODERATE
```

Compared to the one-dimensional vectors, factors, and lists, a data frame has two dimensions and is displayed in matrix format. This particular data frame has one column for each vector of patient data and one row for each patient. In machine learning terms, the data frame's columns are the features or attributes and the rows are the examples.

To extract entire columns (vectors) of data, we can take advantage of the fact that a data frame is simply a list of vectors. Similar to lists, the most direct way to extract a single element is by referring to it by name. For example, to obtain the `subject_name` vector, type:

```
> pt_data$subject_name
[1] "John Doe"       "Jane Doe"       "Steve Graves"
```

Also similar to lists, a vector of names can be used to extract multiple columns from a data frame:

```
> pt_data[c("temperature", "flu_status")]
  temperature flu_status
1        98.1      FALSE
2        98.6      FALSE
3       101.4       TRUE
```

When we request columns in the data frame by name, the result is a data frame containing all rows of data for the specified columns. The command `pt_data[2:3]` will also extract the `temperature` and `flu_status` columns. However, referring to the columns by name results in clear and easy-to-maintain R code that will not break if the data frame is later reordered.

To extract specific values from the data frame, methods like those for accessing values in vectors are used. However, there is an important distinction — because the data frame is two-dimensional, both the desired rows and columns must be specified. Rows are specified first, followed by a comma, followed by the columns in a format like this: [rows, columns]. As with vectors, rows and columns are counted beginning at one.

For instance, to extract the value in the first row and second column of the patient data frame, use the following command:

```
> pt_data[1, 2]
[1] 98.1
```

If you would like more than a single row or column of data, specify vectors indicating the desired rows and columns. The following statement will pull data from the first and third rows and the second and fourth columns:

```
> pt_data[c(1, 3), c(2, 4)]
  temperature gender
1        98.1   MALE
3       101.4   MALE
```

To refer to every row or every column, simply leave the row or column portion blank. For example, to extract all rows of the first column:

```
> pt_data[, 1]
[1] "John Doe"     "Jane Doe"     "Steve Graves"
```

To extract all columns for the first row:

```
> pt_data[1, ]
  subject_name temperature flu_status gender blood symptoms
1     John Doe        98.1      FALSE   MALE     O   SEVERE
```

And to extract everything:

```
> pt_data[ , ]
  subject_name temperature flu_status gender blood symptoms
1     John Doe        98.1      FALSE   MALE     O   SEVERE
2     Jane Doe        98.6      FALSE FEMALE    AB     MILD
3 Steve Graves       101.4       TRUE   MALE     A MODERATE
```

Of course, columns are better accessed by name rather than position, and negative signs can be used to exclude rows or columns of data. Therefore, the output of the command:

```
> pt_data[c(1, 3), c("temperature", "gender")]
  temperature gender
1        98.1   MALE
3       101.4   MALE
```

is equivalent to:

```
> pt_data[-2, c(-1, -3, -5, -6)]
  temperature gender
1        98.1   MALE
3       101.4   MALE
```

Sometimes it is necessary to create new columns in data frames—perhaps, for instance, as a function of existing columns. For example, we may need to convert the Fahrenheit temperature readings in the patient data frame to the Celsius scale. To do this, we simply use the assignment operator to assign the result of the conversion calculation to a new column name as follows:

```
> pt_data$temp_c <- (pt_data$temperature - 32) * (5 / 9)
```

To confirm the calculation worked, let's compare the new Celsius-based temp_c column to the previous Fahrenheit-scale temperature column:

```
> pt_data[c("temperature", "temp_c")]
  temperature    temp_c
1        98.1  36.72222
2        98.6  37.00000
3       101.4  38.55556
```

Seeing these side by side, we can confirm that the calculation has worked correctly.

To become more familiar with data frames, try practicing similar operations with the patient dataset, or even better, use data from one of your own projects. These types of operations are crucial for much of the work we will do in upcoming chapters.

Matrices and arrays

In addition to data frames, R provides other structures that store values in tabular form. A **matrix** is a data structure that represents a two-dimensional table with rows and columns of data. Like vectors, R matrices can contain only one type of data, although they are most often used for mathematical operations and therefore typically store only numbers.

To create a matrix, simply supply a vector of data to the matrix() function, along with a parameter specifying the number of rows (nrow) or number of columns (ncol). For example, to create a 2x2 matrix storing the numbers one to four, we can use the nrow parameter to request the data to be divided into two rows:

```
> m <- matrix(c(1, 2, 3, 4), nrow = 2)
> m
     [,1] [,2]
[1,]    1    3
[2,]    2    4
```

This is equivalent to the matrix produced using ncol = 2:

```
> m <- matrix(c(1, 2, 3, 4), ncol = 2)
> m
     [,1] [,2]
[1,]    1    3
[2,]    2    4
```

You will notice that R loaded the first column of the matrix first before loading the second column. This is called **column-major order**, which is R's default method for loading matrices.

To override this default setting and load a matrix by rows, set the parameter byrow = TRUE when creating the matrix.

To illustrate this further, let's see what happens if we add more values to the matrix.

With six values, requesting two rows creates a matrix with three columns:

```
> m <- matrix(c(1, 2, 3, 4, 5, 6), nrow = 2)
> m
     [,1] [,2] [,3]
[1,]    1    3    5
[2,]    2    4    6
```

Requesting two columns creates a matrix with three rows:

```
> m <- matrix(c(1, 2, 3, 4, 5, 6), ncol = 2)
> m
     [,1] [,2]
[1,]    1    4
[2,]    2    5
[3,]    3    6
```

As with data frames, values in matrices can be extracted using `[row, column]` notation. For instance, `m[1, 1]` will return the value 1 while `m[3, 2]` will extract 6 from the m matrix. Additionally, entire rows or columns can be requested:

```
> m[1, ]
[1] 1 4
> m[, 1]
[1] 1 2 3
```

Closely related to the matrix structure is the **array**, which is a multidimensional table of data. Where a matrix has rows and columns of values, an array has rows, columns, and a number of additional layers of values. Although we will occasionally use matrices in later chapters, the use of arrays is unnecessary within the scope of this book.

Managing data with R

One of the challenges faced while working with massive datasets involves gathering, preparing, and otherwise managing data from a variety of sources. Although we will cover data preparation, data cleaning, and data management in depth by working on real-world machine learning tasks in later chapters, this section highlights the basic functionality for getting data in and out of R.

Saving, loading, and removing R data structures

When you have spent a lot of time getting a data frame into the desired form, you shouldn't need to recreate your work each time you restart your R session. To save a data structure to a file that can be reloaded later or transferred to another system, use the `save()` function. The `save()` function writes one or more R data structures to the location specified by the `file` parameter. R data files have an `.RData` extension.

Suppose you had three objects named x, y, and z that you would like to save to a permanent file. Regardless of whether they are vectors, factors, lists, or data frames, they can be saved to a file named `mydata.RData` using the following command:

```
> save(x, y, z, file = "mydata.RData")
```

The `load()` command can recreate any data structures that have been saved to an `.RData` file. To load the `mydata.RData` file created in the preceding code, simply type:

```
> load("mydata.RData")
```

This will recreate the x, y, and z data structures in your R environment.

Be careful what you are loading! All data structures stored in the file you are importing with the load() command will be added to your workspace, even if they overwrite something else you are working on.

If you need to wrap up your R session in a hurry, the save.image() command will write your entire session to a file simply called .RData. By default, R will look for this file the next time you start R, and your session will be recreated just as you had left it.

After you've been working in an R session for some time, you may have accumulated a number of data structures. The listing function ls() returns a vector of all data structures currently in memory. For example, if you've been following along with the code in this chapter, the ls() function returns the following:

```
> ls()
[1] "blood"        "flu_status" "gender"       "m"
[5] "subject_name" "subject1"   "symptoms"
[9] "temperature"
```

R automatically clears all data structures from memory upon quitting the session, but for large objects, you may want to free up the memory sooner. The remove function rm() can be used for this purpose. For example, to eliminate the m and subject1 objects, simply type:

```
> rm(m, subject1)
```

The rm() function can also be supplied with a character vector of object names to remove. This works with the ls() function to clear the entire R session:

```
> rm(list = ls())
```

Be very careful when executing the preceding code, as you will not be prompted before your objects are removed!

Importing and saving data from CSV files

It is very common for public datasets to be stored in text files. Text files can be read on virtually any computer or operating system, which makes the format nearly universal. They can also be exported and imported from and to programs such as Microsoft Excel, providing a quick and easy way to work with spreadsheet data.

A **tabular** (as in "table") data file is structured in matrix form, such that each line of text reflects one example, and each example has the same number of features. The feature values on each line are separated by a predefined symbol known as a **delimiter**. Often, the first line of a tabular data file lists the names of the data columns. This is called a **header** line.

Perhaps the most common tabular text file format is the **comma-separated values** (**CSV**) file, which, as the name suggests, uses the comma as a delimiter. CSV files can be imported to and exported from many common applications. A CSV file representing the medical dataset constructed previously could be stored as:

```
subject_name,temperature,flu_status,gender,blood_type
John Doe,98.1,FALSE,MALE,O
Jane Doe,98.6,FALSE,FEMALE,AB
Steve Graves,101.4,TRUE, MALE,A
```

Given a patient data file named `pt_data.csv` located in the R working directory, the `read.csv()` function can be used as follows to load the file into R:

```
> pt_data <- read.csv("pt_data.csv", stringsAsFactors = FALSE)
```

This will read the CSV file into a data frame titled `pt_data`. Just as we had done previously when constructing a data frame, we need to use the `stringsAsFactors = FALSE` parameter to prevent R from converting all text variables to factors. Unless you are certain that every column in the CSV file is truly a factor, this step is better left to you, not R, to perform.

> If your dataset resides outside the R working directory, the full path to the CSV file (for example, `"/path/to/mydata.csv"`) can be used when calling the `read.csv()` function.

By default, R assumes that the CSV file includes a header line listing the names of the features in the dataset. If a CSV file does not have a header, specify the option `header = FALSE` as shown in the following command, and R will assign default feature names by numbering them as V1, V2, and so on:

```
> mydata <- read.csv("mydata.csv", stringsAsFactors = FALSE,
                   header = FALSE)
```

The `read.csv()` function is a special case of the `read.table()` function, which can read tabular data in many different forms, including other delimited formats such as **tab-separated values** (**TSV**). For more detailed information on the `read.table()` family of functions, refer to the R help page using the `?read.table` command.

To save a data frame to a CSV file, use the `write.csv()` function. If your data frame is named `pt_data`, simply enter:

```
> write.csv(pt_data, file = "pt_data.csv", row.names = FALSE)
```

This will write a CSV file with the name `pt_data.csv` to the R working folder. The `row.names` parameter overrides R's default setting, which is to output row names in the CSV file. Generally, this output is unnecessary and will simply inflate the size of the resulting file.

Exploring and understanding data

After collecting data and loading it into R data structures, the next step in the machine learning process involves examining the data in detail. It is during this step that you will begin to explore the data's features and examples, and realize the peculiarities that make your data unique. The better you understand your data, the better you will be able to match a machine learning model to your learning problem.

The best way to learn the process of data exploration is by example. In this section, we will explore the `usedcars.csv` dataset, which contains actual data about used cars advertised for sale on a popular US website in the year 2012.

> The `usedcars.csv` dataset is available for download on the Packt Publishing support page for this book. If you are following along with the examples, be sure that this file has been downloaded and saved to your R working directory.

Since the dataset is stored in CSV form, we can use the `read.csv()` function to load the data into an R data frame:

```
> usedcars <- read.csv("usedcars.csv", stringsAsFactors = FALSE)
```

Given the `usedcars` data frame, we will now assume the role of a data scientist who has the task of understanding the used car data. Although data exploration is a fluid process, the steps can be imagined as a sort of investigation in which questions about the data are answered. The exact questions may vary across projects, but the types of questions are always similar. You should be able to adapt the basic steps of this investigation to any dataset you like, whether large or small.

Exploring the structure of data

The first questions to ask in an investigation of a new dataset should be about how the dataset is organized. If you are fortunate, your source will provide a **data dictionary**, a document that describes the dataset's features. In our case, the used car data does not come with this documentation, so we'll need to create our own.

The `str()` function provides a method for displaying the structure of R objects, such as data frames, vectors, or lists. It can be used to create the basic outline for our data dictionary:

```
> str(usedcars)
'data.frame':    150 obs. of 6 variables:
 $ year        : int   2011 2011 2011 2011 ...
 $ model       : chr   "SEL" "SEL" "SEL" "SEL" ...
 $ price       : int   21992 20995 19995 17809 ...
 $ mileage     : int   7413 10926 7351 11613 ...
 $ color       : chr   "Yellow" "Gray" "Silver" "Gray" ...
 $ transmission: chr   "AUTO" "AUTO" "AUTO" "AUTO" ...
```

For such a simple command, we learn a wealth of information about the dataset. The statement `150 obs` informs us that the data includes 150 **observations**, which is just another way of saying that the dataset contains 150 records or examples. The number of observations is often simply abbreviated as n. Since we know that the data describes used cars, we can now presume that we have examples of $n = 150$ automobiles for sale.

The `6 variables` statement refers to the six features that were recorded in the data. These features are listed by name on separate lines. Looking at the line for the feature called `color`, we note some additional details:

```
 $ color       : chr   "Yellow" "Gray" "Silver" "Gray" ...
```

After the variable's name, the `chr` label tells us that the feature is `character` type. In this dataset, three of the variables are character while three are noted as `int`, which refers to the `integer` type. Although the `usedcars` dataset includes only character and integer variables, you are also likely to encounter `num`, or `numeric` type when using non-integer data. Any factors would be listed as `factor` type. Following each variable's type, R presents a sequence of the first few feature values. The values `"Yellow" "Gray" "Silver" "Gray"` are the first four values of the `color` feature.

Applying a bit of subject-area knowledge to the feature names and values allows us to make some assumptions about what the variables represent. The `year` variable could refer to the year the vehicle was manufactured, or it could specify the year the advertisement was posted. We'll have to investigate this feature later in more detail, since the four example values (`2011 2011 2011 2011`) could be used to argue for either possibility. The `model`, `price`, `mileage`, `color`, and `transmission` variables most likely refer to the characteristics of the car for sale.

Although our data appears to have been given meaningful variable names, this is not always the case. Sometimes datasets have features with nonsensical names or codes, like v1. In these cases, it may be necessary to do additional sleuthing to determine what a feature actually represents. Still, even with helpful feature names, it is always prudent to be skeptical about the provided labels. Let's investigate further.

Exploring numeric variables

To investigate the numeric variables in the used car data, we will employ a common set of measurements for describing values known as **summary statistics**. The summary() function displays several common summary statistics. Let's take a look at a single feature, year:

```
> summary(usedcars$year)
   Min. 1st Qu.  Median    Mean 3rd Qu.    Max.
   2000    2008    2009    2009    2010    2012
```

Ignoring the meaning of the values for now, the fact that we see numbers such as 2000, 2008, and 2009 leads us to believe that the year variable indicates the year of manufacture rather than the year the advertisement was posted, since we know the vehicle listings were obtained in 2012.

By supplying a vector of column names, we can also use the summary() function to obtain summary statistics for several numeric variables at the same time:

```
> summary(usedcars[c("price", "mileage")])
    price             mileage
 Min.   : 3800   Min.   :  4867
 1st Qu.:10995   1st Qu.: 27200
 Median :13592   Median : 36385
 Mean   :12962   Mean   : 44261
 3rd Qu.:14904   3rd Qu.: 55125
 Max.   :21992   Max.   :151479
```

The six summary statistics that the summary() function provides are simple, yet powerful tools for investigating data. They can be divided into two types: measures of center and measures of spread.

Measuring the central tendency – mean and median

Measures of **central tendency** are a class of statistics used to identify a value that falls in the middle of a set of data. You are most likely already familiar with one common measure of center: the average. In common use, when something is deemed average, it falls somewhere between the extreme ends of the scale.

An average student might have marks falling in the middle of his or her classmates. An average weight is neither unusually light nor heavy. In general, an average item is typical and not too unlike the others in its group. You might think of it as an exemplar by which all the others are judged.

In statistics, the average is also known as the **mean**, which is a measurement defined as the sum of all values divided by the number of values. For example, to calculate the mean income in a group of three people with incomes of $36,000, $44,000, and $56,000 we could type:

```
> (36000 + 44000 + 56000) / 3
[1] 45333.33
```

R also provides a mean() function, which calculates the mean for a vector of numbers:

```
> mean(c(36000, 44000, 56000))
[1] 45333.33
```

The mean income of this group of people is about $45,333. Conceptually, this can be imagined as the income each person would have if the total amount of income was divided equally across every person.

Recall that the preceding summary() output listed mean values for the price and mileage variables. These values suggest that the typical used car in this dataset was listed at a price of $12,962 and had an odometer reading of 44,261. What does this tell us about our data? We can note that because the average price is relatively low, we might expect that the dataset contains economy-class cars. Of course, the data can also include late-model luxury cars with high mileage, but the relatively low mean mileage statistic doesn't provide evidence to support this hypothesis. On the other hand, it doesn't provide evidence to ignore the possibility either. We'll need to keep this in mind as we examine the data further.

Although the mean is by far the most commonly cited statistic for measuring the center of a dataset, it is not always the most appropriate one. Another commonly used measure of central tendency is the **median**, which is the value that occurs at the midpoint of an ordered list of values. As with the mean, R provides a median() function, which we can apply to our salary data as shown in the following example:

```
> median(c(36000, 44000, 56000))
[1] 44000
```

So, because the middle value is 44000, the median income is $44,000.

If a dataset has an even number of values, there is no middle value. In this case, the median is commonly calculated as the average of the two values at the center of the ordered list. For example, the median of the values 1, 2, 3, and 4 is 2.5.

At first glance, it seems like the median and mean are very similar measures. Certainly, the mean value of $45,333 and the median value of $44,000 are not very far apart. Why have two measures of central tendency? The reason is due to the fact that the mean and median are affected differently by values falling at far ends of the range. In particular, the mean is highly sensitive to **outliers**, or values that are atypically high or low relative to the majority of data. So, because the mean is more responsive to outliers, it is more likely to be shifted higher or lower by a small number of extreme values.

Recall again the reported median values in the summary() output for the used car dataset. Although the mean and median for price are fairly similar (differing by approximately five percent), there is a much larger difference between the mean and median for mileage. For mileage, the mean of 44,261 is more than 20 percent larger than the median of 36,385. Since the mean is more sensitive to extreme values than the median, the fact that the mean is much higher than the median might lead us to suspect that there are some used cars in the dataset with extremely high mileage values. To investigate this further, we'll need to add additional summary statistics to our analysis.

Measuring spread – quartiles and the five-number summary

The mean and median provide ways to quickly summarize values, but these measures of center tell us little about whether or not there is diversity in the measurements. To measure the diversity, we need to employ another type of summary statistics concerned with the **spread** of the data, or how tightly or loosely the values are spaced. Knowing about the spread provides a sense of the data's highs and lows, and whether most values are like or unlike the mean and median.

The **five-number summary** is a set of five statistics that roughly depict the spread of a feature's values. All five of the statistics are included in the output of the summary() function. Written in order, they are:

1. Minimum (Min.)
2. First quartile, or Q1 (1st Qu.)
3. Median, or Q2 (Median)

4. Third quartile, or Q3 (`3rd Qu.`)

5. Maximum (`Max.`)

As you would expect, the minimum and maximum are the most extreme feature values, indicating the smallest and largest values respectively. R provides the `min()` and `max()` functions to calculate these for a vector.

The span between the minimum and maximum value is known as the **range**. In R, the `range()` function returns both the minimum and maximum value:

```
> range(usedcars$price)
[1]  3800 21992
```

Combining `range()` with the `diff()` difference function allows you to compute the range statistic with a single line of code:

```
> diff(range(usedcars$price))
[1] 18192
```

The first and third quartiles, Q1 and Q3, refer to the value below and above which one quarter of the values are found. Along with the median (Q2), the **quartiles** divide a dataset into four portions, each with the same number of values.

Quartiles are a special case of a type of statistic called **quantiles**, which are numbers that divide data into equally sized quantities. In addition to quartiles, commonly used quantiles include **tertiles** (three parts), **quintiles** (five parts), **deciles** (10 parts), and **percentiles** (100 parts).

 Percentiles are often used to describe the ranking of a value; for instance, a student whose test score was ranked at the 99th percentile performed better than or equal to 99 percent of the other test takers.

The middle 50 percent of data, found between the first and third quartiles, is of particular interest because it is a simple measure of spread. The difference between Q1 and Q3 is known as the **interquartile range (IQR)**, and can be calculated with the `IQR()` function:

```
> IQR(usedcars$price)
[1] 3909.5
```

We could have also calculated this value by hand from the `summary()` output for the `usedcars$price` variable by computing *14904 – 10995 = 3909*. The small difference between our calculation and the `IQR()` output is due to the fact that R automatically rounds the `summary()` output.

The `quantile()` function provides a versatile tool for identifying quantiles for a set of values. By default, the `quantile()` function returns the five-number summary. Applying the function to the `usedcars$price` variable results in the same summary statistics as before:

```
> quantile(usedcars$price)
     0%      25%      50%      75%     100%
 3800.0  10995.0  13591.5  14904.5  21992.0
```

 When computing quantiles, there are many methods for handling ties among sets of values with no single middle value. The `quantile()` function allows you to specify among nine different tie-breaking algorithms by specifying the `type` parameter. If your project requires a precisely defined quantile, it is important to read the function documentation using the `?quantile` command.

By specifying an additional `probs` parameter using a vector denoting cut points, we can obtain arbitrary quantiles, such as the 1st and 99th percentiles:

```
> quantile(usedcars$price, probs = c(0.01, 0.99))
      1%      99%
 5428.69 20505.00
```

The sequence function `seq()` generates vectors of evenly-spaced values. This makes it easy to obtain other slices of data, such as the quintiles (five groups) shown in the following command:

```
> quantile(usedcars$price, seq(from = 0, to = 1, by = 0.20))
     0%      20%      40%      60%      80%     100%
 3800.0  10759.4  12993.8  13992.0  14999.0  21992.0
```

Equipped with an understanding of the five-number summary, we can re-examine the used car `summary()` output. On the `price` variable, the minimum was $3,800 and the maximum was $21,992. Interestingly, the difference between the minimum and Q1 is about $7,000, as is the difference between Q3 and the maximum; yet, the difference from Q1 to the median to Q3 is roughly $2,000. This suggests that the lower and upper 25 percent of values are more widely dispersed than the middle 50 percent of values, which seem to be more tightly grouped around the center. We also see a similar trend with the `mileage` variable. As you will learn later in this chapter, this pattern of spread is common enough that it has been called a "normal" distribution of data.

The spread of the `mileage` variable also exhibits another interesting property — the difference between Q3 and the maximum is far greater than that between the minimum and Q1. In other words, the larger values are far more spread out than the smaller values.

This finding helps explain why the mean value is much greater than the median. Because the mean is sensitive to extreme values, it is pulled higher, while the median stays in relatively the same place. This is an important property, which becomes more apparent when the data is presented visually.

Visualizing numeric variables – boxplots

Visualizing numeric variables can be helpful for diagnosing data problems. A common visualization of the five-number summary is a **boxplot**, also known as a **box-and-whisker** plot. The boxplot displays the center and spread of a numeric variable in a format that allows you to quickly obtain a sense of the range and skew of a variable or compare it to other variables.

Let's take a look at a boxplot for the used car price and mileage data. To obtain a boxplot for a variable, we will use the `boxplot()` function. We will also specify a pair of extra parameters, `main` and `ylab`, to add a title to the figure and label the *y* axis (the vertical axis), respectively. The commands to create the `price` and `mileage` boxplots are:

```
> boxplot(usedcars$price, main = "Boxplot of Used Car Prices",
          ylab = "Price ($)")
> boxplot(usedcars$mileage, main = "Boxplot of Used Car Mileage",
          ylab = "Odometer (mi.)")
```

R will produce figures as follows:

Figure 2.1: Boxplots of used car price and mileage data

A boxplot depicts the five-number summary using horizontal lines and dots. The horizontal lines forming the box in the middle of each figure represent Q1, Q2 (the median), and Q3 when reading the plot from bottom to top. The median is denoted by the dark line, which lines up with $13,592 on the vertical axis for price and 36,385 mi. on the vertical axis for mileage.

In simple boxplots, such as those in the preceding diagram, the box width is arbitrary and does not illustrate any characteristic of the data. For more sophisticated analyses, it is possible to use the shape and size of the boxes to facilitate comparisons of the data across several groups. To learn more about such features, begin by examining the notch and varwidth options in the R boxplot() documentation by typing the ?boxplot command.

The minimum and maximum values can be illustrated using the whiskers that extend below and above the box; however, a widely used convention only allows the whiskers to extend to a minimum or maximum of 1.5 times the IQR below Q1 or above Q3. Any values that fall beyond this threshold are considered outliers and are denoted as circles or dots. For example, recall that the IQR for the price variable was 3,909 with Q1 of 10,995 and Q3 of 14,904. An outlier is therefore any value that is less than *10995 - 1.5 * 3909 = 5131.5* or greater than *14904 + 1.5 * 3909 = 20767.5*.

The price boxplot shows two outliers on both the high and low ends. On the mileage boxplot, there are no outliers on the low end and thus the bottom whisker extends to the minimum value of 4,867. On the high end, we see several outliers beyond the 100,000 mile mark. These outliers are responsible for our earlier finding, which noted that the mean value was much greater than the median.

Visualizing numeric variables – histograms

A **histogram** is another way to visualize the spread of a numeric variable. It is similar to a boxplot in that it divides the variable's values into a predefined number of portions or **bins** that act as containers for values. Their similarities end there, however. Where a boxplot creates four portions containing the same number of values but varying in range, a histogram uses a larger number of portions of identical range and allows the bins to contain different numbers of values.

We can create a histogram for the used car price and mileage data using the hist() function. As we did with the boxplot, we will specify a title for the figure using the main parameter and label the *x* axis with the xlab parameter. The commands to create the histograms are:

```
> hist(usedcars$price, main = "Histogram of Used Car Prices",
        xlab = "Price ($)")
> hist(usedcars$mileage, main = "Histogram of Used Car Mileage",
        xlab = "Odometer (mi.)")
```

This produces the following diagrams:

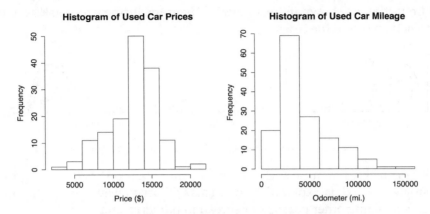

Figure 2.2: Histograms of used car price and mileage data

The histogram is composed of a series of bars with heights indicating the count, or **frequency**, of values falling within each of the equal-width bins partitioning the values. The vertical lines that separate the bars, as labeled on the horizontal axis, indicate the start and end points of the range of values falling within the bin.

You may have noticed that the preceding histograms have differing numbers of bins. This is because the hist() function attempts to identify the optimal number of bins for the variable's range. If you'd like to override this default, use the breaks parameter. Supplying an integer such as breaks = 10 creates exactly 10 bins of equal width, while supplying a vector such as c(5000, 10000, 15000, 20000) creates bins that break at the specified values.

On the price histogram, each of the 10 bars spans an interval of $2,000, beginning at $2,000 and ending at $22,000. The tallest bar in the center of the figure covers the range from $12,000 to $14,000 and has a frequency of 50. Since we know our data includes 150 cars, we know that one-third of all the cars are priced from $12,000 to $14,000. Nearly 90 cars—more than half—are priced from $12,000 to $16,000.

The mileage histogram includes eight bars representing bins of 20,000 miles each, beginning at 0 and ending at 160,000 miles. Unlike the price histogram, the tallest bar is not in the center of the data, but on the left-hand side of the diagram. The 70 cars contained in this bin have odometer readings from 20,000 to 40,000 miles.

You might also notice that the shape of the two histograms is somewhat different. It seems that the used car prices tend to be evenly divided on both sides of the middle, while the car mileages stretch further to the right.

This characteristic is known as **skew**, or more specifically *right* skew, because the values on the high end (right side) are far more spread out than the values on the low end (left side). As shown in the following diagram, histograms of skewed data look stretched on one of the sides:

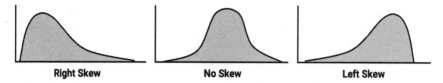

Figure 2.3: Three skew patterns visualized with idealized histograms

The ability to quickly diagnose such patterns in our data is one of the strengths of the histogram as a data exploration tool. This will become even more important as we start examining other patterns of spread in numeric data.

Understanding numeric data – uniform and normal distributions

Histograms, boxplots, and statistics describing the center and spread provide ways to examine the distribution of a variable's values. A variable's **distribution** describes how likely a value is to fall within various ranges.

If all values are equally likely to occur—say, for instance, in a dataset recording the values rolled on a fair six-sided die—the distribution is said to be **uniform**. A uniform distribution is easy to detect with a histogram because the bars are approximately the same height. The histogram may look something like the following diagram:

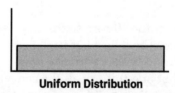

Figure 2.4: A uniform distribution visualized with an idealized histogram

It's important to note that not all random events are uniform. For instance, rolling a weighted six-sided trick die would result in some numbers coming up more often than others. While each roll of the die results in a randomly selected number, they are not equally likely.

Take, for instance, the used car `price` and `mileage` data. This is clearly not uniform, since some values are seemingly far more likely to occur than others. In fact, on the `price` histogram, it seems that values become less likely to occur as they are further away from both sides of the center bar, which results in a bell-shaped distribution of data. This characteristic is so common in real-world data that it is the hallmark of the so-called **normal distribution**. The stereotypical bell-shaped curve of the normal distribution is shown in the following diagram:

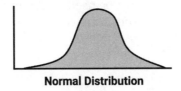

Normal Distribution

Figure 2.5: A normal distribution visualized with an idealized histogram

Although there are numerous types of non-normal distributions, many real-world phenomena generate data that can be described by the normal distribution. Therefore, the normal distribution's properties have been studied in great detail.

Measuring spread – variance and standard deviation

Distributions allow us to characterize a large number of values using a smaller number of parameters. The normal distribution, which describes many types of real-world data, can be defined with just two: center and spread. The center of the normal distribution is defined by its mean value, which we have used before. The spread is measured by a statistic called the **standard deviation**.

In order to calculate the standard deviation, we must first obtain the **variance**, which is defined as the average of the squared differences between each value and the mean value. In mathematical notation, the variance of a set of n values in a set named x is defined by the following formula. The Greek letter *mu* (similar in appearance to an *m* or *u*) denotes the mean of the values, and the variance itself is denoted by the Greek letter *sigma* squared (similar to a *b* turned sideways):

$$\text{Var}(X) = \sigma^2 = \frac{1}{n}\sum_{i=1}^{n}(x_i - \mu)^2$$

The standard deviation is the square root of the variance, and is denoted by *sigma* as shown in the following formula:

$$\text{StdDev}(X) = \sigma = \sqrt{\frac{1}{n}\sum_{i=1}^{n}(x_i - \mu)^2}$$

In R, the `var()` and `sd()` functions can be used to obtain the variance and standard deviation. For example, computing the variance and standard deviation on the `price` and `mileage` vectors, we find:

```
> var(usedcars$price)
[1] 9749892
> sd(usedcars$price)
[1] 3122.482
> var(usedcars$mileage)
[1] 728033954
> sd(usedcars$mileage)
[1] 26982.1
```

When interpreting the variance, larger numbers indicate that the data is spread more widely around the mean. The standard deviation indicates, on average, how much each value differs from the mean.

If you compute these statistics by hand using the formulas in the preceding diagrams, you will obtain a slightly different result than the built-in R functions. This is because the preceding formulas use the population variance (which divides by *n*), while R uses the sample variance (which divides by *n* - 1). Except for very small datasets, the distinction is minor.

The standard deviation can be used to quickly estimate how extreme a given value is under the assumption that it came from a normal distribution. The **68–95–99.7 rule** states that 68 percent of values in a normal distribution fall within one standard deviation of the mean, while 95 percent and 99.7 percent of values fall within two and three standard deviations, respectively. This is illustrated in the following diagram:

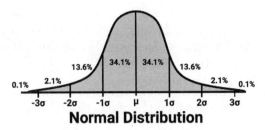

Normal Distribution

Figure 2.6: The percent of values within one, two, and three
standard deviations of a normal distribution's mean

Applying this information to the used car data, we know that the mean and
standard deviation of `price` were $12,962 and $3,122 respectively. Therefore,
by assuming that the prices are normally distributed, approximately 68 percent
of cars in our data were advertised at prices between *$12,962 - $3,122 = $9,840* and
$12,962 + $3,122 = $16,804.

Although, strictly speaking, the 68–95–99.7 rule only applies to normal
distributions, the basic principle applies to any data; values more than
three standard deviations away from the mean are exceedingly rare events.

Exploring categorical variables

If you recall, the used car dataset contains three categorical variables: `model`, `color`,
and `transmission`. R has stored these as `character` (`chr`) vectors rather than
`factor` type because we used the `stringsAsFactors = FALSE` parameter when
loading the data. Additionally, although the `year` variable is stored as a numeric
(`int`) vector, each year can be imagined as a category applied to multiple cars.
We may therefore consider also treating it as categorical.

In contrast to numeric data, categorical data is typically examined using tables rather
than summary statistics. A table that presents a single categorical variable is known
as a **one-way table**. The `table()` function can be used to generate one-way tables
for the used car data:

```
> table(usedcars$year)

2000 2001 2002 2003 2004 2005 2006 2007 2008 2009 2010 2011 2012
   3    1    1    1    3    2    6   11   14   42   49   16    1
> table(usedcars$model)
 SE SEL SES
78  23  49
> table(usedcars$color)
 Black   Blue   Gold   Gray  Green    Red Silver  White Yellow
    35     17      1     16      5     25     32     16      3
```

The `table()` output lists the categories of the nominal variable and a count of the number of values falling into each category. Since we know there are 150 used cars in the dataset, we can determine that roughly one-third of all the cars were manufactured in 2010, given that *49 / 150 = 0.327*.

R can also perform the calculation of table proportions directly, by using the `prop.table()` command on a table produced by the `table()` function:

```
> model_table <- table(usedcars$model)
> prop.table(model_table)

       SE        SEL       SES
0.5200000 0.1533333 0.3266667
```

The results of `prop.table()` can be combined with other R functions to transform the output. Suppose we would like to display the results in percentages with a single decimal place. We can do this by multiplying the proportions by 100, then using the `round()` function while specifying `digits = 1`, as shown in the following example:

```
> color_table <- table(usedcars$color)
> color_pct <- prop.table(color_table) * 100
> round(color_pct, digits = 1)
Black   Blue   Gold   Gray  Green    Red Silver  White Yellow
 23.3   11.3    0.7   10.7    3.3   16.7   21.3   10.7    2.0
```

Although this includes the same information as the default `prop.table()` output, the changes make it easier to read. The results show that black is the most common color, with nearly a quarter (23.3 percent) of all advertised cars. Silver is a close second with 21.3 percent, and red is third with 16.7 percent.

Measuring the central tendency – the mode

In statistics terminology, the **mode** of a feature is the value occurring most often. Like the mean and median, the mode is another measure of central tendency. It is typically used for categorical data, since the mean and median are not defined for nominal variables.

For example, in the used car data, the mode of the `year` variable is 2010, while the modes for the `model` and `color` variables are SE and Black, respectively. A variable may have more than one mode; a variable with a single mode is **unimodal**, while a variable with two modes is **bimodal**. Data with multiple modes is more generally called **multimodal**.

Although you might suspect that you could use the `mode()` function, R uses this to obtain the type of variable (as in numeric, list, and so on) rather than the statistical mode. Instead, to find the statistical mode, simply look at the `table()` output for the category with the greatest number of values.

The mode or modes are used in a qualitative sense to gain an understanding of important values. Even so, it would be dangerous to place too much emphasis on the mode since the most common value is not necessarily a majority. For instance, although black was the single most common car color, it was only about a quarter of all advertised cars.

It is best to think about the modes in relation to the other categories. Is there one category that dominates all others, or are there several? Thinking about modes this way may help to generate testable hypotheses by raising questions about what makes certain values more common than others. If black and silver are common used car colors, we might believe that the data represents luxury cars, which tend to be sold in more conservative colors. Alternatively, these colors could indicate economy cars, which are sold with fewer color options. We will keep these questions in mind as we continue to examine this data.

Thinking about the modes as common values allows us to apply the concept of the statistical mode to numeric data. Strictly speaking, it would be unlikely to have a mode for a continuous variable, since no two values are likely to repeat. However, if we think about modes as the highest bars on a histogram, we can discuss the modes of variables such as `price` and `mileage`. It can be helpful to consider the mode when exploring numeric data, particularly to examine whether or not the data is multimodal.

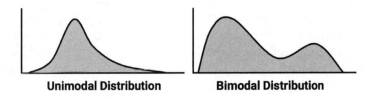

Unimodal Distribution   **Bimodal Distribution**

Figure 2.7: Hypothetical distributions of numeric data with one and two modes

Exploring relationships between variables

So far, we have examined variables one at a time, calculating only **univariate** statistics. During our investigation, we raised questions that we were unable to answer before:

- Does the `price` and `mileage` data imply that we are examining only economy-class cars, or are there luxury cars with high mileage?

- Do relationships between `model` and `color` provide insight into the types of cars we are examining?

These types of questions can be addressed by looking at **bivariate** relationships, which consider the relationship between two variables. Relationships of more than two variables are called **multivariate** relationships. Let's begin with the bivariate case.

Visualizing relationships – scatterplots

A **scatterplot** is a diagram that visualizes a bivariate relationship between numeric features. It is a two-dimensional figure in which dots are drawn on a coordinate plane using the values of one feature to provide the horizontal *x* coordinates, and the values of another feature to provide the vertical *y* coordinates. Patterns in the placement of dots reveal underlying associations between the two features.

To answer our question about the relationship between `price` and `mileage`, we will examine a scatterplot. We'll use the `plot()` function, along with the `main`, `xlab`, and `ylab` parameters used previously to label the diagram.

To use `plot()`, we need to specify `x` and `y` vectors containing the values used to position the dots on the figure. Although the conclusions would be the same regardless of the variable used to supply the *x* and *y* coordinates, convention dictates that the *y* variable is the one that is presumed to depend on the other (and is therefore known as the **dependent variable**). Since a seller cannot modify a car's odometer reading, mileage is unlikely to be dependent on the car's price. Instead, our hypothesis is that a car's price depends on the odometer mileage. Therefore, we will select `price` as the dependent *y* variable.

The full command to create our scatterplot is:

```
> plot(x = usedcars$mileage, y = usedcars$price,
       main = "Scatterplot of Price vs. Mileage",
       xlab = "Used Car Odometer (mi.)",
       ylab = "Used Car Price ($)")
```

This results in the following scatterplot:

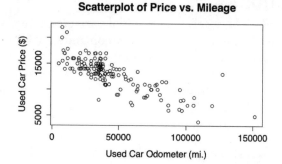

Figure 2.8: The relationship between used car price and mileage

Using the scatterplot, we notice a clear relationship between the price of a used car and the odometer reading. To read the plot, examine how the values of the y axis variable change as the values on the x axis increase. In this case, car prices tend to be lower as the mileage increases. If you have ever sold or shopped for a used car, this is not a profound insight.

Perhaps a more interesting finding is the fact that there are very few cars that have both high price and high mileage, aside from a lone outlier at about 125,000 miles and $14,000. The absence of more points like this provides evidence to support a conclusion that our dataset is unlikely to include any high-mileage luxury cars. All of the most expensive cars in the data, particularly those above $17,500, seem to have extraordinarily low mileage, which implies that we could be looking at a single type of car that retails for a price around $20,000 when new.

The relationship we've observed between car prices and mileage is known as a negative association because it forms a pattern of dots in a line sloping downward. A positive association would appear to form a line sloping upward. A flat line, or a seemingly random scattering of dots, is evidence that the two variables are not associated at all. The strength of a linear association between two variables is measured by a statistic known as **correlation**. Correlations are discussed in detail in *Chapter 6, Forecasting Numeric Data – Regression Methods*, which covers methods for modeling linear relationships.

 Keep in mind that not all associations form straight lines. Sometimes the dots form a U shape or V shape, while sometimes the pattern seems to be weaker or stronger for increasing values of the x or y variable. Such patterns imply that the relationship between the two variables is not linear.

Examining relationships – two-way cross-tabulations

To examine a relationship between two nominal variables, a **two-way cross-tabulation** is used (also known as a **crosstab** or **contingency table**). A cross-tabulation is similar to a scatterplot in that it allows you to examine how the values of one variable vary by the values of another. The format is a table in which the rows are the levels of one variable, while the columns are the levels of another. Counts in each of the table's cells indicate the number of values falling into the particular row and column combination.

To answer our earlier question about whether there is a relationship between `model` and `color`, we will examine a crosstab. There are several functions to produce two-way tables in R, including `table()`, which we used for one-way tables. The `CrossTable()` function in the `gmodels` package by Gregory R. Warnes is perhaps the most user-friendly, as it presents the row, column, and margin percentages in a single table, saving us the trouble of computing them ourselves. To install the `gmodels` package, type:

```
> install.packages("gmodels")
```

After the package installs, type `library(gmodels)` to load the package. You will need to do this during each R session in which you plan to use the `CrossTable()` function.

Before proceeding with our analysis, let's simplify our project by reducing the number of levels in the `color` variable. This variable has nine levels, but we don't really need this much detail. What we are actually interested in is whether or not the car's color is conservative. Toward this end, we'll divide the nine colors into two groups—the first group will include the conservative colors `Black`, `Gray`, `Silver`, and `White`; the second group will include `Blue`, `Gold`, `Green`, `Red`, and `Yellow`. We will create a binary indicator variable (often called a **dummy variable**), indicating whether or not the car's color is conservative by our definition. Its value will be `1` if true and `0` otherwise:

```
> usedcars$conservative <-
    usedcars$color %in% c("Black", "Gray", "Silver", "White")
```

You may have noticed a new command here: the `%in%` operator returns `TRUE` or `FALSE` for each value in the vector on the left-hand side of the operator, indicating whether the value is found in the vector on the right-hand side. In simple terms, you can translate this line as "is the used car color in the set of black, gray, silver, and white?"

Examining the `table()` output for our newly created variable, we see that about two-thirds of the cars have conservative colors while one-third does not:

```
> table(usedcars$conservative)
FALSE   TRUE
   51     99
```

Now, let's look at a cross-tabulation to see how the proportion of conservatively-colored cars varies by model. Since we're assuming that the model of car dictates the choice of color, we'll treat the conservative color indicator as the dependent (y) variable. The `CrossTable()` command is therefore:

```
> CrossTable(x = usedcars$model, y = usedcars$conservative)
```

This results in the following table:

```
   Cell Contents
|-------------------------|
|                       N |
| Chi-square contribution |
|           N / Row Total |
|           N / Col Total |
|         N / Table Total |
|-------------------------|

Total Observations in Table:  150

              | usedcars$conservative
usedcars$model |     FALSE |      TRUE | Row Total |
--------------|-----------|-----------|-----------|
           SE |        27 |        51 |        78 |
              |     0.009 |     0.004 |           |
              |     0.346 |     0.654 |     0.520 |
              |     0.529 |     0.515 |           |
              |     0.180 |     0.340 |           |
--------------|-----------|-----------|-----------|
          SEL |         7 |        16 |        23 |
              |     0.086 |     0.044 |           |
              |     0.304 |     0.696 |     0.153 |
              |     0.137 |     0.162 |           |
              |     0.047 |     0.107 |           |
--------------|-----------|-----------|-----------|
          SES |        17 |        32 |        49 |
              |     0.007 |     0.004 |           |
              |     0.347 |     0.653 |     0.327 |
              |     0.333 |     0.323 |           |
              |     0.113 |     0.213 |           |
--------------|-----------|-----------|-----------|
 Column Total |        51 |        99 |       150 |
              |     0.340 |     0.660 |           |
--------------|-----------|-----------|-----------|
```

There is a wealth of data in the `CrossTable()` output. The legend at the top (labeled `Cell Contents`) indicates how to interpret each value. The table rows indicate the three models of used cars: SE, SEL, and SES (plus an additional row for the total across all models). The columns indicate whether or not the car's color is conservative (plus a column totaling across both types of color).

The first value in each cell indicates the number of cars with that combination of model and color. The proportions indicate each cell's contribution to the chi-square statistic, the row total, the column total, and the table's overall total.

What we are most interested in is the proportion of conservative cars for each model. The row proportions tell us that 0.654 (65 percent) of SE cars are colored conservatively, in comparison to 0.696 (70 percent) of SEL cars, and 0.653 (65 percent) of SES. These differences are relatively small, which suggests that there are no substantial differences in the types of colors chosen for each model of car.

The chi-square values refer to the cell's contribution in the **Pearson's chi-squared test for independence** between two variables. This test measures how likely it is that the difference in cell counts in the table is due to chance alone. If the probability is very low, it provides strong evidence that the two variables are associated.

You can obtain the chi-squared test results by adding an additional parameter specifying chisq = TRUE when calling the CrossTable() function. In this case, the probability is about 93 percent, suggesting that it is very likely that the variations in cell count are due to chance alone, and not due to a true association between model and color.

Summary

In this chapter, we learned about the basics of managing data in R. We started by taking an in-depth look at the structures used for storing various types of data. The foundational R data structure is the vector, which is extended and combined into more complex data types, such as lists and data frames. The data frame is an R data structure that corresponds to the notion of a dataset having both features and examples. R provides functions for reading and writing data frames to spreadsheet-like tabular data files.

We then explored a real-world dataset containing prices of used cars. We examined numeric variables using common summary statistics of center and spread, and visualized relationships between prices and odometer readings with a scatterplot. Next, we examined nominal variables using tables. In examining the used car data, we followed an exploratory process that can be used to understand any dataset. These skills will be required for the other projects throughout this book.

Now that we have spent some time understanding the basics of data management with R, you are ready to begin using machine learning to solve real-world problems. In the next chapter, we will tackle our first classification task using nearest neighbor methods.

3

Lazy Learning – Classification Using Nearest Neighbors

A curious type of dining experience has appeared in cities around the world. Patrons are served in a completely darkened restaurant by waiters who move via memorized routes, using only their senses of touch and sound. The allure of these establishments is the belief that depriving oneself of sight will enhance the senses of taste and smell, and foods will be experienced in new ways. Each bite provides a sense of wonder while discovering the flavors the chef has prepared.

Can you imagine how a diner experiences the unseen food? Upon first bite, the senses are overwhelmed. What are the dominant flavors? Does the food taste savory or sweet? Does it taste similar to something eaten previously? Personally, I imagine this process of discovery in terms of a slightly modified adage—if it smells like a duck and tastes like a duck, then you are probably eating duck.

This illustrates an idea that can be used for machine learning—as does another maxim involving poultry—birds of a feather flock together. Stated differently, things that are alike are likely to have properties that are alike. Machine learning uses this principle to classify data by placing it in the same category as similar, or "nearest" neighbors. This chapter is devoted to classifiers that use this approach. You will learn:

- The key concepts that define nearest neighbor classifiers and why they are considered "lazy" learners

- Methods to measure the similarity of two examples using distance

- How to apply a popular nearest neighbor classifier called k-NN

If all of this talk about food is making you hungry, our first task will be to understand the k-NN approach by putting it to use while we settle a long-running culinary debate.

Understanding nearest neighbor classification

In a single sentence, **nearest neighbor** classifiers are defined by their characteristic of classifying unlabeled examples by assigning them the class of similar labeled examples. This is analogous to the dining experience described in the chapter introduction, in which a person identifies new foods through comparison to those previously encountered. With nearest neighbor classification, computers apply a human-like ability to recall past experiences to make conclusions about current circumstances. Despite the simplicity of this idea, nearest neighbor methods are extremely powerful. They have been used successfully for:

- Computer vision applications, including optical character recognition and facial recognition in both still images and video

- Recommendation systems that predict whether a person will enjoy a movie or song

- Identifying patterns in genetic data to detect specific proteins or diseases

In general, nearest neighbor classifiers are well suited for classification tasks where relationships among the features and the target classes are numerous, complicated, or otherwise extremely difficult to understand, yet the items of similar class type tend to be fairly homogeneous. Another way of putting it would be to say that if a concept is difficult to define, but you know it when you see it, then nearest neighbors might be appropriate. On the other hand, if the data is noisy and thus no clear distinction exists among the groups, nearest neighbor algorithms may struggle to identify the class boundaries.

The k-NN algorithm

The nearest neighbors approach to classification is exemplified by the **k-nearest neighbors** algorithm (**k-NN**). Although this is perhaps one of the simplest machine learning algorithms, it is still used widely.

The strengths and weaknesses of this algorithm are as follows:

Strengths	Weaknesses
• Simple and effective • Makes no assumptions about the underlying data distribution • Fast training phase	• Does not produce a model, limiting the ability to understand how the features are related to the class • Requires selection of an appropriate k • Slow classification phase • Nominal features and missing data require additional processing

The k-NN algorithm gets its name from the fact that it uses information about an example's k nearest neighbors to classify unlabeled examples. The letter k is a variable term implying that any number of nearest neighbors could be used. After choosing k, the algorithm requires a training dataset made up of examples that have been classified into several categories, as labeled by a nominal variable. Then, for each unlabeled record in the test dataset, k-NN identifies the k records in the training data that are the "nearest" in similarity. The unlabeled test instance is assigned the class representing the majority of the k nearest neighbors.

To illustrate this process, let's revisit the blind tasting experience described in the introduction. Suppose that prior to eating the mystery meal, we had created a dataset in which we recorded our impressions of a number of previously-tasted ingredients. To keep things simple, we rated only two features of each ingredient. The first is a measure from 1 to 10 of how crunchy the ingredient is, and the second is a 1 to 10 score of how sweet the ingredient tastes. We then labeled each ingredient as one of three types of food: fruits, vegetables, or proteins, ignoring other foods such as grains and fats.

The first few rows of such a dataset might be structured as follows:

Ingredient	Sweetness	Crunchiness	Food type
Apple	10	9	Fruit
Bacon	1	4	Protein
Banana	10	1	Fruit
Carrot	7	10	Vegetable
Celery	3	10	Vegetable
Cheese	1	1	Protein

The k-NN algorithm treats the features as coordinates in a multidimensional **feature space**. Because the ingredient dataset includes only two features, its feature space is two-dimensional. We can plot two-dimensional data on a scatterplot, with the *x* dimension indicating the ingredient's sweetness and the *y* dimension indicating the crunchiness. After adding a few more ingredients to the taste dataset, the scatterplot might look like this:

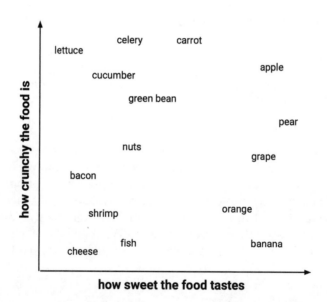

Figure 3.1: A scatterplot of selected foods' crunchiness versus sweetness

Do you notice a pattern? Similar types of food tend to be grouped closely together. As illustrated in the next figure, vegetables tend to be crunchy but not sweet; fruits tend to be sweet and either crunchy or not crunchy; and proteins tend to be neither crunchy nor sweet:

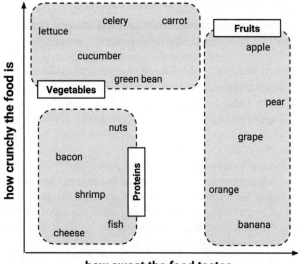

Figure 3.2: Foods that are similarly classified tend to have similar attributes

Suppose that after constructing this dataset, we decide to use it to settle the age-old question: is a tomato a fruit or a vegetable? We can use the nearest neighbor approach to determine which class is a better fit, as shown in the following diagram:

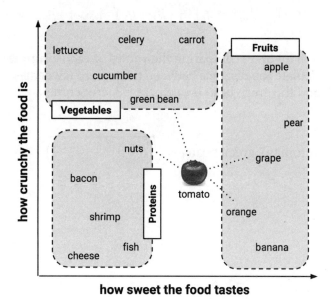

Figure 3.3: The tomato's nearest neighbors provide insight into whether it is a fruit or vegetable

Measuring similarity with distance

Locating the tomato's nearest neighbors requires a **distance function**, which is a formula that measures the similarity between two instances.

There are many different ways to calculate distance. Traditionally, the k-NN algorithm uses **Euclidean distance**, which is the distance one would measure if it were possible to use a ruler to connect two points. This is illustrated in the previous figure by the dotted lines connecting the tomato to its neighbors.

Euclidean distance is measured "as the crow flies," implying the shortest direct route. Another common distance measure is **Manhattan distance**, which is based on the paths a pedestrian would take by walking city blocks. If you are interested in learning more about other distance measures, you can read the documentation for R's distance function using the `?dist` command.

Euclidean distance is specified by the following formula, where p and q are the examples to be compared, each having n features. The term p_1 refers to the value of the first feature of example p, while q_1 refers to the value of the first feature of example q:

$$\text{dist}(p,q) = \sqrt{(p_1 - q_1)^2 + (p_2 - q_2)^2 + \ldots + (p_n - q_n)^2}$$

The distance formula involves comparing the values of each example's features. For example, to calculate the distance between the tomato (sweetness = 6, crunchiness = 4), and the green bean (sweetness = 3, crunchiness = 7), we can use the formula as follows:

$$\text{dist}(\text{tomato}, \text{green bean}) = \sqrt{(6-3)^2 + (4-7)^2} = 4.2$$

In a similar vein, we can calculate the distance between the tomato and several of its closest neighbors as follows:

Ingredient	Sweetness	Crunchiness	Food type	Distance to the tomato
Grape	8	5	Fruit	sqrt((6 - 8)^2 + (4 - 5)^2) = 2.2
Green bean	3	7	Vegetable	sqrt((6 - 3)^2 + (4 - 7)^2) = 4.2
Nuts	3	6	Protein	sqrt((6 - 3)^2 + (4 - 6)^2) = 3.6
Orange	7	3	Fruit	sqrt((6 - 7)^2 + (4 - 3)^2) = 1.4

To classify the tomato as a vegetable, protein, or fruit, we'll begin by assigning the tomato the food type of its single nearest neighbor. This is called 1-NN classification because $k = 1$. The orange is the single nearest neighbor to the tomato, with a distance of 1.4. As orange is a fruit, the 1-NN algorithm would classify a tomato as a fruit.

If we use the k-NN algorithm with $k = 3$ instead, it performs a vote among the three nearest neighbors: orange, grape, and nuts. Now, because the majority class among these neighbors is fruit (two of the three votes), the tomato again is classified as a fruit.

Choosing an appropriate k

The decision of how many neighbors to use for k-NN determines how well the model will generalize to future data. The balance between overfitting and underfitting the training data is a problem known as the **bias-variance tradeoff**. Choosing a large k reduces the impact or variance caused by noisy data but can bias the learner such that it runs the risk of ignoring small, but important patterns.

Suppose we took the extreme stance of setting a very large k, as large as the total number of observations in the training data. With every training instance represented in the final vote, the most common class always has a majority of the voters. The model would consequently always predict the majority class, regardless of the nearest neighbors.

On the opposite extreme, using a single nearest neighbor allows noisy data and outliers to unduly influence the classification of examples. For example, suppose some of the training examples were accidentally mislabeled. Any unlabeled example that happens to be nearest to the incorrectly labeled neighbor will be predicted to have the incorrect class, even if nine other nearby neighbors would have voted differently.

Obviously, the best k value is somewhere between these two extremes.

The following figure illustrates, more generally, how the decision boundary (depicted by a dashed line) is affected by larger or smaller k values. Smaller values allow more complex decision boundaries that more carefully fit the training data. The problem is that we do not know whether the straight boundary or the curved boundary better represents the true underlying concept to be learned.

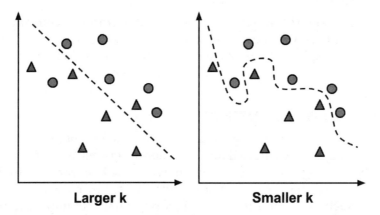

Larger k   **Smaller k**

Figure 3.4: A larger k has higher bias and lower variance than a smaller k

In practice, the choice of k depends on the difficulty of the concept to be learned and the number of records in the training data. One common practice is to begin with k equal to the square root of the number of training examples. In the food classifier that we developed previously, we might set $k = 4$ because there were 15 example ingredients in the training data and the square root of 15 is 3.87.

However, such rules may not always result in the single best k. An alternative approach is to test several k values on a variety of test datasets and choose the one that delivers the best classification performance. That said, unless the data is very noisy, a large training dataset can make the choice of k less important. This is because even subtle concepts will have a sufficiently large pool of examples to vote as nearest neighbors.

A less common, but interesting solution to this problem is to choose a larger *k* and use a **weighted voting** process in which the vote of closer neighbors is considered more authoritative than the vote of neighbors that are far away. Some k-NN implementations offer this option.

Preparing data for use with k-NN

Features are typically transformed to a standard range prior to applying the k-NN algorithm. The rationale for this step is that the distance formula is highly dependent on how features are measured. In particular, if certain features have a much larger range of values than others, the distance measurements will be strongly dominated by the features with larger ranges. This wasn't a problem for the food tasting example, as both sweetness and crunchiness were measured on a scale from 1 to 10.

However, suppose that we added an additional feature to the dataset to represent a food's spiciness, which was measured using the Scoville scale. If you are not familiar with this metric, it is a standardized measure of spice heat, ranging from zero (not at all spicy) to over a million (for the hottest chili peppers). Since the difference between spicy and non-spicy foods can be over a million, while the difference between sweet and non-sweet, or crunchy and non-crunchy, foods is at most 10, the difference in scale allows the spice level to impact the distance function much more than the other two factors. Without adjusting our data, we might find that our distance measures only differentiate foods by their spiciness; the impact of crunchiness and sweetness would be dwarfed by the contribution of spiciness.

The solution is to rescale the features by shrinking or expanding their range such that each one contributes relatively equally to the distance formula. For example, if sweetness and crunchiness are both measured on a scale from 1 to 10, we would also like spiciness to be measured on a scale from 1 to 10. There are several common ways to accomplish such scaling.

The traditional method of rescaling features for k-NN is **min-max normalization**. This process transforms a feature such that all of its values fall in a range between 0 and 1. The formula for normalizing a feature is as follows:

$$X_{new} = \frac{X - \min(X)}{\max(X) - \min(X)}$$

Essentially, for each value of feature X, the formula subtracts the minimum X value and divides by the range of X.

The resulting normalized feature values can be interpreted as indicating how far, from 0 percent to 100 percent, the original value fell along the range between the original minimum and maximum.

Another common transformation is called **z-score standardization**. The following formula subtracts the mean value of feature X, and divides the result by the standard deviation of X:

$$X_{new} = \frac{X - \mu}{\sigma} = \frac{X - \text{Mean}(X)}{\text{StdDev}(X)}$$

This formula, which is based on properties of the normal distribution covered in *Chapter 2, Managing and Understanding Data*, rescales each of a feature's values in terms of how many standard deviations they fall above or below the mean. The resulting value is called a **z-score**. The z-scores fall in an unbounded range of negative and positive numbers. Unlike the normalized values, they have no predefined minimum and maximum.

The same rescaling method used on the k-NN training dataset must also be applied to the test examples that the algorithm will later classify. This can lead to a tricky situation for min-max normalization, as the minimum or maximum of future cases might be outside the range of values observed in the training data. If you know the plausible minimum or maximum value ahead of time, you can use these constants rather than the observed values. Alternatively, you can use z-score standardization under the assumption that the future examples will have similar mean and standard deviation as the training examples.

The Euclidean distance formula is not defined for nominal data. Therefore, to calculate the distance between nominal features, we need to convert them into a numeric format. A typical solution utilizes **dummy coding**, where a value of *1* indicates one category, and *0* indicates the other. For instance, dummy coding for a gender variable could be constructed as:

$$\text{male} = \begin{cases} 1 & \text{if } x = \text{male} \\ 0 & \text{otherwise} \end{cases}$$

Notice how dummy coding of the two-category (binary) gender variable results in a single new feature named male. There is no need to construct a separate feature for female. Since the two sexes are mutually exclusive, knowing one or the other is enough.

This is true more generally as well. An *n*-category nominal feature can be dummy coded by creating binary indicator variables for *n* - *1* levels of the feature. For example, dummy coding for a three-category temperature variable (for example, hot, medium, or cold) could be set up as *(3 - 1) = 2* features, as shown here:

$$\text{hot} = \begin{cases} 1 & \text{if } x = \text{hot} \\ 0 & \text{otherwise} \end{cases}$$

$$\text{medium} = \begin{cases} 1 & \text{if } x = \text{medium} \\ 0 & \text{otherwise} \end{cases}$$

Knowing that hot and medium are both *0* provides enough information to know that the temperature is cold. Therefore, we do not need a third feature for the cold category. Since only one attribute is coded as *1* and the others must all be *0*, dummy coding is also known as **one-hot encoding**.

A convenient aspect of dummy coding is that the distance between dummy-coded features is always one or zero, and thus, the values fall on the same scale as min-max normalized numeric data. No additional transformation is necessary.

If a nominal feature is ordinal (one could make such an argument for temperature) an alternative to dummy coding is to number the categories and apply normalization. For instance, cold, warm, and hot could be numbered as 1, 2, and 3, which normalizes to 0, 0.5, and 1. A caveat to this approach is that it should only be used if the steps between categories are equivalent. For instance, although income categories for poor, middle class, and wealthy are ordered, the difference between poor and middle class may be different than the difference between middle class and wealthy. Since the steps between groups are not equal, dummy coding is a safer approach.

Why is the k-NN algorithm lazy?

Classification algorithms based on nearest neighbor methods are considered **lazy learning** algorithms because, technically speaking, no abstraction occurs. The abstraction and generalization processes are skipped altogether, which undermines the definition of learning proposed in *Chapter 1, Introducing Machine Learning*.

Under the strict definition of learning, a lazy learner is not really learning anything. Instead, it merely stores the training data verbatim. This allows the training phase, which is not actually training anything, to occur very rapidly.

Of course, the downside is that the process of making predictions tends to be relatively slow by comparison. Due to the heavy reliance on the training instances rather than an abstracted model, lazy learning is also known as **instance-based learning** or **rote learning**.

As instance-based learners do not build a model, the method is said to be in a class of **non-parametric** learning methods — no parameters are learned about the data. Without generating theories about the underlying data, non-parametric methods limit our ability to understand how the classifier is using the data. On the other hand, this allows the learner to find natural patterns, rather than trying to fit the data into a preconceived and potentially biased functional form.

Figure 3.5: Machine learning algorithms have different biases and may come to different conclusions!

Although k-NN classifiers may be considered lazy, they are still quite powerful. As you will soon see, the simple principles of nearest neighbor learning can be used to automate the process of screening for cancer.

Example – diagnosing breast cancer with the k-NN algorithm

Routine breast cancer screening allows the disease to be diagnosed and treated prior to it causing noticeable symptoms. The process of early detection involves examining the breast tissue for abnormal lumps or masses. If a lump is found, a fine-needle aspiration biopsy is performed, which uses a hollow needle to extract a small sample of cells from the mass. A clinician then examines the cells under a microscope to determine whether the mass is likely to be malignant or benign.

If machine learning could automate the identification of cancerous cells, it would provide considerable benefit to the health system. Automated processes are likely to improve the efficiency of the detection process, allowing physicians to spend less time diagnosing and more time treating the disease. An automated screening system might also provide greater detection accuracy by removing the inherently subjective human component from the process.

We will investigate the utility of machine learning for detecting cancer by applying the k-NN algorithm to measurements of biopsied cells from women with abnormal breast masses.

Step 1 – collecting data

We will utilize the Breast Cancer Wisconsin (Diagnostic) dataset from the UCI Machine Learning Repository at `http://archive.ics.uci.edu/ml`. This data was donated by researchers of the University of Wisconsin and includes measurements from digitized images of fine-needle aspirate of a breast mass. The values represent characteristics of the cell nuclei present in the digital image.

> To read more about this dataset, refer to *Breast Cancer Diagnosis and Prognosis via Linear Programming, Mangasarian OL, Street WN, Wolberg WH, Operations Research, 1995, Vol. 43, pp. 570-577.*

The breast cancer data includes 569 examples of cancer biopsies, each with 32 features. One feature is an identification number, another is the cancer diagnosis, and 30 are numeric-valued laboratory measurements. The diagnosis is coded as "M" to indicate malignant or "B" to indicate benign.

The 30 numeric measurements comprise the mean, standard error, and worst (that is, largest) value for 10 different characteristics of the digitized cell nuclei. These include:

- Radius
- Texture
- Perimeter
- Area
- Smoothness
- Compactness
- Concavity
- Concave points
- Symmetry
- Fractal dimension

Based on these names, all features seem to relate to the shape and size of the cell nuclei. Unless you are an oncologist, you are unlikely to know how each relates to benign or malignant masses. These patterns will be revealed as we continue in the machine learning process.

Step 2 – exploring and preparing the data

Let's explore the data and see if we can shine some light on the relationships. In doing so, we will prepare the data for use with the k-NN learning method.

> If you plan on following along, download the `wisc_bc_data.csv` file from the Packt website and save it to your R working directory. For this book, the dataset was modified very slightly from its original form. In particular, a header line was added and the rows of data were randomly ordered.

We'll begin by importing the CSV data file as we have done in previous chapters, saving the Wisconsin breast cancer data to the `wbcd` data frame:

```
> wbcd <- read.csv("wisc_bc_data.csv", stringsAsFactors = FALSE)
```

Using the command `str(wbcd)`, we can confirm that the data is structured with 569 examples and 32 features, as we expected. The first several lines of output are as follows:

```
'data.frame':   569 obs. of   32 variables:
 $ id             : int  87139402 8910251 905520 ...
 $ diagnosis      : chr  "B" "B" "B" "B" ...
 $ radius_mean    : num  12.3 10.6 11 11.3 15.2 ...
 $ texture_mean   : num  12.4 18.9 16.8 13.4 13.2 ...
 $ perimeter_mean : num  78.8 69.3 70.9 73 97.7 ...
 $ area_mean      : num  464 346 373 385 712 ...
```

The first variable is an integer variable named `id`. As this is simply a unique identifier (ID) for each patient in the data, it does not provide useful information and we will need to exclude it from the model.

> Regardless of the machine learning method, ID variables should always be excluded. Neglecting to do so can lead to erroneous findings because the ID can be used to correctly predict each example. Therefore, a model that includes an ID column will almost definitely suffer from overfitting and generalize poorly to future data.

Let's drop the `id` feature altogether. As it is located in the first column, we can exclude it by making a copy of the wbcd data frame without column 1:

```
> wbcd <- wbcd[-1]
```

The next variable, `diagnosis`, is of particular interest as it is the outcome we hope to predict. This feature indicates whether the example is from a benign or malignant mass. The `table()` output indicates that 357 masses are benign, while 212 are malignant:

```
> table(wbcd$diagnosis)

  B   M
357 212
```

Many R machine learning classifiers require the target feature to be coded as a factor, so we will need to recode the `diagnosis` variable. We will also take this opportunity to give the "B" and "M" values more informative labels using the `labels` parameter:

```
> wbcd$diagnosis <- factor(wbcd$diagnosis, levels = c("B", "M"),
    labels = c("Benign", "Malignant"))
```

When we look at the `prop.table()` output, we now find that the values have been labeled `Benign` and `Malignant`, with 62.7 percent and 37.3 percent of the masses, respectively:

```
> round(prop.table(table(wbcd$diagnosis)) * 100, digits = 1)

   Benign Malignant
     62.7      37.3
```

The remaining 30 features are all numeric and, as expected, consist of three different measurements of 10 characteristics. For illustrative purposes, we will only take a closer look at three of these features:

```
> summary(wbcd[c("radius_mean", "area_mean", "smoothness_mean")])
  radius_mean       area_mean      smoothness_mean
 Min.   : 6.981   Min.   : 143.5   Min.   :0.05263
 1st Qu.:11.700   1st Qu.: 420.3   1st Qu.:0.08637
 Median :13.370   Median : 551.1   Median :0.09587
 Mean   :14.127   Mean   : 654.9   Mean   :0.09636
 3rd Qu.:15.780   3rd Qu.: 782.7   3rd Qu.:0.10530
 Max.   :28.110   Max.   :2501.0   Max.   :0.16340
```

Looking at the three side-by-side, do you notice anything problematic about the values? Recall that the distance calculation for k-NN is heavily dependent upon the measurement scale of the input features. Since smoothness ranges from 0.05 to 0.16, while area ranges from 143.5 to 2501.0, the impact of area is going to be much greater than smoothness in the distance calculation. This could potentially cause problems for our classifier, so let's apply normalization to rescale the features to a standard range of values.

Transformation – normalizing numeric data

To normalize these features, we need to create a `normalize()` function in R. This function takes a vector x of numeric values, and for each value in x, subtracts the minimum x value and divides by the range of x values. Lastly, the resulting vector is returned. The code for the function is as follows:

```
> normalize <- function(x) {
      return ((x - min(x)) / (max(x) - min(x)))
   }
```

After executing the previous code, the `normalize()` function is available for use in R. Let's test the function on a couple of vectors:

```
> normalize(c(1, 2, 3, 4, 5))
[1] 0.00 0.25 0.50 0.75 1.00
> normalize(c(10, 20, 30, 40, 50))
[1] 0.00 0.25 0.50 0.75 1.00
```

The function appears to be working correctly. Despite the fact that the values in the second vector are 10 times larger than the first vector, after normalization, they both appear exactly the same.

We can now apply the `normalize()` function to the numeric features in our data frame. Rather than normalizing each of the 30 numeric variables individually, we will use one of R's functions to automate the process.

The `lapply()` function takes a list and applies a specified function to each list element. As a data frame is a list of equal-length vectors, we can use `lapply()` to apply `normalize()` to each feature in the data frame. The final step is to convert the list returned by `lapply()` to a data frame using the `as.data.frame()` function. The full process looks like this:

```
> wbcd_n <- as.data.frame(lapply(wbcd[2:31], normalize))
```

In plain English, this command applies the `normalize()` function to columns 2 to 31 in the `wbcd` data frame, converts the resulting list to a data frame, and assigns it the name `wbcd_n`. The `_n` suffix is used here as a reminder that the values in `wbcd` have been normalized.

To confirm that the transformation was applied correctly, let's look at one variable's summary statistics:

```
> summary(wbcd_n$area_mean)
Min.    1st Qu. Median  Mean    3rd Qu. Max.
0.0000  0.1174  0.1729  0.2169  0.2711  1.0000
```

As expected, the `area_mean` variable, which originally ranged from 143.5 to 2501.0, now ranges from 0 to 1.

Data preparation – creating training and test datasets

Although all 569 biopsies are labeled with a benign or malignant status, it is not very interesting to predict what we already know. Additionally, any performance measures we obtain during training may be misleading, as we do not know the extent to which the data has been overfitted or how well the learner will generalize to new cases. For these reasons, a more interesting question is how well our learner performs on a dataset of unseen data. If we had access to a laboratory, we could apply our learner to measurements taken from the next 100 masses of unknown cancer status and see how well the machine learner's predictions compare to diagnoses obtained using conventional methods.

In the absence of such data, we can simulate this scenario by dividing our data into two portions: a training dataset that will be used to build the k-NN model and a test dataset that will be used to estimate the predictive accuracy of the model. We will use the first 469 records for the training dataset and the remaining 100 to simulate new patients.

Using the data extraction methods presented in *Chapter 2, Managing and Understanding Data*, we will split the `wbcd_n` data frame into `wbcd_train` and `wbcd_test`:

```
> wbcd_train <- wbcd_n[1:469, ]
> wbcd_test <- wbcd_n[470:569, ]
```

If the previous commands are confusing, remember that data is extracted from data frames using the `[row, column]` syntax. A blank value for the row or column value indicates that all rows or columns should be included. Hence, the first line of code requests rows 1 to 469 and all columns, and the second line requests 100 rows from 470 to 569 and all columns.

When constructing training and test datasets, it is important that each dataset is a representative subset of the full set of data. The wbcd records were already randomly ordered, so we could simply extract 100 consecutive records to create a test dataset. This would not be appropriate if the data was ordered chronologically or in groups of similar values. In these cases, random sampling methods would be needed. Random sampling will be discussed in *Chapter 5, Divide and Conquer – Classification Using Decision Trees and Rules*.

When we constructed our normalized training and test datasets, we excluded the target variable, diagnosis. For training the k-NN model, we will need to store these class labels in factor vectors, split between the training and test datasets:

```
> wbcd_train_labels <- wbcd[1:469, 1]
> wbcd_test_labels <- wbcd[470:569, 1]
```

This code takes the diagnosis factor in the first column of the wbcd data frame and creates the vectors wbcd_train_labels and wbcd_test_labels. We will use these in the next steps of training and evaluating our classifier.

Step 3 – training a model on the data

Equipped with our training data and vector of labels, we are now ready to classify our test records. For the k-NN algorithm, the training phase actually involves no model building; the process of training a lazy learner like k-NN simply involves storing the input data in a structured format.

To classify our test instances, we will use a k-NN implementation from the class package, which provides a set of basic R functions for classification. If this package is not already installed on your system, you can install it by typing:

```
> install.packages("class")
```

To load the package during any session in which you wish to use the functions, simply enter the library(class) command.

The knn() function in the class package provides a standard, classic implementation of the k-NN algorithm. For each instance in the test data, the function will identify the *k* nearest neighbors, using Euclidean distance, where *k* is a user-specified number. The test instance is classified by taking a "vote" among the *k* nearest neighbors—specifically, this involves assigning the class of the majority of the neighbors. A tie vote is broken at random.

There are several other k-NN functions in other R packages that provide more sophisticated or more efficient implementations. If you run into limitations with knn (), search for k-NN at the Comprehensive R Archive Network (CRAN).

Training and classification using the knn () function is performed in a single command with four parameters, as shown in the following table:

---

kNN classification syntax

using the **knn ()** function in the **class** package

Building the classifier and making predictions:

```
p <- knn(train, test, class, k)
```

- **train** is a data frame containing numeric training data
- **test** is a data frame containing numeric test data
- **class** is a factor vector with the class for each row in the training data
- **k** is an integer indicating the number of nearest neighbors

The function returns a factor vector of predicted classes for each row in the test data frame.

Example:

```
wbcd_pred <- knn(train = wbcd_train, test = wbcd_test,
                 cl = wbcd_train_labels, k = 3)
```

We now have nearly everything we need to apply the k-NN algorithm to this data. We've split our data into training and test datasets, each with exactly the same numeric features. The labels for the training data are stored in a separate factor vector. The only remaining parameter is k, which specifies the number of neighbors to include in the vote.

As our training data includes 469 instances, we might try k = 21, an odd number roughly equal to the square root of 469. With a two-category outcome, using an odd number eliminates the chance of ending with a tie vote.

Now we can use the knn () function to classify the test data:

```
> wbcd_test_pred <- knn(train = wbcd_train, test = wbcd_test,
                        cl = wbcd_train_labels, k = 21)
```

The knn () function returns a factor vector of predicted labels for each of the examples in the wbcd_test dataset. We have assigned these predictions to wbcd_test_pred.

Step 4 – evaluating model performance

The next step of the process is to evaluate how well the predicted classes in the wbcd_test_pred vector match the actual values in the wbcd_test_labels vector. To do this, we can use the CrossTable() function in the gmodels package, which was introduced in *Chapter 2, Managing and Understanding Data*. If you haven't done so already, please install this package using the install. packages("gmodels") command.

After loading the package with the library(gmodels) command, we can create a cross tabulation indicating the agreement between the predicted and actual label vectors. Specifying prop.chisq = FALSE will remove the unnecessary chi-square values from the output:

```
> CrossTable(x = wbcd_test_labels, y = wbcd_test_pred,
             prop.chisq = FALSE)
```

The resulting table looks like this:

```
                 | wbcd_test_pred
wbcd_test_labels |    Benign | Malignant | Row Total |
-----------------|-----------|-----------|-----------|
          Benign |        61 |         0 |        61 |
                 |     1.000 |     0.000 |     0.610 |
                 |     0.968 |     0.000 |           |
                 |     0.610 |     0.000 |           |
-----------------|-----------|-----------|-----------|
       Malignant |         2 |        37 |        39 |
                 |     0.051 |     0.949 |     0.390 |
                 |     0.032 |     1.000 |           |
                 |     0.020 |     0.370 |           |
-----------------|-----------|-----------|-----------|
    Column Total |        63 |        37 |       100 |
                 |     0.630 |     0.370 |           |
-----------------|-----------|-----------|-----------|
```

The cell percentages in the table indicate the proportion of values that fall into four categories. The top-left cell indicates the **true negative** results. These 61 of 100 values are cases where the mass was benign and the k-NN algorithm correctly identified it as such. The bottom-right cell indicates the **true positive** results, where the classifier and the clinically determined label agree that the mass is malignant. A total of 37 of 100 predictions were true positives.

The cells falling on the other diagonal contain counts of examples where the k-NN prediction disagreed with the true label. The two examples in the lower-left cell are **false negative** results; in this case, the predicted value was benign, but the tumor was actually malignant. Errors in this direction could be extremely costly, as they might lead a patient to believe that she is cancer-free, but in reality, the disease may continue to spread.

The top-right cell would contain the **false positive** results, if there were any. These values occur when the model has classified a mass as malignant when in reality it was benign. Although such errors are less dangerous than a false negative result, they should also be avoided, as they could lead to additional financial burden on the health care system or stress for the patient, as unnecessary tests or treatment may be provided.

 If we desired, we could totally eliminate false negatives by classifying every mass as malignant. Obviously, this is not a realistic strategy. Still, it illustrates the fact that prediction involves striking a balance between the false positive rate and the false negative rate. In *Chapter 10, Evaluating Model Performance*, you will learn methods for evaluating predictive accuracy that can be used to optimize performance to the costs of each type of error.

A total of 2 out of 100, or 2 percent of masses were incorrectly classified by the k-NN approach. While 98 percent accuracy seems impressive for a few lines of R code, we might try another iteration of the model to see if we can improve the performance and reduce the number of values that have been incorrectly classified, especially because the errors were dangerous false negatives.

Step 5 – improving model performance

We will attempt two simple variations on our previous classifier. First, we will employ an alternative method for rescaling our numeric features. Second, we will try several different *k* values.

Transformation – z-score standardization

Although normalization is commonly used for k-NN classification, z-score standardization may be a more appropriate way to rescale the features in a cancer dataset. Since z-score standardized values have no predefined minimum and maximum, extreme values are not compressed towards the center. Even in the absence of formal medical-domain training, one might suspect that a malignant tumor might lead to extreme outliers as tumors grow uncontrollably. With this in mind, it might be reasonable to allow the outliers to be weighted more heavily in the distance calculation. Let's see whether z-score standardization improves our predictive accuracy.

To standardize a vector, we can use R's built-in `scale()` function, which by default rescales values using the z-score standardization. The `scale()` function can be applied directly to a data frame, so there is no need to use the `lapply()` function. To create a z-score standardized version of the `wbcd` data, we can use the following command:

```
> wbcd_z <- as.data.frame(scale(wbcd[-1]))
```

This rescales all features with the exception of `diagnosis` in the first column and stores the result as the `wbcd_z` data frame. The _z suffix is a reminder that the values were z-score transformed.

To confirm that the transformation was applied correctly, we can look at the summary statistics:

```
> summary(wbcd_z$area_mean)
```

Min.	1st Qu.	Median	Mean	3rd Qu.	Max.
-1.4530	-0.6666	-0.2949	0.0000	0.3632	5.2460

The mean of a z-score standardized variable should always be zero, and the range should be fairly compact. A z-score greater than 3 or less than -3 indicates an extremely rare value. Examining the summary statistics with these criteria in mind, the transformation seems to have worked.

As we have done before, we need to divide the z-score-transformed data into training and test sets, and classify the test instances using the `knn()` function. We'll then compare the predicted labels to the actual labels using `CrossTable()`:

```
> wbcd_train <- wbcd_z[1:469, ]
> wbcd_test <- wbcd_z[470:569, ]
> wbcd_train_labels <- wbcd[1:469, 1]
> wbcd_test_labels <- wbcd[470:569, 1]
> wbcd_test_pred <- knn(train = wbcd_train, test = wbcd_test,
                        cl = wbcd_train_labels, k = 21)
> CrossTable(x = wbcd_test_labels, y = wbcd_test_pred,
             prop.chisq = FALSE)
```

Unfortunately, in the following table, the results of our new transformation show a slight decline in accuracy. Using the same instances in which we had previously classified 98 percent of examples correctly, we now classified only 95 percent correctly. Making matters worse, we did no better at classifying the dangerous false negatives.

```
                      | wbcd_test_pred
    wbcd_test_labels  |   Benign | Malignant | Row Total |
    ------------------|----------|-----------|-----------|
              Benign  |       61 |         0 |        61 |
                      |    1.000 |     0.000 |     0.610 |
                      |    0.924 |     0.000 |           |
                      |    0.610 |     0.000 |           |
    ------------------|----------|-----------|-----------|
           Malignant  |        5 |        34 |        39 |
                      |    0.128 |     0.872 |     0.390 |
                      |    0.076 |     1.000 |           |
                      |    0.050 |     0.340 |           |
    ------------------|----------|-----------|-----------|
        Column Total  |       66 |        34 |       100 |
                      |    0.660 |     0.340 |           |
    ------------------|----------|-----------|-----------|
```

Testing alternative values of k

We may be able to optimize the performance of the k-NN model by examining its performance across various *k* values. Using the normalized training and test datasets, the same 100 records were classified using several different choices of *k*. The number of false negatives and false positives are shown for each iteration:

k value	False negatives	False positives	Percent classified incorrectly
1	1	3	4 percent
5	2	0	2 percent
11	3	0	3 percent
15	3	0	3 percent
21	2	0	2 percent
27	4	0	4 percent

Although the classifier was never perfect, the 1-NN approach was able to avoid some of the false negatives at the expense of adding false positives. It is important to keep in mind, however, that it would be unwise to tailor our approach too closely to our test data; after all, a different set of 100 patient records is likely to be somewhat different from those used to measure our performance.

If you need to be certain that a learner will generalize to future data, you might create several sets of 100 patients at random and repeatedly retest the result. Such methods to carefully evaluate the performance of machine learning models will be discussed further in *Chapter 10, Evaluating Model Performance*.

Summary

In this chapter, we learned about classification using k-NN. Unlike many classification algorithms, k-nearest neighbors does not do any learning. It simply stores the training data verbatim. Unlabeled test examples are then matched to the most similar records in the training set using a distance function, and the unlabeled example is assigned the label of its neighbors.

In spite of the fact that k-NN is a very simple algorithm, it is capable of tackling extremely complex tasks such as the identification of cancerous masses. In a few simple lines of R code, we were able to correctly identify whether a mass was malignant or benign 98 percent of the time.

In the next chapter, we will examine a classification method that uses probability to estimate the likelihood that an observation falls into certain categories. It will be interesting to compare how this approach differs from k-NN. Later on, in *Chapter 9, Finding Groups of Data – Clustering with k-means*, we will learn about a close relative to k-NN, which uses distance measures for a completely different learning task.

4

Probabilistic Learning – Classification Using Naive Bayes

When a meteorologist provides a weather forecast, precipitation is typically described with phrases like "70 percent chance of rain." Such forecasts are known as probability of precipitation reports. Have you ever considered how they are calculated? It is a puzzling question because in reality, either it will rain or not.

Weather estimates are based on probabilistic methods, which are those concerned with describing uncertainty. They use data on past events to extrapolate future events. In the case of the weather, the chance of rain describes the proportion of prior days with similar atmospheric conditions in which precipitation occurred. A 70 percent chance of rain implies that in seven out of 10 past cases with similar conditions, precipitation occurred somewhere in the area.

This chapter covers the Naive Bayes algorithm, which uses probabilities in much the same way as a weather forecast. While studying this method, you will learn:

- Basic principles of probability
- The specialized methods and data structures needed to analyze text data with R
- How to employ Naive Bayes to build an SMS junk message filter

If you've taken a statistics class before, some of the material in this chapter may be a review. Even so, it may be helpful to refresh your knowledge of probability, as these principles are the basis of how Naive Bayes got such a strange name.

Understanding Naive Bayes

The basic statistical ideas necessary to understand the Naive Bayes algorithm have existed for centuries. The technique descended from the work of the 18th century mathematician Thomas Bayes, who developed foundational principles for describing the probability of events and how probabilities should be revised in light of additional information. These principles formed the foundation for what are now known as **Bayesian methods**.

We will cover these methods in greater detail later on. For now, it suffices to say that a probability is a number between zero and one (that is, from zero to 100 percent), which captures the chance that an event will occur in light of the available evidence. The lower the probability, the less likely the event is to occur. A probability of zero indicates that the event will definitely not occur, while a probability of one indicates that the event will occur with absolute certainty.

Classifiers based on Bayesian methods utilize training data to calculate the probability of each outcome based on the evidence provided by feature values. When the classifier is later applied to unlabeled data, it uses these calculated probabilities to predict the most likely class for the new example. It's a simple idea, but it results in a method that can have results on a par with more sophisticated algorithms. In fact, Bayesian classifiers have been used for:

- Text classification, such as junk email (spam) filtering
- Intrusion or anomaly detection in computer networks
- Diagnosing medical conditions given a set of observed symptoms

Typically, Bayesian classifiers are best applied to problems in which the information from numerous attributes should be considered simultaneously in order to estimate the overall probability of an outcome. While many machine learning algorithms ignore features that have weak effects, Bayesian methods utilize all available evidence to subtly change the predictions. This implies that even if a large number of features have relatively minor effects, their combined impact in a Bayesian model could be quite large.

Basic concepts of Bayesian methods

Before jumping into the Naive Bayes algorithm, it's worth spending some time defining the concepts that are used across Bayesian methods. Summarized in a single sentence, Bayesian probability theory is rooted in the idea that the estimated likelihood of an **event**, or potential outcome, should be based on the evidence at hand across multiple **trials**, or opportunities for the event to occur.

The following table illustrates events and trials for several real-world outcomes:

Event	Trial
Heads result	Coin flip
Rainy weather	A single day
Message is spam	Incoming email message
Candidate becomes president	Presidential election
Win the lottery	Lottery ticket

Bayesian methods provide insights into how the probability of these events can be estimated from observed data. To see how, we'll need to formalize our understanding of probability.

Understanding probability

The probability of an event is estimated from observed data by dividing the number of trials in which the event occurred by the total number of trials. For instance, if it rained three out of 10 days with similar conditions as today, the probability of rain today can be estimated as $3/10 = 0.30$ or 30 percent. Similarly, if 10 out of 50 prior email messages were spam, then the probability of any incoming message being spam can be estimated as $10/50 = 0.20$ or 20 percent.

To denote these probabilities, we use notation in the form $P(A)$, which signifies the probability of event A. For example, $P(rain) = 0.30$ and $P(spam) = 0.20$.

The probability of all possible outcomes of a trial must always sum to one, because a trial always results in some outcome happening. Thus, if the trial has two outcomes that cannot occur simultaneously, such as rainy versus sunny, or spam versus ham (not spam), then knowing the probability of either outcome reveals the probability of the other. For example, given the value $P(spam) = 0.20$, we can calculate $P(ham)$ $= 1 - 0.20 = 0.80$. This works because spam and ham are **mutually exclusive and exhaustive** events, which implies that they cannot occur at the same time and are the only possible outcomes.

Now, because an event cannot happen and not happen simultaneously, an event is always mutually exclusive and exhaustive with its **complement**, or the event comprising the outcomes in which the event of interest does not happen. The complement of event A is typically denoted A^c or A'. Additionally, the shorthand notation $P(\neg A)$ can be used to denote the probability of event A not occurring, as in $P(\neg spam) = 0.80$. This notation is equivalent to $P(A^c)$.

To illustrate events and their complements, it is often helpful to imagine a two-dimensional space that is partitioned into probabilities for each event. In the following diagram, the rectangle represents the possible outcomes for an email message. The circle represents the 20 percent probability that the message is spam. The remaining 80 percent represents the complement *P(¬spam)*, or the messages that are not spam:

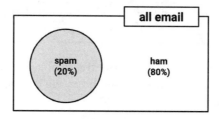

Figure 4.1: The probability space of all email can be visualized as partitions of spam and ham

Understanding joint probability

Often, we are interested in monitoring several non-mutually exclusive events for the same trial. If certain events occur concurrently with the event of interest, we may be able to use them to make predictions. Consider, for instance, a second event based on the outcome that an email message contains the word "Viagra." The preceding diagram, updated for this second event, might appear as shown in the following diagram:

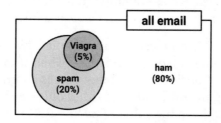

Figure 4.2: Non-mutually exclusive events are depicted as overlapping partitions

Notice in the diagram that the Viagra circle does not completely fill the spam circle, nor is it completely contained by the spam circle. This implies that not all spam messages contain the word "Viagra" and not every email with the word "Viagra" is spam. However, because this word appears very little outside spam, its presence in a new incoming message would be strong evidence that the message is spam.

To zoom in for a closer look at the overlap between these circles, we'll employ a visualization known as a **Venn diagram**. First used in the late 19th century by mathematician John Venn, the diagram uses circles to illustrate the overlap between sets of items. As is the case here, in most Venn diagrams, the size of the circles and the degree of the overlap is not meaningful.

Instead, it is used as a reminder to allocate probability to all combinations of events. A Venn diagram for spam and Viagra might be depicted as follows:

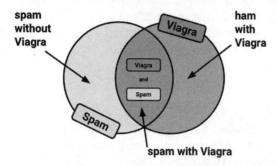

Figure 4.3: A Venn diagram illustrates the overlap of the spam and Viagra events

We know that 20 percent of all messages were spam (the left circle), and 5 percent of all messages contained the word "Viagra" (the right circle). We would like to quantify the degree of overlap between these two proportions. In other words, we hope to estimate the probability that both *P(spam)* and *P(Viagra)* occur, which can be written as *P(spam ∩ Viagra)*. The upside-down "U" symbol signifies the **intersection** of the two events; the notation *A ∩ B* refers to the event in which both *A* and *B* occur.

Calculating *P(spam ∩ Viagra)* depends on the **joint probability** of the two events, or how the probability of one event is related to the probability of the other. If the two events are totally unrelated, they are called **independent events**. This is not to say that independent events cannot occur at the same time; event independence simply implies that knowing the outcome of one event does not provide any information about the outcome of the other. For instance, the outcome of a heads result on a coin flip is independent from whether the weather is rainy or sunny on any given day.

If all events were independent, it would be impossible to predict one event by observing another. In other words, **dependent events** are the basis of predictive modeling. Just as the presence of clouds is predictive of a rainy day, the appearance of the word "Viagra" is predictive of a spam email.

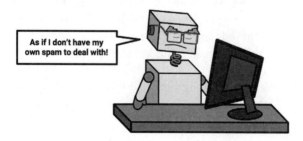

Figure 4.4: Dependent events are required for machines to learn to identify useful patterns

Calculating the probability of dependent events is a bit more complex than for independent events. If *P(spam)* and *P(Viagra)* were independent, we could easily calculate *P(spam ∩ Viagra)*, the probability of both events happening at the same time. Because 20 percent of all messages are spam, and 5 percent of all emails contain the word "Viagra," we could assume that 1 percent of all messages are spam with the term "Viagra." This is because *0.05 * 0.20 = 0.01*. More generally, for independent events *A* and *B*, the probability of both happening can be computed as *P(A ∩ B) = P(A) * P(B)*.

That said, we know that *P(spam)* and *P(Viagra)* are likely to be highly dependent, which means that this calculation is incorrect. To obtain a more reasonable estimate, we need to use a more careful formulation of the relationship between these two events, which is based on more advanced Bayesian methods.

Computing conditional probability with Bayes' theorem

The relationships between dependent events can be described using **Bayes' theorem**, which provides a way of thinking about how to revise an estimate of the probability of one event in light of the evidence provided by another. One formulation is as follows:

$$P(A \mid B) = \frac{P(A \cap B)}{P(B)}$$

The notation *P(A | B)* is read as the probability of event *A* given that event *B* occurred. This is known as **conditional probability**, since the probability of *A* is dependent (that is, conditional) on what happened with event *B*. Bayes' theorem tells us that our estimate of *P(A | B)* should be based on *P(A ∩ B)*, a measure of how often *A* and *B* are observed to occur together, and *P(B)*, a measure of how often *B* is observed to occur in general.

Bayes' theorem states that the best estimate of *P(A | B)* is the proportion of trials in which *A* occurred with *B*, out of all the trials in which *B* occurred. This implies that the probability of event *A* is higher if *A* and *B* often occur together each time *B* is observed. Note that this formula adjusts *P(A ∩ B)* for the probability of *B* occurring. If *B* is extremely rare, *P(B)* and *P(A ∩ B)* will always be small; however, if *A* and *B* almost always happen together, *P(A | B)* will be high regardless of the probability of *B*.

By definition, $P(A \cap B) = P(A \mid B) * P(B)$, a fact which can be easily derived by applying a bit of algebra to the previous formula. Rearranging this formula once more with the knowledge that $P(A \cap B) = P(B \cap A)$ results in the conclusion that $P(A \cap B) = P(B \mid A) * P(A)$, which we can then use in the following formulation of Bayes' theorem:

$$P(A \mid B) = \frac{P(A \cap B)}{P(B)} = \frac{P(B \mid A)P(A)}{P(B)}$$

In fact, this is the traditional formulation of Bayes' theorem, for reasons that will become clear as we apply it to machine learning. First, to better understand how Bayes' theorem works in practice, let's revisit our hypothetical spam filter.

Without knowledge of an incoming message's content, the best estimate of its spam status would be *P(spam)*, the probability that any prior message was spam. This estimate is known as the **prior probability**. We found this previously to be 20 percent.

Suppose that you obtained additional evidence by looking more carefully at the set of previously received messages and examining the frequency that the term "Viagra" appeared. The probability that the word "Viagra" was used in previous spam messages, or *P(Viagra | spam)*, is called the **likelihood**. The probability that "Viagra" appeared in any message at all, or *P(Viagra)*, is known as the **marginal likelihood**.

By applying Bayes' theorem to this evidence, we can compute a **posterior probability** that measures how likely a message is to be spam. If the posterior probability is greater than 50 percent, the message is more likely to be spam than ham, and it should perhaps be filtered. The following formula shows how Bayes' theorem is applied to the evidence provided by previous email messages:

$$\underset{\text{posterior probability}}{P(\text{spam}|\text{Viagra})} = \frac{\overset{\text{likelihood}}{P(\text{Viagra}|\text{spam})}\overset{\text{prior probability}}{P(\text{spam})}}{\underset{\text{marginal likelihood}}{P(\text{Viagra})}}$$

To calculate the components of Bayes' theorem, it helps to construct a **frequency table** (shown on the left in the following diagram) that records the number of times "Viagra" appeared in spam and ham messages. Just like a two-way cross-tabulation, one dimension of the table indicates levels of the class variable (spam or ham), while the other dimension indicates levels for features (Viagra: yes or no). The cells then indicate the number of instances that have the particular combination of class value and feature value.

The frequency table can then be used to construct a **likelihood table**, as shown on the right in the following diagram. The rows of the likelihood table indicate the conditional probabilities for "Viagra" (yes/no), given that an email was either spam or ham.

Frequency	Viagra		Total
	Yes	No	
spam	4	16	20
ham	1	79	80
Total	5	95	100

Likelihood	Viagra		Total
	Yes	No	
spam	4 / 20	16 / 20	20
ham	1 / 80	79 / 80	80
Total	5 / 100	95 / 100	100

Figure 4.5: Frequency and likelihood tables are the basis for computing the posterior probability of spam

The likelihood table reveals that $P(Viagra=Yes \mid spam) = 4 / 20 = 0.20$, indicating that there is a 20 percent probability a message contains the term "Viagra" given that the message is spam. Additionally, since $P(A \cap B) = P(B \mid A) * P(A)$, we can calculate $P(spam \cap Viagra)$ as $P(Viagra \mid spam) * P(spam) = (4 / 20) * (20 / 100) = 0.04$. The same result can be found in the frequency table, which notes that four out of 100 messages were spam with the term "Viagra." Either way, this is four times greater than the previous estimate of 0.01 we calculated as $P(A \cap B) = P(A) * P(B)$ under the false assumption of independence. This, of course, illustrates the importance of Bayes' theorem for estimating joint probability.

To compute the posterior probability, $P(spam \mid Viagra)$, we simply take $P(Viagra \mid spam) * P(spam) / P(Viagra)$, or $(4 / 20) * (20 / 100) / (5 / 100) = 0.80$. Therefore, the probability is 80 percent that a message is spam, given that it contains the word "Viagra." In light of this result, any message containing this term should probably be filtered.

This is very much how commercial spam filters work, although they consider a much larger number of words simultaneously when computing the frequency and likelihood tables. In the next section, we'll see how this method is put to use when additional features are involved.

The Naive Bayes algorithm

The **Naive Bayes** algorithm defines a simple method to apply Bayes' theorem to classification problems. Although it is not the only machine learning method that utilizes Bayesian methods, it is the most common. It grew in popularity due to its successes in text classification, where it was once the de facto standard. The strengths and weaknesses of this algorithm are as follows:

Strengths	Weaknesses
• Simple, fast, and very effective • Does well with noisy and missing data • Requires relatively few examples for training, but also works well with very large numbers of examples • Easy to obtain the estimated probability for a prediction	• Relies on an often-faulty assumption of equally important and independent features • Not ideal for datasets with many numeric features • Estimated probabilities are less reliable than the predicted classes

The Naive Bayes algorithm is named as such because it makes some "naive" assumptions about the data. In particular, Naive Bayes assumes that all of the features in the dataset are **equally important and independent**. These assumptions are rarely true in most real-world applications.

For example, if you were attempting to identify spam by monitoring email messages, it is almost certainly true that some features will be more important than others. For example, the email sender may be a more important indicator of spam than the message text. Additionally, the words in the message body are not independent from one another, since the appearance of some words is a very good indication that other words are also likely to appear. A message with the word "Viagra" will probably also contain the words "prescription" or "drugs."

However, in most cases, even when these assumptions are violated, Naive Bayes still performs fairly well. This is true even in circumstances where strong dependencies are found among the features. Due to the algorithm's versatility and accuracy across many types of conditions, particularly with smaller training datasets, Naive Bayes is often a reasonable baseline candidate for classification learning tasks.

The exact reason why Naive Bayes works well in spite of its faulty assumptions has been the subject of much speculation. One explanation is that it is not important to obtain a precise estimate of probability so long as the predictions are accurate. For instance, if a spam filter correctly identifies spam, does it matter whether it was 51 percent or 99 percent confident in its prediction? For one discussion of this topic, refer to *On the Optimality of the Simple Bayesian Classifier under Zero-One Loss*, Domingos P and *Pazzani M, Machine Learning, 1997, Vol. 29, pp. 103-130.*

Classification with Naive Bayes

Let's extend our spam filter by adding a few additional terms to be monitored in addition to the term "Viagra": "money," "groceries," and "unsubscribe." The Naive Bayes learner is trained by constructing a likelihood table for the appearance of these four words (labeled W_1, W_2, W_3, and W_4), as shown in the following diagram for 100 emails:

Likelihood	Viagra (W_1)		Money (W_2)		Groceries (W_3)		Unsubscribe (W_4)		Total
	Yes	No	Yes	No	Yes	No	Yes	No	
spam	4 / 20	16 / 20	10 / 20	10 / 20	0 / 20	20 / 20	12 / 20	8 / 20	20
ham	1 / 80	79 / 80	14 / 80	66 / 80	8 / 80	71 / 80	23 / 80	57 / 80	80
Total	5 / 100	95 / 100	24 / 100	76 / 100	8 / 100	91 / 100	35 / 100	65 / 100	100

Figure 4.6: An expanded table adds likelihoods for additional terms in spam and ham messages

As new messages are received, we need to calculate the posterior probability to determine whether they are more likely spam or ham, given the likelihood of the words being found in the message text. For example, suppose that a message contains the terms "Viagra" and "unsubscribe" but does not contain either "money" or "groceries."

Using Bayes' theorem, we can define the problem as shown in the following formula. This computes the probability that a message is spam, given that *Viagra = Yes, Money = No, Groceries = No,* and *Unsubscribe = Yes*:

$$P\left(\text{spam} \mid W_1 \cap \neg W_2 \cap \neg W_3 \cap W_4\right) = \frac{P\left(W_1 \cap \neg W_2 \cap \neg W_3 \cap W_4 \mid \text{spam}\right) P\left(\text{spam}\right)}{P\left(W_1 \cap \neg W_2 \cap \neg W_3 \cap W_4\right)}$$

For a number of reasons, this formula is computationally difficult to solve. As additional features are added, tremendous amounts of memory are needed to store probabilities for all of the possible intersecting events. Imagine the complexity of a Venn diagram for the events for four words, let alone for hundreds or more. Many of these potential intersections will never have been observed in past data, which would lead to a joint probability of zero and problems that will become clear later.

The computation becomes more reasonable if we exploit the fact that Naive Bayes makes the naive assumption of independence among events. Specifically, it assumes **class-conditional independence**, which means that events are independent so long as they are conditioned on the same class value. The conditional independence assumption allows us to use the probability rule for independent events, which states that *P(A ∩ B) = P(A) * P(B)*. This simplifies the numerator by allowing us to multiply the individual conditional probabilities rather than computing a complex conditional joint probability.

Lastly, because the denominator does not depend on the target class (spam or ham), it is treated as a constant value and can be ignored for the time being. This means that the conditional probability of spam can be expressed as:

$$P(\text{spam} \mid W_1 \cap \neg W_2 \cap \neg W_3 \cap W_4) \propto P(W_1 \mid \text{spam}) P(\neg W_2 \mid \text{spam}) P(\neg W_3 \mid \text{spam}) P(W_4 \mid \text{spam}) P(\text{spam})$$

And the probability that the message is ham can be expressed as:

$$P(\text{ham} \mid W_1 \cap \neg W_2 \cap \neg W_3 \cap W_4) \propto P(W_1 \mid \text{ham}) P(\neg W_2 \mid \text{ham}) P(\neg W_3 \mid \text{ham}) P(W_4 \mid \text{ham}) P(\text{ham})$$

Note that the equals symbol has been replaced by the proportional-to symbol (similar to a sideways, open-ended "8") to indicate the fact that the denominator has been omitted.

Using the values in the likelihood table, we can start filling numbers in these equations. The overall likelihood of spam is then:

*(4/20) * (10/20) * (20/20) * (12/20) * (20/100) = 0.012*

While the likelihood of ham is:

*(1/80) * (66/80) * (71/80) * (23/80) * (80/100) = 0.002*

Because *0.012/0.002 = 6*, we can say that this message is six times more likely to be spam than ham. However, to convert these numbers to probabilities, we need one last step to reintroduce the denominator that had been excluded. Essentially, we must re-scale the likelihood of each outcome by dividing it by the total likelihood across all possible outcomes.

In this way, the probability of spam is equal to the likelihood that the message is spam divided by the likelihood that the message is either spam or ham:

0.012 / (0.012 + 0.002) = 0.857

Similarly, the probability of ham is equal to the likelihood that the message is ham divided by the likelihood that the message is either spam or ham:

0.002 / (0.012 + 0.002) = 0.143

Given the pattern of words found in this message, we expect that the message is spam with 85.7 percent probability, and ham with 14.3 percent probability. Because these are mutually exclusive and exhaustive events, the probabilities sum up to one.

The Naive Bayes classification algorithm used in the preceding example can be summarized by the following formula. The probability of level L for class C, given the evidence provided by features F_1 through F_n, is equal to the product of the probabilities of each piece of evidence conditioned on the class level, the prior probability of the class level, and a scaling factor $1/Z$, which converts the likelihood values to probabilities. This is formulated as:

$$P(C_L \mid F_1, \ldots, F_n) = \frac{1}{Z} p(C_L) \prod_{i=1}^{n} p(F_i \mid C_L)$$

Although this equation seems intimidating, as the spam filtering example illustrated, the series of steps is fairly straightforward. Begin by building a frequency table, use this to build a likelihood table, and multiply out the conditional probabilities with the "naive" assumption of independence. Finally, divide by the total likelihood to transform each class likelihood into a probability. After attempting this calculation a few times by hand, it will become second nature.

The Laplace estimator

Before we employ Naive Bayes on more complex problems, there are some nuances to consider. Suppose we received another message, this time containing all four terms: "Viagra," "groceries," "money," and "unsubscribe." Using the Naive Bayes algorithm as before, we can compute the likelihood of spam as:

$$(4/20) * (10/20) * (0/20) * (12/20) * (20/100) = 0$$

And the likelihood of ham is:

$$(1/80) * (14/80) * (8/80) * (23/80) * (80/100) = 0.00005$$

Therefore, the probability of spam is:

$$0/(0 + 0.00005) = 0$$

And the probability of ham is:

$$0.00005/(0 + 0.\,0.00005) = 1$$

These results suggest that the message is spam with zero percent probability and ham with 100 percent probability. Does this prediction make sense? Probably not. The message contains several words usually associated with spam, including "Viagra," which is rarely used in legitimate messages. It is therefore very likely that the message has been incorrectly classified.

This problem arises if an event never occurs for one or more levels of the class and therefore their joint probability is zero. For instance, the term "groceries" had never previously appeared in a spam message. Consequently, *P(spam | groceries) = 0%*.

Now, because probabilities in the Naive Bayes formula are multiplied in a chain, this zero percent value causes the posterior probability of spam to be zero, giving the word "groceries" the ability to effectively nullify and overrule all of the other evidence. Even if the email was otherwise overwhelmingly expected to be spam, the absence of the word "groceries" in spam will always veto the other evidence and result in the probability of spam being zero.

A solution to this problem involves using something called the **Laplace estimator**, which is named after the French mathematician Pierre-Simon Laplace. The Laplace estimator adds a small number to each of the counts in the frequency table, which ensures that each feature has a non-zero probability of occurring with each class. Typically, the Laplace estimator is set to one, which ensures that each class-feature combination is found in the data at least once.

The Laplace estimator can be set to any value and does not necessarily even have to be the same for each of the features. If you were a devoted Bayesian, you could use a Laplace estimator to reflect a presumed prior probability of how the feature relates to the class. In practice, given a large enough training dataset, this is excessive. Consequently, the value of one is almost always used.

Let's see how this affects our prediction for this message. Using a Laplace value of one, we add one to each numerator in the likelihood function. Then, we need to add four to each conditional probability denominator to compensate for the four additional values added to the numerator. The likelihood of spam is therefore:

$$(5/24) * (11/24) * (1/24) * (13/24) * (20/100) = 0.0004$$

And the likelihood of ham is:

$$(2/84) * (15/84) * (9/84) * (24/84) * (80/100) = 0.0001$$

By computing *0.0004 / (0.0004 + 0.0001)*, we find that the probability of spam is 80 percent and therefore the probability of ham is about 20 percent. This is a more plausible result than the *P(spam) = 0* computed when the term "groceries" alone determined the result.

Although the Laplace estimator was added to the numerator and denominator of the likelihoods, it was not added to the prior probabilities—the values of *20/100* and *80/100*. This is because our best estimate of the overall probability of spam and ham remains 20% and 80% given what was observed in the data.

Using numeric features with Naive Bayes

Naive Bayes uses frequency tables for learning the data, which means that each feature must be categorical in order to create the combinations of class and feature values comprising the matrix. Since numeric features do not have categories of values, the preceding algorithm does not work directly with numeric data. There are, however, ways that this can be addressed.

One easy and effective solution is to **discretize** numeric features, which simply means that the numbers are put into categories known as **bins**. For this reason, discretization is also sometimes called **binning**. This method works best when there are large amounts of training data.

There are several different ways to discretize a numeric feature. Perhaps the most common is to explore the data for natural categories or **cut points** in the distribution. For example, suppose that you added a feature to the spam dataset that recorded the time of day or night the email was sent, from zero to 24 hours past midnight. Depicted using a histogram, the time data might look something like the following diagram:

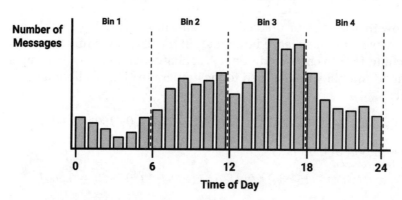

Figure 4.7: A histogram visualizing the distribution of the time emails were received

In the early hours of the morning, message frequency is low. Activity picks up during business hours and tapers off in the evening. This creates four natural bins of activity, as partitioned by the dashed lines. These indicate places where the numeric data could be divided into levels to create a new categorical feature, which could then be used with Naive Bayes.

The choice of four bins was based on the natural distribution of data and a hunch about how the proportion of spam might change throughout the day. We might expect that spammers operate in the late hours of the night, or they may operate during the day, when people are likely to check their email. That said, to capture these trends, we could have just as easily used three bins or twelve.

 If there are no obvious cut points, one option is to discretize the feature using quantiles. You could divide the data into three bins with tertiles, four bins with quartiles, or five bins with quintiles.

One thing to keep in mind is that discretizing a numeric feature always results in a reduction of information, as the feature's original granularity is reduced to a smaller number of categories. It is important to strike a balance. Too few bins can result in important trends being obscured. Too many bins can result in small counts in the Naive Bayes frequency table, which can increase the algorithm's sensitivity to noisy data.

Example – filtering mobile phone spam with the Naive Bayes algorithm

As the worldwide use of mobile phones has grown, a new avenue for electronic junk mail has opened for disreputable marketers. These advertisers utilize short message service (SMS) text messages to target potential consumers with unwanted advertising known as SMS spam. This type of spam is troublesome because, unlike email spam, an SMS message is particularly disruptive, due to the omnipresence of one's mobile phone. Developing a classification algorithm that could filter SMS spam would provide a useful tool for cellular phone providers.

Since Naive Bayes has been used successfully for email spam filtering, it seems likely that it could also be applied to SMS spam. However, relative to email spam, SMS spam poses additional challenges for automated filters. SMS messages are often limited to 160 characters, reducing the amount of text that can be used to identify whether a message is junk. The limit, combined with small mobile phone keyboards, has led many to adopt a form of SMS shorthand lingo, which further blurs the line between legitimate messages and spam. Let's see how a simple Naive Bayes classifier handles these challenges.

Step 1 – collecting data

To develop the Naive Bayes classifier, we will use data adapted from the SMS Spam Collection at `http://www.dt.fee.unicamp.br/~tiago/smsspamcollection/`.

> To read more about how the SMS Spam Collection was developed, refer to *On the Validity of a New SMS Spam Collection, Gómez JM, Almeida TA, and Yamakami A, Proceedings of the 11th IEEE International Conference on Machine Learning and Applications, 2012.*

This dataset includes the text of SMS messages, along with a label indicating whether the message is unwanted. Junk messages are labeled spam, while legitimate messages are labeled ham. Some examples of spam and ham are shown in the following table:

Sample SMS ham	Sample SMS spam
• Better. Made up for Friday and stuffed myself like a pig yesterday. Now I feel bleh. But at least its not writhing pain kind of bleh. • If he started searching he will get job in few days. he have great potential and talent. • I got another job! The one at the hospital doing data analysis or something, starts on monday! Not sure when my thesis will got finished	• Congratulations ur awarded 500 of CD vouchers or 125gift guaranteed & Free entry 2 100 wkly draw txt MUSIC to 87066 • December only! Had your mobile 11mths+? You are entitled to update to the latest colour camera mobile for Free! Call The Mobile Update Co FREE on 08002986906 • Valentines Day Special! Win over £1000 in our quiz and take your partner on the trip of a lifetime! Send GO to 83600 now. 150p/msg rcvd.

Looking at the preceding messages, do you notice any distinguishing characteristics of spam? One notable characteristic is that two of the three spam messages use the word "free," yet this word does not appear in any of the ham messages. On the other hand, two of the ham messages cite specific days of the week, as compared to zero in spam messages.

Our Naive Bayes classifier will take advantage of such patterns in the word frequency to determine whether the SMS messages seem to better fit the profile of spam or ham. While it's not inconceivable that the word "free" would appear outside of a spam SMS, a legitimate message is likely to provide additional words giving context. For instance, a ham message might ask, "Are you free on Sunday?" whereas a spam message might use the phrase "free ringtones." The classifier will compute the probability of spam and ham given the evidence provided by all the words in the message.

Step 2 – exploring and preparing the data

The first step towards constructing our classifier involves processing the raw data for analysis. Text data is challenging to prepare because it is necessary to transform the words and sentences into a form that a computer can understand. We will transform our data into a representation known as **bag-of-words**, which ignores word order and simply provides a variable indicating whether the word appears at all.

> The dataset used here has been modified slightly from the original in order to make it easier to work with in R. If you plan on following along with the example, download the sms_spam.csv file from the Packt website and save it to your R working directory.

We'll begin by importing the CSV data and saving it to a data frame:

```
> sms_raw <- read.csv("sms_spam.csv", stringsAsFactors = FALSE)
```

Using the str() function, we see that the sms_raw data frame includes 5,559 total SMS messages with two features: type and text. The SMS type has been coded as either ham or spam. The text element stores the full raw SMS message text.

```
> str(sms_raw)
'data.frame':    5559 obs. of  2 variables:
 $ type: chr  "ham" "ham" "ham" "spam" ...
 $ text: chr  "Hope you are having a good week. Just checking in"
"K..give back my thanks." "Am also doing in cbe only. But have to
pay." "complimentary 4 STAR Ibiza Holiday or £10,000 cash needs your
URGENT collection. 09066364349 NOW from Landline not to lose out"|
__truncated__  ...
```

The type element is currently a character vector. Since this is a categorical variable, it would be better to convert it to a factor, as shown in the following code:

```
> sms_raw$type <- factor(sms_raw$type)
```

Examining this with the str() and table() functions, we see that type has now been appropriately recoded as a factor. Additionally, we see that 747 (about 13 percent) of SMS messages in our data were labeled as spam, while the others were labeled as ham:

```
> str(sms_raw$type)
 Factor w/ 2 levels "ham","spam": 1 1 1 2 2 1 1 1 2 1 ...
> table(sms_raw$type)
 ham spam
4812  747
```

For now, we will leave the message text alone. As you will learn in the next section, processing the raw SMS messages will require the use of a new set of powerful tools designed specifically to process text data.

Data preparation – cleaning and standardizing text data

SMS messages are strings of text composed of words, spaces, numbers, and punctuation. Handling this type of complex data takes a large amount of thought and effort. One needs to consider how to remove numbers and punctuation; handle uninteresting words, such as *and*, *but*, and *or*; and how to break apart sentences into individual words. Thankfully, this functionality has been provided by members of the R community in a text mining package titled `tm`.

> The `tm` package was originally created by Ingo Feinerer as a dissertation project at the Vienna University of Economics and Business. To learn more, see *Text Mining Infrastructure in R, Feinerer I, Hornik K*, and *Meyer D, Journal of Statistical Software, 2008, Vol. 25, pp. 1-54.*

The `tm` package can be installed via the `install.packages("tm")` command and loaded with the `library(tm)` command. Even if you already have it installed, it may be worth re-running the `install` command to ensure that your version is up-to-date, as the `tm` package is still under active development. This occasionally results in changes to its functionality.

> This chapter was tested using `tm` version 0.7-6, which was current as of February 2019. If you see differences in the output or if the code does not work, you may be using a different version. The Packt support page for this book, as well as its GitHub repository, will post solutions for future `tm` packages if significant changes are noted.

The first step in processing text data involves creating a **corpus**, which is a collection of text documents. The documents can be short or long, from individual news articles, pages in a book, pages from the web, or even entire books. In our case, the corpus will be a collection of SMS messages.

To create a corpus, we'll use the `VCorpus()` function in the `tm` package, which refers to a volatile corpus – volatile as it is stored in memory as opposed to being stored on disk (the `PCorpus()` function is used to access a permanent corpus stored in a database). This function requires us to specify the source of documents for the corpus, which could be a computer's file system, a database, the web, or elsewhere.

Since we already loaded the SMS message text into R, we'll use the VectorSource()
reader function to create a source object from the existing sms_raw$text vector,
which can then be supplied to VCorpus() as follows:

```
> sms_corpus <- VCorpus(VectorSource(sms_raw$text))
```

The resulting corpus object is saved with the name sms_corpus.

> By specifying an optional readerControl parameter, the VCorpus()
> function can be used to import text from sources such as PDFs and
> Microsoft Word files. To learn more, examine the *Data Import* section
> in the tm package vignette using the vignette("tm") command.

By printing the corpus, we see that it contains documents for each of the 5,559 SMS
messages in the training data:

```
> print(sms_corpus)
<<VCorpus>>
Metadata:  corpus specific: 0, document level (indexed): 0
Content:  documents: 5559
```

Now, because the tm corpus is essentially a complex list, we can use list operations
to select documents in the corpus. The inspect() function shows a summary of the
result. For example, the following command will view a summary of the first and
second SMS messages in the corpus:

```
> inspect(sms_corpus[1:2])
<<VCorpus>>
Metadata:  corpus specific: 0, document level (indexed): 0
Content:  documents: 2

[[1]]
<<PlainTextDocument>>
Metadata:  7
Content:  chars: 49

[[2]]
<<PlainTextDocument>>
Metadata:  7
Content:  chars: 23
```

To view the actual message text, the `as.character()` function must be applied to the desired messages. To view one message, use the `as.character()` function on a single list element, noting that the double-bracket notation is required:

```
> as.character(sms_corpus[[1]])
[1] "Hope you are having a good week. Just checking in"
```

To view multiple documents, we'll need to apply `as.character()` to several items in the `sms_corpus` object. For this, we'll use the `lapply()` function, which is part of a family of R functions that applies a procedure to each element of an R data structure. These functions, which include `apply()` and `sapply()` among others, are one of the key idioms of the R language. Experienced R coders use these much like the way `for` or `while` loops are used in other programming languages, as they result in more readable (and sometimes more efficient) code. The `lapply()` function for applying `as.character()` to a subset of corpus elements is as follows:

```
> lapply(sms_corpus[1:2], as.character)
$'1'
[1] "Hope you are having a good week. Just checking in"


$'2'
[1] "K..give back my thanks."
```

As noted earlier, the corpus contains the raw text of 5,559 text messages. To perform our analysis, we need to divide these messages into individual words. First, we need to clean the text to standardize the words and remove punctuation characters that clutter the result. For example, we would like the strings *Hello!*, *HELLO*, and *hello* to be counted as instances of the same word.

The `tm_map()` function provides a method to apply a transformation (also known as a mapping) to a `tm` corpus. We will use this function to clean up our corpus using a series of transformations and save the result in a new object called `corpus_clean`.

Our first transformation will standardize the messages to use only lowercase characters. To this end, R provides a `tolower()` function that returns a lowercase version of text strings. In order to apply this function to the corpus, we need to use the `tm` wrapper function `content_transformer()` to treat `tolower()` as a transformation function that can be used to access the corpus. The full command is as follows:

```
> sms_corpus_clean <- tm_map(sms_corpus,
    content_transformer(tolower))
```

To check whether the command worked as expected, let's inspect the first message in the original corpus and compare it to the same in the transformed corpus:

```
> as.character(sms_corpus[[1]])
[1] "Hope you are having a good week. Just checking in"
> as.character(sms_corpus_clean[[1]])
[1] "hope you are having a good week. just checking in"
```

As expected, uppercase letters in the clean corpus have been replaced by lowercase versions of the same.

> The content_transformer() function can be used to apply more sophisticated text processing and cleanup processes like grep pattern matching and replacement. Simply write a custom function and wrap it before applying the tm_map() function.

Let's continue our cleanup by removing numbers from the SMS messages. Although some numbers may provide useful information, the majority are likely to be unique to individual senders and thus will not provide useful patterns across all messages. With this in mind, we'll strip all numbers from the corpus as follows:

```
> sms_corpus_clean <- tm_map(sms_corpus_clean, removeNumbers)
```

> Note that the preceding code did not use the content_transformer() function. This is because removeNumbers() is built into tm along with several other mapping functions that do not need to be wrapped. To see the other built-in transformations, simply type getTransformations().

Our next task is to remove filler words such as *to, and, but,* and *or* from the SMS messages. These terms are known as **stop words** and are typically removed prior to text mining. This is due to the fact that although they appear very frequently, they do not provide much useful information for machine learning.

Rather than define a list of stop words ourselves, we'll use the stopwords() function provided by the tm package. This function allows us to access sets of stop words from various languages. By default, common English language stop words are used. To see the default list, type stopwords() at the R command prompt. To see the other languages and options available, type ?stopwords for the documentation page.

Even within a single language, there is no single definitive list of stop words. For example, the default English list in tm includes about 174 words, while another option includes 571 words. You can even specify your own list of stop words. Regardless of the list you choose, keep in mind the goal of this transformation, which is to eliminate useless data while keeping as much useful information as possible.

The stop words alone are not a transformation. What we need is a way to remove any words that appear in the stop words list. The solution lies in the removeWords() function, which is a transformation included with the tm package. As we have done before, we'll use the tm_map() function to apply this mapping to the data, providing the stopwords() function as a parameter to indicate exactly the words we would like to remove. The full command is as follows:

```
> sms_corpus_clean <- tm_map(sms_corpus_clean,
    removeWords, stopwords())
```

Since stopwords() simply returns a vector of stop words, if we had so chosen, we could have replaced this function call with our own vector of words to remove. In this way, we could expand or reduce the list of stop words to our liking or remove a different set of words entirely.

Continuing our cleanup process, we can also eliminate any punctuation from the text messages using the built-in removePunctuation() transformation:

```
> sms_corpus_clean <- tm_map(sms_corpus_clean, removePunctuation)
```

The removePunctuation() transformation completely strips punctuation characters from the text, which can lead to unintended consequences. For example, consider what happens when it is applied as follows:

```
> removePunctuation("hello...world")
[1] "helloworld"
```

As shown, the lack of a blank space after the ellipses caused the words *hello* and *world* to be joined as a single word. While this is not a substantial problem right now, it is worth noting for the future.

To work around the default behavior of removePunctuation(), create a custom function that replaces rather than removes punctuation characters:

```
> replacePunctuation <- function(x) {
    gsub("[[:punct:]]+", " ", x)
}
```

This uses R's gsub() function to substitute any punctuation characters in x with a blank space. This replacePunctuation() function can then be used with tm_map() as with other transformations.

Another common standardization for text data involves reducing words to their root form in a process called **stemming**. The stemming process takes words like *learned, learning,* and *learns,* and strips the suffix in order to transform them into the base form, *learn.* This allows machine learning algorithms to treat the related terms as a single concept rather than attempting to learn a pattern for each variant.

The tm package provides stemming functionality via integration with the SnowballC package. At the time of writing, SnowballC is not installed by default with tm, so do so with install.packages("SnowballC") if you have not done so already.

> The SnowballC package is maintained by Milan Bouchet-Valat and provides an R interface to the C-based libstemmer library, itself based on M.F. Porter's "Snowball" word stemming algorithm, a widely used open-source stemming method. For more detail, see http:// snowballstem.org.

The SnowballC package provides a wordStem() function, which for a character vector, returns the same vector of terms in its root form. For example, the function correctly stems the variants of the word *learn* as described previously:

```
> library(SnowballC)
> wordStem(c("learn", "learned", "learning", "learns"))
[1] "learn"   "learn"   "learn"   "learn"
```

To apply the wordStem() function to an entire corpus of text documents, the tm package includes a stemDocument() transformation. We apply this to our corpus with the tm_map() function exactly as before:

```
> sms_corpus_clean <- tm_map(sms_corpus_clean, stemDocument)
```

> If you receive an error message when applying the stemDocument() transformation, please confirm that you have the SnowballC package installed. With the package installed, if you encounter a message about *all scheduled cores encountered errors* you might also try forcing the tm_map() command to a single core by adding an additional parameter specifying mc.cores = 1.

After removing numbers, stop words, and punctuation, and also performing stemming, the text messages are left with the blank spaces that once separated the now-missing pieces. Therefore, the final step in our text cleanup process is to remove additional whitespace using the built-in stripWhitespace() transformation:

```
> sms_corpus_clean <- tm_map(sms_corpus_clean, stripWhitespace)
```

The following table shows the first three messages in the SMS corpus before and after the cleaning process. The messages have been limited to the most interesting words, and punctuation and capitalization have been removed:

SMS messages before cleaning	SMS messages after cleaning

`> as.character(sms_corpus[1:3])`
 	
`[[1]] Hope you are having a good week. Just checking in`
	`> as.character(sms_corpus_clean[1:3])`
	
	`[[1]] hope good week just check`

`[[2]] K..give back my thanks.`	`[[2]] kgive back thank`
`[[3]] Am also doing in cbe only. But have to pay.`	`[[3]] also cbe pay`

Data preparation – splitting text documents into words

Now that the data are processed to our liking, the final step is to split the messages into individual terms through a process called **tokenization**. A token is a single element of a text string; in this case, the tokens are words.

As you might assume, the `tm` package provides functionality to tokenize the SMS message corpus. The `DocumentTermMatrix()` function takes a corpus and creates a data structure called a **document-term matrix (DTM)** in which rows indicate documents (SMS messages) and columns indicate terms (words).

The `tm` package also provides a data structure for a **term-document matrix (TDM)**, which is simply a transposed DTM in which the rows are terms and the columns are documents. Why the need for both? Sometimes it is more convenient to work with one or the other. For example, if the number of documents is small, while the word list is large, it may make sense to use a TDM because it is usually easier to display many rows than to display many columns. That said, the two are generally interchangeable.

Each cell in the matrix stores a number indicating a count of the times the word represented by the column appears in the document represented by the row. The following illustration depicts only a small portion of the DTM for the SMS corpus, as the complete matrix has 5,559 rows and over 7,000 columns:

message #	balloon	balls	bam	bambling	band
1	0	0	0	0	0
2	0	0	0	0	0
3	0	0	0	0	0
4	0	0	0	0	0
5	0	0	0	0	0

Figure 4.8: The DTM for the SMS messages is filled with mostly zeros

The fact that each cell in the table is zero implies that none of the words listed at the top of the columns appear in any of the first five messages in the corpus. This highlights the reason why this data structure is called a **sparse matrix**; the vast majority of cells in the matrix are filled with zeros. Stated in real-world terms, although each message must contain at least one word, the probability of any one word appearing in a given message is small.

Creating a DTM sparse matrix from a tm corpus involves a single command:

```
> sms_dtm <- DocumentTermMatrix(sms_corpus_clean)
```

This will create an sms_dtm object that contains the tokenized corpus using the default settings, which applies minimal processing. The default settings are appropriate because we have already prepared the corpus manually.

On the other hand, if we hadn't already performed the preprocessing, we could do so here by providing a list of control parameter options to override the defaults. For example, to create a DTM directly from the raw, unprocessed SMS corpus, we can use the following command:

```
> sms_dtm2 <- DocumentTermMatrix(sms_corpus, control = list(
    tolower = TRUE,
    removeNumbers = TRUE,
    stopwords = TRUE,
    removePunctuation = TRUE,
    stemming = TRUE
))
```

This applies the same preprocessing steps to the SMS corpus in the same order as done earlier. However, comparing sms_dtm to the sms_dtm2, we see a slight difference in the number of terms in the matrix:

```
> sms_dtm
<<DocumentTermMatrix (documents: 5559, terms: 6559)>>
Non-/sparse entries: 42147/36419334
Sparsity           : 100%
```

```
Maximal term length: 40
Weighting          : term frequency (tf)


> sms_dtm2
<<DocumentTermMatrix (documents: 5559, terms: 6961)>>
Non-/sparse entries: 43221/38652978
Sparsity            : 100%
Maximal term length: 40
Weighting          : term frequency (tf)
```

The reason for this discrepancy has to do with a minor difference in the ordering of the preprocessing steps. The `DocumentTermMatrix()` function applies its cleanup functions to the text strings only after they have been split apart into words. Thus, it uses a slightly different stop words removal function. Consequently, some words are split differently than when they are cleaned before tokenization.

> To force the two prior DTMs to be identical, we can override the default stop words function with our own that uses the original replacement function. Simply replace `stopwords = TRUE` with the following:
>
> `stopwords = function(x) { removeWords(x, stopwords()) }`

The differences between these two cases illustrate an important principle of cleaning text data: the order of operations matters. With this in mind, it is very important to think through how early steps in the process are going to affect later ones. The order presented here will work in many cases, but when the process is tailored more carefully to specific datasets and use cases, it may require rethinking. For example, if there are certain terms you hope to exclude from the matrix, consider whether to search for them before or after stemming. Also consider how the removal of punctuation—and whether the punctuation is eliminated or replaced by blank space—affects these steps.

Data preparation – creating training and test datasets

With our data prepared for analysis, we now need to split the data into training and test datasets, so that after our spam classifier is built, it can be evaluated on data it had not previously seen. However, even though we need to keep the classifier blinded as to the contents of the test dataset, it is important that the split occurs after the data have been cleaned and processed. We need exactly the same preparation steps to have occurred on both the training and test datasets.

We'll divide the data into two portions: 75 percent for training and 25 percent for testing. Since the SMS messages are sorted in a random order, we can simply take the first 4,169 for training and leave the remaining 1,390 for testing. Thankfully, the DTM object acts very much like a data frame and can be split using the standard [row, col] operations. As our DTM stores SMS messages as rows and words as columns, we must request a specific range of rows and all columns for each:

```
> sms_dtm_train <- sms_dtm[1:4169, ]
> sms_dtm_test  <- sms_dtm[4170:5559, ]
```

For convenience later on, it is also helpful to save a pair of vectors with the labels for each of the rows in the training and testing matrices. These labels are not stored in the DTM, so we need to pull them from the original sms_raw data frame:

```
> sms_train_labels <- sms_raw[1:4169, ]$type
> sms_test_labels  <- sms_raw[4170:5559, ]$type
```

To confirm that the subsets are representative of the complete set of SMS data, let's compare the proportion of spam in the training and test data frames:

```
> prop.table(table(sms_train_labels))

      ham      spam
0.8647158 0.1352842

> prop.table(table(sms_test_labels))

      ham      spam
0.8683453 0.1316547
```

Both the training data and test data contain about 13 percent spam. This suggests that the spam messages were divided evenly between the two datasets.

Visualizing text data – word clouds

A **word cloud** is a way to visually depict the frequency at which words appear in text data. The cloud is composed of words scattered somewhat randomly around the figure. Words appearing more often in the text are shown in a larger font, while less common terms are shown in smaller fonts. This type of figure grew in popularity as a way to observe trending topics on social media websites.

The wordcloud package provides a simple R function to create this type of diagram. We'll use it to visualize the words in SMS messages. Comparing the clouds for spam and ham messages will help us gauge whether our Naive Bayes spam filter is likely to be successful. If you haven't already done so, install and load the package by typing install.packages("wordcloud") and library(wordcloud) at the R command-line.

The `wordcloud` package was written by Ian Fellows. For more information about this package, visit his blog at: `http://blog.fellstat.com/?cat=11`.

A word cloud can be created directly from a `tm` corpus object using the syntax:

```
> wordcloud(sms_corpus_clean, min.freq = 50, random.order = FALSE)
```

This will create a word cloud from our prepared SMS corpus. Since we specified `random.order = FALSE`, the cloud will be arranged in non-random order, with higher-frequency words placed closer to the center. If we do not specify `random.order`, the cloud will be arranged randomly by default.

The `min.freq` parameter specifies the number of times a word must appear in the corpus before it will be displayed in the cloud. Since a frequency of 50 is about one percent of the corpus, this means that a word must be found in at least one percent of the SMS messages to be included in the cloud.

 You might get a warning message noting that R was unable to fit all of the words on the figure. If so, try increasing the `min.freq` to reduce the number of words in the cloud. It might also help to use the `scale` parameter to reduce the font size.

The resulting word cloud should appear similar to the following:

Figure 4.9: A word cloud depicting words appearing in all SMS messages

A perhaps more interesting visualization involves comparing the clouds for SMS spam and ham. Since we did not construct separate corpora for spam and ham, this is an appropriate time to note a very helpful feature of the `wordcloud()` function. Given a vector of raw text strings, it will automatically apply common text preparation processes before displaying the cloud.

Let's use R's `subset()` function to take a subset of the `sms_raw` data by the SMS type. First, we'll create a subset where the `type` is `spam`:

```
> spam <- subset(sms_raw, type == "spam")
```

Next, we'll do the same thing for the `ham` subset:

```
> ham <- subset(sms_raw, type == "ham")
```

 Be careful to note the double equal sign. Like many programming languages, R uses `==` to test equality. If you accidently use a single equals sign, you'll end up with a subset much larger than you expected!

We now have two data frames, `spam` and `ham`, each with a `text` feature containing the raw text strings for SMS messages. Creating word clouds is as simple as before. This time, we'll use the `max.words` parameter to look at the 40 most common words in each of the two sets. The `scale` parameter allows us to adjust the maximum and minimum font size for words in the cloud. Feel free to adjust these parameters as you see fit. This is illustrated in the following code:

```
> wordcloud(spam$text, max.words = 40, scale = c(3, 0.5))
> wordcloud(ham$text, max.words = 40, scale = c(3, 0.5))
```

The resulting word clouds should appear similar to the following:

Figure 4.10: Side-by-side word clouds depicting SMS spam and ham messages

Do you have a hunch which one is the spam cloud and which represents ham?

> Because of the randomization process, each word cloud may look slightly different. Running the `wordcloud()` function several times allows you to choose the cloud that is the most visually appealing for presentation purposes.

As you probably guessed, the spam cloud is on the left. Spam messages include words such as *urgent, free, mobile, claim,* and *stop*; these terms do not appear in the ham cloud at all. Instead, ham messages use words such as *can, sorry, need,* and *time*. These stark differences suggest that our Naive Bayes model will have some strong key words to differentiate between the classes.

Data preparation – creating indicator features for frequent words

The final step in the data preparation process is to transform the sparse matrix into a data structure that can be used to train a Naive Bayes classifier. Currently, the sparse matrix includes over 6,500 features; this is a feature for every word that appears in at least one SMS message. It's unlikely that all of these are useful for classification. To reduce the number of features, we'll eliminate any word that appears in less than five messages, or in less than about 0.1 percent of records in the training data.

Finding frequent words requires the use of the `findFreqTerms()` function in the `tm` package. This function takes a DTM and returns a character vector containing the words that appear at least a minimum number of times. For instance, the following command displays the words appearing at least five times in the `sms_dtm_train` matrix:

```
> findFreqTerms(sms_dtm_train, 5)
```

The result of the function is a character vector, so let's save our frequent words for later:

```
> sms_freq_words <- findFreqTerms(sms_dtm_train, 5)
```

A peek into the contents of the vector shows us that there are 1,139 terms appearing in at least five SMS messages:

```
> str(sms_freq_words)
 chr [1:1139] "£wk" "€~m" "€~s" "abiola" "abl" "abt" "accept" "access"
"account" "across" "act" "activ" ...
```

We now need to filter our DTM to include only the terms appearing in the frequent word vector. As before, we'll use data frame style [row, col] operations to request specific sections of the DTM, noting that the columns are named after the words the DTM contains. We can take advantage of this fact to limit the DTM to specific words. Since we want all rows but only the columns representing the words in the sms_ freq_words vector, our commands are:

```
> sms_dtm_freq_train <- sms_dtm_train[ , sms_freq_words]
> sms_dtm_freq_test <- sms_dtm_test[ , sms_freq_words]
```

The training and test datasets now include 1,139 features, which correspond to words appearing in at least five messages.

The Naive Bayes classifier is usually trained on data with categorical features. This poses a problem, since the cells in the sparse matrix are numeric and measure the number of times a word appears in a message. We need to change this to a categorical variable that simply indicates yes or no, depending on whether the word appears at all.

The following defines a convert_counts() function to convert counts to Yes or No strings:

```
> convert_counts <- function(x) {
    x <- ifelse(x > 0, "Yes", "No")
  }
```

By now, some of the pieces of the preceding function should look familiar. The first line defines the function. The statement ifelse(x > 0, "Yes", "No") transforms the values in x such that if the value is greater than 0, then it will be replaced with "Yes", otherwise it will be replaced by a "No" string. Lastly, the newly transformed vector x is returned.

We now need to apply convert_counts() to each of the columns in our sparse matrix. You may be able to guess the R function to do exactly this. The function is simply called apply() and is used much like lapply() was used previously.

The apply() function allows a function to be used on each of the rows or columns in a matrix. It uses a MARGIN parameter to specify either rows or columns. Here, we'll use MARGIN = 2 since we're interested in the columns (MARGIN = 1 is used for rows). The commands to convert the training and test matrices are as follows:

```
> sms_train <- apply(sms_dtm_freq_train, MARGIN = 2,
    convert_counts)
> sms_test  <- apply(sms_dtm_freq_test, MARGIN = 2,
    convert_counts)
```

The result will be two character-type matrices, each with cells indicating "Yes" or "No" for whether the word represented by the column appears at any point in the message represented by the row.

Step 3 – training a model on the data

Now that we have transformed the raw SMS messages into a format that can be represented by a statistical model, it is time to apply the Naive Bayes algorithm. The algorithm will use the presence or absence of words to estimate the probability that a given SMS message is spam.

The Naive Bayes implementation we will employ is in the e1071 package. This package was developed at the statistics department at the Vienna University of Technology (TU Wien) and includes a variety of functions for machine learning. If you have not done so already, be sure to install and load the package using the install.packages("e1071") and library(e1071) commands before continuing.

> Many machine learning approaches are implemented in more than one R package, and Naive Bayes is no exception. One other option is NaiveBayes() in the klaR package, which is nearly identical to the one in the e1071 package. Feel free to use whichever you prefer.

Unlike the k-NN algorithm we used for classification in the previous chapter, training a Naive Bayes learner and using it for classification occur in separate stages. Still, as shown in the following table, these steps are fairly straightforward:

Naive Bayes classification syntax
using the **naiveBayes()** function in the **e1071** package
Building the classifier:
`m <- naiveBayes(train, class, laplace = 0)` • **train** is a data frame or matrix containing training data • **class** is a factor vector with the class for each row in the training data • **laplace** is a number to control the Laplace estimator (by default, 0) The function will return a naive Bayes model object that can be used to make predictions.
Making predictions:
`p <- predict(m, test, type = "class")` • **m** is a model trained by the **naiveBayes()** function • **test** is a data frame or matrix containing test data with the same features as the training data used to build the classifier • **type** is either **"class"** or **"raw"** and specifies whether the predictions should be the most likely class value or the raw predicted probabilities The function will return a vector of predicted class values or raw predicted probabilities depending upon the value of the **type** parameter.
Example:
`sms_classifier <- naiveBayes(sms_train, sms_type)` `sms_predictions <- predict(sms_classifier, sms_test)`

To build our model on the `sms_train` matrix, we'll use the following command:

```
> sms_classifier <- naiveBayes(sms_train, sms_train_labels)
```

The `sms_classifier` variable now contains a `naiveBayes` classifier object that can be used to make predictions.

Step 4 – evaluating model performance

To evaluate the SMS classifier, we need to test its predictions on the unseen messages in the test data. Recall that the unseen message features are stored in a matrix named `sms_test`, while the class labels (spam or ham) are stored in a vector named `sms_test_labels`. The classifier that we trained has been named `sms_classifier`. We will use this classifier to generate predictions and then compare the predicted values to the true values.

The `predict()` function is used to make the predictions. We will store these in a vector named `sms_test_pred`. We simply supply this function with the names of our classifier and test dataset as shown:

```
> sms_test_pred <- predict(sms_classifier, sms_test)
```

To compare the predictions to the true values, we'll use the `CrossTable()` function in the `gmodels` package, which we used previously. This time, we'll add some additional parameters to eliminate unnecessary cell proportions, and use the `dnn` parameter (dimension names) to relabel the rows and columns as shown in the following code:

```
> library(gmodels)
> CrossTable(sms_test_pred, sms_test_labels,
    prop.chisq = FALSE, prop.c = FALSE, prop.r = FALSE,
    dnn = c('predicted', 'actual'))
```

This produces the following table:

```
         Total Observations in Table:  1390

             | actual
   predicted |       ham |      spam | Row Total |
-------------|-----------|-----------|-----------|
         ham |      1201 |        30 |      1231 |
             |     0.864 |     0.022 |           |
-------------|-----------|-----------|-----------|
        spam |         6 |       153 |       159 |
             |     0.004 |     0.110 |           |
-------------|-----------|-----------|-----------|
Column Total |      1207 |       183 |      1390 |
-------------|-----------|-----------|-----------|
```

Looking at the table, we can see that a total of only *6 + 30 = 36* of 1,390 SMS messages were incorrectly classified (2.6 percent). Among the errors were six out of 1,207 ham messages that were misidentified as spam, and 30 of 183 spam messages that were incorrectly labeled as ham. Considering the little effort that we put into the project, this level of performance seems quite impressive. This case study exemplifies the reason why Naive Bayes is so often used for text classification: directly out of the box, it performs surprisingly well.

On the other hand, the six legitimate messages that were incorrectly classified as spam could cause significant problems for the deployment of our filtering algorithm, because the filter could cause a person to miss an important text message. We should investigate to see whether we can slightly tweak the model to achieve better performance.

Step 5 – improving model performance

You may have noticed that we didn't set a value for the Laplace estimator when training our model. This allows words that appeared in zero spam or zero ham messages to have an indisputable say in the classification process. Just because the word "ringtone" only appeared in spam messages in the training data, it does not mean that every message with this word should be classified as spam.

We'll build a Naive Bayes model as before, but this time set `laplace = 1`:

```
> sms_classifier2 <- naiveBayes(sms_train, sms_train_labels,
    laplace = 1)
```

Next, we'll make predictions:

```
> sms_test_pred2 <- predict(sms_classifier2, sms_test)
```

Finally, we'll compare the predicted classes to the actual classifications using a cross tabulation:

```
> CrossTable(sms_test_pred2, sms_test_labels,
    prop.chisq = FALSE, prop.c = FALSE, prop.r = FALSE,
    dnn = c('predicted', 'actual'))
```

This results in the following table:

```
Total Observations in Table:  1390

              | actual
    predicted |      ham |     spam | Row Total |
--------------|----------|----------|-----------|
          ham |     1202 |       28 |      1230 |
              |    0.996 |    0.153 |           |
--------------|----------|----------|-----------|
         spam |        5 |      155 |       160 |
              |    0.004 |    0.847 |           |
--------------|----------|----------|-----------|
 Column Total |     1207 |      183 |      1390 |
              |    0.868 |    0.132 |           |
--------------|----------|----------|-----------|
```

Adding the Laplace estimator reduced the number of false positives (ham messages erroneously classified as spam) from six to five, and the number of false negatives from 30 to 28. Although this seems like a small change, it's substantial considering that the model's accuracy was already quite impressive. We'd need to be careful before tweaking the model too much more, as it is important to maintain a balance between being overly aggressive and overly passive when filtering spam. Users would prefer that a small number of spam messages slip through the filter rather than an alternative in which ham messages are filtered too aggressively.

Summary

In this chapter, we learned about classification using Naive Bayes. This algorithm constructs tables of probabilities that are used to estimate the likelihood that new examples belong to various classes. The probabilities are calculated using a formula known as Bayes' theorem, which specifies how dependent events are related. Although Bayes' theorem can be computationally expensive, a simplified version that makes so-called "naive" assumptions about the independence of features is capable of handling much larger datasets.

The Naive Bayes classifier is often used for text classification. To illustrate its effectiveness, we employed Naive Bayes on a classification task involving spam SMS messages. Preparing the text data for analysis required the use of specialized R packages for text processing and visualization. Ultimately, the model was able to classify over 97 percent of all the SMS messages correctly as spam or ham.

In the next chapter, we will examine two more machine learning methods. Each performs classification by partitioning data into groups of similar values.

5

Divide and Conquer – Classification Using Decision Trees and Rules

When deciding between job offers with various levels of pay and benefits, many people begin by making lists of pros and cons, then eliminate options using simple rules. For instance, saying "If I must commute more than an hour, I will be unhappy," or "If I make less than $50K, I won't be able to support my family." In this way, the complex and difficult decision of predicting one's future happiness can be reduced to a series of simple decisions.

This chapter covers decision trees and rule learners—two machine learning methods that also make complex decisions from sets of simple choices. These methods present their knowledge in the form of logical structures that can be understood with no statistical knowledge. This aspect makes these models particularly useful for business strategy and process improvement.

By the end of this chapter, you will learn:

- How trees and rules "greedily" partition data into interesting segments
- The most common decision tree and classification rule learners, including the C5.0, 1R, and RIPPER algorithms
- How to use these algorithms for performing real-world classification tasks, such as identifying risky bank loans and poisonous mushrooms

We will begin by examining decision trees and follow with a look at classification rules. Then, we will summarize what we've learned by previewing later chapters, which discuss methods that use trees and rules as a foundation for more advanced machine learning techniques.

Understanding decision trees

Decision tree learners are powerful classifiers that utilize a **tree structure** to model the relationships among the features and the potential outcomes. As illustrated in the following figure, this structure earned its name due to the fact that it mirrors the way a literal tree begins at a wide trunk and splits into narrower and narrower branches as it is followed upward. In much the same way, a decision tree classifier uses a structure of branching decisions that channel examples into a final predicted class value.

To better understand how this works in practice, let's consider the following tree, which predicts whether a job offer should be accepted. A job offer under consideration begins at the **root node**, where it is then passed through **decision nodes** that require choices to be made based on the attributes of the job. These choices split the data across **branches** that indicate potential outcomes of a decision. They are depicted here as yes or no outcomes, but in other cases, there may be more than two possibilities.

If a final decision can be made, the tree is terminated by **leaf nodes** (also known as **terminal nodes**) that denote the action to be taken as the result of the series of decisions. In the case of a predictive model, the leaf nodes provide the expected result given the series of events in the tree.

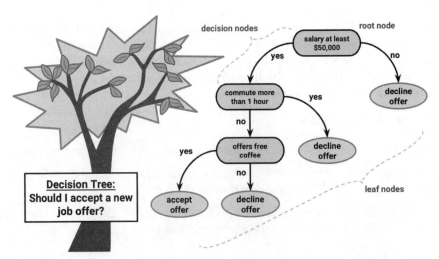

Figure 5.1: A decision tree depicting the process of determining whether to accept a new job offer

A great benefit of decision tree algorithms is that the flowchart-like tree structure is not only for the machine's internal use. After the model is created, many decision tree algorithms output the resulting structure in a human-readable format. This provides insight into how and why the model works or doesn't work well for a particular task. This also makes decision trees particularly appropriate for applications in which the classification mechanism needs to be transparent for legal reasons, or if the results need to be shared with others to inform future business practices. With this in mind, some potential uses include:

- Credit scoring models in which the criteria that cause an applicant to be rejected need to be clearly documented and free from bias

- Marketing studies of customer behavior, such as satisfaction or churn, which will be shared with management or advertising agencies

- Diagnosis of medical conditions based on laboratory measurements, symptoms, or rate of disease progression

Although the previous applications illustrate the value of trees for informing decision processes, this is not to suggest that their utility ends here. In fact, decision trees are perhaps the single most widely used machine learning technique, and can be applied for modeling almost any type of data—often with excellent out-of-the-box performance.

That said, in spite of their wide applicability, it is worth noting that there are some scenarios where trees may not be an ideal fit. This includes tasks where the data has a large number of nominal features with many levels or a large number of numeric features. These cases may result in a very large number of decisions and an overly complex tree. They may also contribute to the tendency of decision trees to overfit data, though as we will soon see, even this weakness can be overcome by adjusting some simple parameters.

Divide and conquer

Decision trees are built using a heuristic called **recursive partitioning**. This approach is also commonly known as **divide and conquer** because it splits the data into subsets, which are then split repeatedly into even smaller subsets, and so on and so forth until the process stops when the algorithm determines the data within the subsets are sufficiently homogenous, or another stopping criterion has been met.

To see how splitting a dataset can create a decision tree, imagine a bare root node that will grow into a mature tree. At first, the root node represents the entire dataset, since no splitting has transpired. Here, the decision tree algorithm must choose a feature to split upon; ideally, it chooses the feature most predictive of the target class. The examples are then partitioned into groups according to the distinct values of this feature, and the first set of tree branches is formed.

Working down each branch, the algorithm continues to divide and conquer the data, choosing the best candidate feature each time to create another decision node until a stopping criterion is reached. Divide and conquer might stop at a node if:

- All (or nearly all) of the examples at the node have the same class
- There are no remaining features to distinguish among examples
- The tree has grown to a predefined size limit

To illustrate the tree-building process, let's consider a simple example. Imagine that you work for a Hollywood studio, where your role is to decide whether the studio should move forward with producing the screenplays pitched by promising new authors. After returning from a vacation, your desk is piled high with proposals. Without the time to read each proposal cover-to-cover, you decide to develop a decision tree algorithm to predict whether a potential movie would fall into one of three categories: **Critical Success**, **Mainstream Hit**, or **Box Office Bust**.

To build the decision tree, you turn to the studio archives to examine the factors leading to the success and failure of the company's 30 most recent releases. You quickly notice a relationship between the film's estimated shooting budget, the number of A-list celebrities lined up for starring roles, and the level of success. Excited about this finding, you produce a scatterplot to illustrate the pattern:

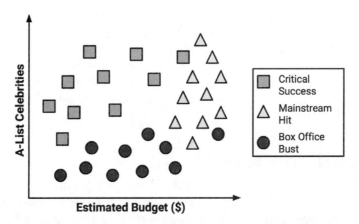

Figure 5.2: A scatterplot depicting the relationship between a movie's budget and celebrity count

Using the divide and conquer strategy, we can build a simple decision tree from this data. First, to create the tree's root node, we split the feature indicating the number of celebrities, partitioning the movies into groups with and without a significant number of A-list stars:

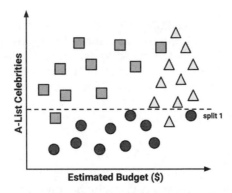

Figure 5.3: The decision tree's first split divides the data into high and low celebrity counts

Next, among the group of movies with a larger number of celebrities, we can make another split between movies with and without a high budget:

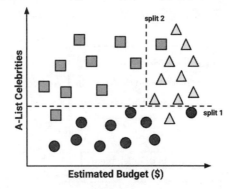

Figure 5.4: The decision tree's second split further divides the films
with a high celebrity count into those with low and high budgets

At this point, we have partitioned the data into three groups. The group at the top-left corner of the diagram is composed entirely of critically acclaimed films. This group is distinguished by a high number of celebrities and a relatively low budget. At the top-right corner, the majority of movies are box office hits with high budgets and a large number of celebrities. The final group, which has little star power but budgets ranging from small to large, contains the flops.

If we wanted, we could continue to divide and conquer the data by splitting it on increasingly specific ranges of budget and celebrity count until each of the currently misclassified values is correctly classified in its own tiny partition. However, it is not advisable to overfit a decision tree in this way. Although there is nothing stopping the algorithm from splitting the data indefinitely, overly specific decisions do not always generalize more broadly. We'll limit the problem of overfitting by stopping the algorithm here, since more than 80 percent of the examples in each group are from a single class. This forms the basis of our stopping criterion.

You might have noticed that diagonal lines might have split the data even more cleanly. This is one limitation of the decision tree's knowledge representation, which uses **axis-parallel splits**. The fact that each split considers one feature at a time prevents the decision tree from forming more complex decision boundaries. For example, a diagonal line could be created by a decision that asks, "Is the number of celebrities greater than the estimated budget?" If so, then "it will be a critical success."

Our model for predicting the future success of movies can be represented in a simple tree, as shown in the following diagram. Each step in the tree shows the fraction of examples falling into each class, which shows how the data becomes more homogeneous as the branches get closer to a leaf. To evaluate a new movie script, follow the branches through each decision until the script's success or failure has been predicted. Using this approach, you will be able to quickly identify the most promising options among the backlog of scripts and get back to more important work, such as writing an Academy Awards acceptance speech!

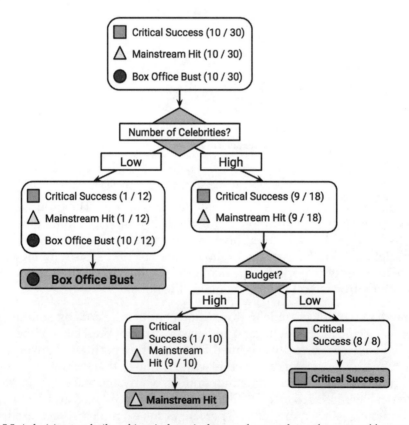

Figure 5.5: A decision tree built on historical movie data can forecast the performance of future movies

Since real-world data contains more than two features, decision trees quickly become far more complex than this, with many more nodes, branches, and leaves. In the next section, you will learn about a popular algorithm to build decision tree models automatically.

The C5.0 decision tree algorithm

There are numerous implementations of decision trees, but one of the most well-known is the **C5.0 algorithm**. This algorithm was developed by computer scientist J. Ross Quinlan as an improved version of his prior algorithm, **C4.5**, which itself is an improvement over his **Iterative Dichotomiser 3 (ID3)** algorithm. Although Quinlan markets C5.0 to commercial clients (see http://www.rulequest.com/ for details), the source code for a single-threaded version of the algorithm was made public, and has therefore been incorporated into programs such as R.

 To further confuse matters, a popular Java-based open-source alternative to C4.5, titled **J48**, is included in R's RWeka package. As the differences among C5.0, C4.5, and J48 are minor, the principles in this chapter will apply to any of these three methods and the algorithms should be considered synonymous.

The C5.0 algorithm has become the industry standard for producing decision trees because it does well for most types of problems directly out of the box. Compared to other advanced machine learning models, such as those described in *Chapter 7, Black Box Methods – Neural Networks and Support Vector Machines*, the decision trees built by C5.0 generally perform nearly as well but are much easier to understand and deploy. Additionally, as shown in the following table, the algorithm's weaknesses are relatively minor and can be largely avoided.

Strengths	Weaknesses
• An all-purpose classifier that does well on many types of problems • Highly automatic learning process, which can handle numeric or nominal features, as well as missing data • Excludes unimportant features • Can be used on both small and large datasets • Results in a model that can be interpreted without a mathematical background (for relatively small trees) • More efficient than other complex models	• Decision tree models are often biased toward splits on features having a large number of levels • It is easy to overfit or underfit the model • Can have trouble modeling some relationships due to reliance on axis-parallel splits • Small changes in training data can result in large changes to decision logic • Large trees can be difficult to interpret and the decisions they make may seem counterintuitive

To keep things simple, our earlier decision tree example ignored the mathematics involved with how a machine would employ a divide and conquer strategy. Let's explore this in more detail to examine how this heuristic works in practice.

Choosing the best split

The first challenge that a decision tree will face is to identify which feature to split upon. In the previous example, we looked for a way to split the data such that the resulting partitions contained examples primarily of a single class. The degree to which a subset of examples contains only a single class is known as **purity**, and any subset composed of only a single class is called **pure**.

There are various measurements of purity that can be used to identify the best decision tree splitting candidate. C5.0 uses **entropy**, a concept borrowed from information theory that quantifies the randomness, or disorder, within a set of class values. Sets with high entropy are very diverse and provide little information about other items that may also belong in the set, as there is no apparent commonality. The decision tree hopes to find splits that reduce entropy, ultimately increasing homogeneity within the groups.

Typically, entropy is measured in **bits**. If there are only two possible classes, entropy values can range from 0 to 1. For *n* classes, entropy ranges from 0 to $log_2(n)$. In each case, the minimum value indicates that the sample is completely homogenous, while the maximum value indicates that the data are as diverse as possible, and no group has even a small plurality.

In mathematical notion, entropy is specified as:

$$\text{Entropy}(S) = \sum_{i=1}^{c} -p_i log_2(p_i)$$

In this formula, for a given segment of data (*S*), the term *c* refers to the number of class levels, and p_i refers to the proportion of values falling into class level *i*. For example, suppose we have a partition of data with two classes: red (60 percent) and white (40 percent). We can calculate the entropy as:

```
> -0.60 * log2(0.60) - 0.40 * log2(0.40)
[1] 0.9709506
```

We can visualize the entropy for all possible two-class arrangements. If we know the proportion of examples in one class is *x*, then the proportion in the other class is *(1 – x)*. Using the curve() function, we can then plot the entropy for all possible values of *x*:

```
> curve(-x * log2(x) - (1 - x) * log2(1 - x),
        col = "red", xlab = "x", ylab = "Entropy", lwd = 4)
```

This results in the following graph:

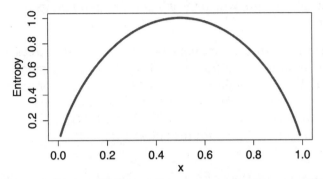

Figure 5.6: The total entropy as the proportion of one class varies in a two-class outcome

As illustrated by the peak in entropy at $x = 0.50$, a 50-50 split results in the maximum entropy. As one class increasingly dominates the other, the entropy reduces to zero.

To use entropy to determine the optimal feature to split upon, the algorithm calculates the change in homogeneity that would result from a split on each possible feature, a measure known as **information gain**. The information gain for a feature F is calculated as the difference between the entropy in the segment before the split (S_1) and the partitions resulting from the split (S_2):

$$\text{InfoGain}(F) = \text{Entropy}(S_1) - \text{Entropy}(S_2)$$

One complication is that after a split, the data is divided into more than one partition. Therefore, the function to calculate *Entropy(S_2)* needs to consider the total entropy across all of the partitions. It does this by weighting each partition's entropy according to the proportion of all records falling into that partition. This can be stated in a formula as:

$$\text{Entropy}(S) = \sum_{i=1}^{n} w_i \, \text{Entropy}(P_i)$$

In simple terms, the total entropy resulting from a split is the sum of entropy of each of the n partitions weighted by the proportion of examples falling in the partition (w_i).

The higher the information gain, the better a feature is at creating homogeneous groups after a split on that feature. If the information gain is zero, there is no reduction in entropy for splitting on this feature. On the other hand, the maximum information gain is equal to the entropy prior to the split. This would imply the entropy after the split is zero, which means that the split results in completely homogeneous groups.

The previous formulas assume nominal features, but decision trees use information gain for splitting on numeric features as well. To do so, a common practice is to test various splits that divide the values into groups greater than or less than a threshold. This reduces the numeric feature into a two-level categorical feature that allows information gain to be calculated as usual. The numeric cut point yielding the largest information gain is chosen for the split.

Though it is used by C5.0, information gain is not the only splitting criterion that can be used to build decision trees. Other commonly used criteria are **Gini index**, **chi-squared statistic**, and **gain ratio**. For a review of these (and many more) criteria, refer to *An Empirical Comparison of Selection Measures for Decision-Tree Induction, Mingers, J, Machine Learning, 1989, Vol. 3, pp. 319-342.*

Pruning the decision tree

As mentioned earlier, a decision tree can continue to grow indefinitely, choosing splitting features and dividing into smaller and smaller partitions until each example is perfectly classified or the algorithm runs out of features to split on. However, if the tree grows overly large, many of the decisions it makes will be overly specific and the model will be overfitted to the training data. The process of **pruning** a decision tree involves reducing its size such that it generalizes better to unseen data.

One solution to this problem is to stop the tree from growing once it reaches a certain number of decisions or when the decision nodes contain only a small number of examples. This is called **early stopping** or **pre-pruning** the decision tree. As the tree avoids doing needless work, this is an appealing strategy. However, one downside to this approach is that there is no way to know whether the tree will miss subtle but important patterns that it would have learned had it grown to a larger size.

An alternative, called **post-pruning**, involves growing a tree that is intentionally too large and pruning leaf nodes to reduce the size of the tree to a more appropriate level. This is often a more effective approach than pre-pruning because it is quite difficult to determine the optimal depth of a decision tree without growing it first. Pruning the tree later on allows the algorithm to be certain that all of the important data structures were discovered.

The implementation details of pruning operations are very technical and beyond the scope of this book. For a comparison of some of the available methods, see *A Comparative Analysis of Methods for Pruning Decision Trees, Esposito, F, Malerba, D, Semeraro, G, IEEE Transactions on Pattern Analysis and Machine Intelligence, 1997, Vol. 19, pp. 476-491.*

One of the benefits of the C5.0 algorithm is that it is opinionated about pruning— it takes care of many of the decisions automatically using fairly reasonable defaults. Its overall strategy is to post-prune the tree. It first grows a large tree that overfits the training data. Later, the nodes and branches that have little effect on the classification errors are removed. In some cases, entire branches are moved further up the tree or replaced by simpler decisions. These processes of grafting branches are known as **subtree raising** and **subtree replacement**, respectively.

Getting the right balance of overfitting and underfitting is a bit of an art, but if model accuracy is vital, it may be worth investing some time with various pruning options to see if it improves the test dataset performance. As you will soon see, one of the strengths of the C5.0 algorithm is that it is very easy to adjust the training options.

Example – identifying risky bank loans using C5.0 decision trees

The global financial crisis of 2007-2008 highlighted the importance of transparency and rigor in banking practices. As the availability of credit was limited, banks tightened their lending systems and turned to machine learning to more accurately identify risky loans.

Decision trees are widely used in the banking industry due to their high accuracy and ability to formulate a statistical model in plain language. Since governments in many countries carefully monitor the fairness of lending practices, executives must be able to explain why one applicant was rejected for a loan while another was approved. This information is also useful for customers hoping to determine why their credit rating is unsatisfactory.

It is likely that automated credit scoring models are used for credit card mailings and instant online approval processes. In this section, we will develop a simple credit approval model using C5.0 decision trees. We will also see how the model results can be tuned to minimize errors that result in a financial loss.

Step 1 – collecting data

The motivation for our credit model is to identify factors that are linked to a higher risk of loan default. To do this, we must obtain data on a large number of past bank loans in addition to information about the loan applicants.

Data with these characteristics are available in a dataset donated to the UCI Machine Learning Repository (http://archive.ics.uci.edu/ml) by Hans Hofmann of the University of Hamburg. The dataset contains information on loans obtained from a credit agency in Germany.

The dataset presented in this chapter has been modified slightly from the original in order to eliminate some preprocessing steps. To follow along with the examples, download the credit.csv file from the Packt website and save it to your R working directory.

The credit dataset includes 1,000 examples of loans, plus a set of numeric and nominal features indicating characteristics of the loan and the loan applicant. A class variable indicates whether the loan went into default. Let's see if we can identify any patterns that predict this outcome.

Step 2 – exploring and preparing the data

As we have done previously, we will import the data using the read.csv() function. Now, because the character data is entirely categorical, we can omit the stringsAsFactors parameter, which will use the default value of TRUE. This creates a credit data frame with a number of factor variables:

```
> credit <- read.csv("credit.csv")
```

We can check the resulting object by examining the first few lines of output from the str() function:

```
> str(credit)
'data.frame':1000 obs. of  17 variables:
 $ checking_balance : Factor w/ 4 levels "< 0 DM","> 200 DM",..
 $ months_loan_duration: int  6 48 12 ...
 $ credit_history      : Factor w/ 5 levels "critical","good",..
 $ purpose             : Factor w/ 6 levels "business","car",..
 $ amount              : int  1169 5951 2096 ...
```

We see the expected 1,000 observations and 17 features, which are a combination of factor and integer data types.

Let's take a look at the `table()` output for a couple of loan features that seem likely to predict a default. The applicant's checking and savings account balance are recorded as categorical variables:

```
> table(credit$checking_balance)
   < 0 DM    > 200 DM 1 - 200 DM     unknown
      274          63         269         394
> table(credit$savings_balance)
   < 100 DM > 1000 DM  100 - 500 DM 500 - 1000 DM    unknown
      603         48           103            63        183
```

The checking and savings account balance may prove to be important predictors of loan default status. Note that since the loan data was obtained from Germany, the values use the Deutsche Mark (DM), which was the currency used in Germany prior to the adoption of the Euro.

Some of the loan's features are numeric, such as its duration and the amount of credit requested:

```
> summary(credit$months_loan_duration)
   Min. 1st Qu.  Median    Mean 3rd Qu.     Max.
    4.0    12.0    18.0    20.9    24.0     72.0
> summary(credit$amount)
   Min. 1st Qu.  Median    Mean 3rd Qu.     Max.
    250    1366    2320    3271    3972    18420
```

The loan amounts ranged from 250 DM to 18,420 DM across terms of four to 72 months. They had a median amount of 2,320 DM and median duration of 18 months.

The `default` vector indicates whether the loan applicant was able to meet the agreed payment terms or if they went into default. A total of 30 percent of the loans in this dataset went into default:

```
> table(credit$default)
 no yes
700 300
```

A high rate of default is undesirable for a bank because it means that the bank is unlikely to fully recover its investment. If we are successful, our model will identify applicants who are at high risk of default, allowing the bank to refuse the credit request before the money is given.

Data preparation – creating random training and test datasets

As we have done in previous chapters, we will split our data into two portions: a training dataset to build the decision tree and a test dataset to evaluate its performance on new data. We will use 90 percent of the data for training and 10 percent for testing, which will provide us with 100 records to simulate new applicants.

As prior chapters used data that had been sorted in a random order, we simply divided the dataset into two portions by taking the first 90 percent of records for training and the remaining 10 percent for testing. In contrast, the credit dataset is not randomly ordered, making the prior approach unwise. Suppose that the bank had sorted the data by the loan amount, with the largest loans at the end of the file. If we used the first 90 percent for training and the remaining 10 percent for testing, we would be training a model on only the small loans and testing the model on the big loans. Obviously, this could be problematic.

We'll solve this problem by training the model on a **random sample** of the credit data. A random sample is simply a process that selects a subset of records at random. In R, the `sample()` function is used to perform random sampling. However, before putting it in action, a common practice is to set a **seed** value, which causes the randomization process to follow a sequence that can be replicated later. It may seem that this defeats the purpose of generating random numbers, but there is a good reason for doing it this way. Providing a seed value via the `set.seed()` function ensures that if the analysis is repeated in the future, an identical result is obtained.

You may wonder how a so-called random process can be seeded to produce an identical result. This is due to the fact that computers use a mathematical function called a **pseudorandom number generator** to create random number sequences that appear to act very random, but are actually quite predictable given knowledge of the previous values in the sequence. In practice, modern pseudorandom number sequences are virtually indistinguishable from true random sequences, but have the benefit that computers can generate them quickly and easily.

The following commands use `sample()` with a seed value. Note that the `set.seed()` function uses the arbitrary value `123`. Omitting this seed will cause your training and testing splits to differ from those shown in the remainder of this chapter. If using R version 3.6.0 or greater, you will also need to request the random number generator from R version 3.5.2 using the `RNGversion("3.5.2")` command. The following commands select 900 values at random out of the sequence of integers from 1 to 1,000:

```
> RNGversion("3.5.2"); set.seed(123)
> train_sample <- sample(1000, 900)
```

As expected, the resulting `train_sample` object is a vector of 900 random integers:

```
> str(train_sample)
 int [1:900] 288 788 409 881 937 46 525 887 548 453 ...
```

By using this vector to select rows from the credit data, we can split it into the 90 percent training and 10 percent test datasets we desired. Recall that the negation operator (the - character) used in the selection of the test records tells R to select records that are not in the specified rows; in other words, the test data includes only the rows that are not in the training sample:

```
> credit_train <- credit[train_sample, ]
> credit_test  <- credit[-train_sample, ]
```

If randomization was done correctly, we should have about 30 percent of loans with default in each of the datasets:

```
> prop.table(table(credit_train$default))
       no       yes
0.7033333 0.2966667
```

```
> prop.table(table(credit_test$default))
  no  yes
0.67 0.33
```

Both the training and test datasets had similar distributions of loan defaults, so we can now build our decision tree. In the case that the proportions differ greatly, we may decide to resample the dataset, or attempt a more sophisticated sampling approach, such as those covered in *Chapter 10, Evaluating Model Performance*.

> If your results do not match exactly, ensure that you ran the commands `RNGversion("3.5.2")` and `set.seed(123)` immediately prior to creating the `train_sample` vector.

Step 3 – training a model on the data

We will use the C5.0 algorithm in the `C50` package for training our decision tree model. If you have not done so already, install the package with `install.packages("C50")` and load it to your R session using `library(C50)`.

The following syntax box lists some of the most common parameters used when building decision trees. Compared to the machine learning approaches we have used previously, the C5.0 algorithm offers many more ways to tailor the model to a particular learning problem.

Once the C50 package has been loaded, the `?C5.0Control` command displays the help page for more details on how to finely-tune the algorithm using these options.

C5.0 decision tree syntax

using the `C5.0()` function in the C50 package

Building the classifier:

```
m <- C5.0(train, class, trials = 1, costs = NULL)
```

- `train` is a data frame containing training data
- `class` is a factor vector with the class for each row in the training data
- `trials` is an optional number to control the number of boosting iterations (set to 1 by default)
- `costs` is an optional matrix specifying costs associated with various types of errors

The function will return a C5.0 model object that can be used to make predictions.

Making predictions:

```
p <- predict(m, test, type = "class")
```

- `m` is a model trained by the `C5.0()` function
- `test` is a data frame containing test data with the same features as the training data used to build the classifier.
- `type` is either `"class"` or `"prob"` and specifies whether the predictions should be the most probable class value or the raw predicted probabilities

The function will return a vector of predicted class values or raw predicted probabilities depending upon the value of the `type` parameter.

Example:

```
credit_model <- C5.0(credit_train, loan_default)
credit_prediction <- predict(credit_model,
  credit_test)
```

For the first iteration of the credit approval model, we'll use the default C5.0 settings, as shown in the following code. Column 17 in `credit_train` is the class variable, `default`, so we need to exclude it from the training data frame and supply it as the target factor vector for classification:

```
> credit_model <- C5.0(credit_train[-17], credit_train$default)
```

The `credit_model` object now contains a C5.0 decision tree. We can see some basic data about the tree by typing its name:

```
> credit_model
```

```
Call:
C5.0.default(x = credit_train[-17], y = credit_train$default)
```

```
Classification Tree
Number of samples: 900
Number of predictors: 16

Tree size: 57

Non-standard options: attempt to group attributes
```

The output shows some simple facts about the tree, including the function call that generated it, the number of features (labeled `predictors`), and examples (labeled `samples`) used to grow the tree. Also listed is the tree size of 57, which indicates that the tree is 57 decisions deep—quite a bit larger than the example trees we've considered so far!

To see the tree's decisions, we can call the `summary()` function on the model:

```
> summary(credit_model)
```

This results in the following output:

```
C5.0 [Release 2.07 GPL Edition]
-------------------------------

Class specified by attribute `outcome'

Read 900 cases (17 attributes) from undefined.data

Decision tree:

checking_balance in {> 200 DM,unknown}: no (412/50)
checking_balance in {< 0 DM,1 - 200 DM}:
:...credit_history in {perfect,very good}: yes (59/18)
    credit_history in {critical,good,poor}:
    :...months_loan_duration <= 22:
        :...credit_history = critical: no (72/14)
        :   credit_history = poor:
        :   :...dependents > 1: no (5)
        :   :   dependents <= 1:
        :   :   :...years_at_residence <= 3: yes (4/1)
        :   :       years_at_residence > 3: no (5/1)
```

The preceding output shows some of the first branches in the decision tree. The first three lines could be represented in plain language as:

1. If the checking account balance is unknown or greater than 200 DM, then classify as "not likely to default."
2. Otherwise, if the checking account balance is less than zero DM or between one and 200 DM...
3. ... and the credit history is perfect or very good, then classify as "likely to default."

The numbers in parentheses indicate the number of examples meeting the criteria for that decision and the number incorrectly classified by the decision. For instance, on the first line, 412/50 indicates, of the 412 examples reaching the decision, 50 were incorrectly classified as "not likely to default." In other words, 50 applicants actually defaulted in spite of the model's prediction to the contrary.

Sometimes a tree results in decisions that make little logical sense. For example, why would an applicant whose credit history is very good be likely to default, while those whose checking balance is unknown are not likely to default? Contradictory rules like this occur sometimes. They might reflect a real pattern in the data, or they may be a statistical anomaly. In either case, it is important to investigate such strange decisions to see whether the tree's logic makes sense for business use.

After the tree, the summary(credit_model) output displays a confusion matrix, which is a cross-tabulation that indicates the model's incorrectly classified records in the training data:

```
Evaluation on training data (900 cases):


    Decision Tree
    ----------------

    Size        Errors
      56   133(14.8%)    <<


     (a)     (b)      <-classified as
    ----    ----
     598      35      (a): class no
      98     169      (b): class yes
```

The Errors heading shows that the model correctly classified all but 133 of the 900 training instances for an error rate of 14.8 percent. A total of 35 actual no values were incorrectly classified as yes (false positives), while 98 yes values were misclassified as no (false negatives).

Given the tendency of decision trees to overfit to the training data, the error rate reported here, which is based on training data performance, may be overly optimistic. Therefore, it is especially important to continue our evaluation by applying our decision tree to a test dataset.

Step 4 – evaluating model performance

To apply our decision tree to the test dataset, we use the `predict()` function as shown in the following line of code:

```
> credit_pred <- predict(credit_model, credit_test)
```

This creates a vector of predicted class values, which we can compare to the actual class values using the `CrossTable()` function in the `gmodels` package. Setting the `prop.c` and `prop.r` parameters to `FALSE` removes the column and row percentages from the table. The remaining percentage (`prop.t`) indicates the proportion of records in the cell out of the total number of records:

```
> library(gmodels)
> CrossTable(credit_test$default, credit_pred,
          prop.chisq = FALSE, prop.c = FALSE, prop.r = FALSE,
          dnn = c('actual default', 'predicted default'))
```

This results in the following table:

```
               | predicted default
actual default |       no  |      yes | Row Total |
---------------|----------|----------|-----------|
            no |       59  |        8 |        67 |
               |    0.590  |    0.080 |           |
---------------|----------|----------|-----------|
           yes |       19  |       14 |        33 |
               |    0.190  |    0.140 |           |
---------------|----------|----------|-----------|
  Column Total |       78  |       22 |       100 |
---------------|----------|----------|-----------|
```

Out of the 100 loan applications in the test set, our model correctly predicted that 59 did not default and 14 did default, resulting in an accuracy of 73 percent and an error rate of 27 percent. This is somewhat worse than its performance on the training data, but not unexpected, given that a model's performance is often worse on unseen data. Also note that the model only correctly predicted 14 of the 33 actual loan defaults in the test data, or 42 percent. Unfortunately, this type of error is potentially a very costly mistake, as the bank loses money on each default. Let's see if we can improve the result with a bit more effort.

Step 5 – improving model performance

Our model's error rate is likely to be too high to deploy it in a real-time credit scoring application. In fact, if the model had predicted "no default" for every test case, it would have been correct 67 percent of the time—a result not much worse than our model but requiring much less effort! Predicting loan defaults from 900 examples seems to be a challenging problem.

Making matters even worse, our model performed especially poorly at identifying applicants who do default on their loans. Luckily, there are a couple of simple ways to adjust the C5.0 algorithm that may help to improve the performance of the model, both overall and for the costlier type of mistakes.

Boosting the accuracy of decision trees

One way the C5.0 algorithm improved upon the C4.5 algorithm was through the addition of **adaptive boosting**. This is a process in which many decision trees are built and the trees vote on the best class for each example.

The idea of boosting is based largely upon research by Rob Schapire and Yoav Freund. For more information, try searching the web for their publications or their textbook *Boosting: Foundations and Algorithms, Cambridge, MA, The MIT Press, 2012.*

As boosting can be applied more generally to any machine learning algorithm, it is covered in more detail later in this book in *Chapter 11, Improving Model Performance.* For now, it suffices to say that boosting is rooted in the notion that by combining a number of weak performing learners, you can create a team that is much stronger than any of the learners alone. Each of the models has a unique set of strengths and weaknesses, and may be better or worse at certain problems. Using a combination of several learners with complementary strengths and weaknesses can therefore dramatically improve the accuracy of a classifier.

The `C5.0()` function makes it easy to add boosting to our decision tree. We simply need to add an additional `trials` parameter indicating the number of separate decision trees to use in the boosted team. The `trials` parameter sets an upper limit; the algorithm will stop adding trees if it recognizes that additional trials do not seem to be improving the accuracy. We'll start with 10 trials, a number that has become the de facto standard, as research suggests that this reduces error rates on test data by about 25 percent. Aside from the new parameter, the command is similar to before:

```
> credit_boost10 <- C5.0(credit_train[-17], credit_train$default,
                         trials = 10)
```

While examining the resulting model, we can see that the output now indicates the addition of boosting:

```
> credit_boost10

Number of boosting iterations: 10

Average tree size: 47.5
```

The new output shows that across the 10 iterations, our tree size shrunk. If you would like, you can see all 10 trees by typing summary(credit_boost10) at the command prompt. The output also shows the tree's performance on the training data:

```
> summary(credit_boost10)


      (a)   (b)      <-classified as

      ----  ----

      629    4       (a): class no

       30   237      (b): class yes
```

The classifier made 34 mistakes on 900 training examples for an error rate of 3.8 percent. This is quite an improvement over the 13.9 percent training error rate we noted before adding boosting! However, it remains to be seen whether we see a similar improvement on the test data. Let's take a look:

```
> credit_boost_pred10 <- predict(credit_boost10, credit_test)
> CrossTable(credit_test$default, credit_boost_pred10,
            prop.chisq = FALSE, prop.c = FALSE, prop.r = FALSE,
            dnn = c('actual default', 'predicted default'))
```

The resulting table is as follows:

actual default	predicted default		
	no	yes	Row Total
no	62	5	67
	0.620	0.050	
yes	13	20	33
	0.130	0.200	
Column Total	75	25	100

Here, we reduced the total error rate from 27 percent prior to boosting to 18 percent in the boosted model. This may not seem like a large gain, but it is in fact greater than the 25 percent reduction we expected. On the other hand, the model is still not doing well at predicting defaults, predicting only *20 / 33 = 61%* correctly. The lack of an even greater improvement may be a function of our relatively small training dataset, or it may just be a very difficult problem to solve.

That said, if boosting can be added this easily, why not apply it by default to every decision tree? The reason is twofold. First, if building a decision tree once takes a great deal of computation time, building many trees may be computationally impractical. Secondly, if the training data is very noisy, then boosting might not result in an improvement at all. Still, if greater accuracy is needed, it's worth giving boosting a try.

Making some mistakes cost more than others

Giving a loan to an applicant who is likely to default can be an expensive mistake. One solution to reduce the number of false negatives may be to reject a larger number of borderline applicants under the assumption that the interest that the bank would earn from a risky loan is far outweighed by the massive loss it would incur if the money is not paid back at all.

The C5.0 algorithm allows us to assign a penalty to different types of errors in order to discourage a tree from making more costly mistakes. The penalties are designated in a **cost matrix**, which specifies how many times more costly each error is relative to any other.

To begin constructing the cost matrix, we need to start by specifying the dimensions. Since the predicted and actual values can both take two values, yes or no, we need to describe a 2x2 matrix using a list of two vectors, each with two values. At the same time, we'll also name the matrix dimensions to avoid confusion later on:

```
> matrix_dimensions <- list(c("no", "yes"), c("no", "yes"))
> names(matrix_dimensions) <- c("predicted", "actual")
```

Examining the new object shows that our dimensions have been set up correctly:

```
> matrix_dimensions
$predicted
[1] "no"  "yes"


$actual
[1] "no"  "yes"
```

Next, we need to assign the penalty for the various types of errors by supplying four values to fill the matrix. Since R fills a matrix by filling columns one by one from top to bottom, we need to supply the values in a specific order:

1. Predicted no, actual no
2. Predicted yes, actual no
3. Predicted no, actual yes
4. Predicted yes, actual yes

Suppose we believe that a loan default costs the bank four times as much as a missed opportunity. Our penalty values then could be defined as:

```
> error_cost <- matrix(c(0, 1, 4, 0), nrow = 2,
                dimnames = matrix_dimensions)
```

This creates the following matrix:

```
> error_cost
          actual
predicted no yes
      no   0   4
      yes  1   0
```

As defined by this matrix, there is no cost assigned when the algorithm classifies a no or yes correctly, but a false negative has a cost of 4 versus a false positive's cost of 1. To see how this impacts classification, let's apply it to our decision tree using the `costs` parameter of the `C5.0()` function. We'll otherwise use the same steps as before:

```
> credit_cost <- C5.0(credit_train[-17], credit_train$default,
                  costs = error_cost)
> credit_cost_pred <- predict(credit_cost, credit_test)
> CrossTable(credit_test$default, credit_cost_pred,
           prop.chisq = FALSE, prop.c = FALSE, prop.r = FALSE,
           dnn = c('actual default', 'predicted default'))
```

This produces the following confusion matrix:

actual default	predicted default no	yes	Row Total
no	37 0.370	30 0.300	67
yes	7 0.070	26 0.260	33
Column Total	44	56	100

Compared to our boosted model, this version makes more mistakes overall: 37 percent error here versus 18 percent in the boosted case. However, the types of mistakes are very different. Where the previous models classified only 42 and 61 percent of defaults correctly, in this model, *26 / 33 = 79%* of the actual defaults were correctly predicted to be defaults. This trade-off resulting in a reduction of false negatives at the expense of increasing false positives may be acceptable if our cost estimates were accurate.

Understanding classification rules

Classification rules represent knowledge in the form of logical if-else statements that assign a class to unlabeled examples. They are specified in terms of an **antecedent** and a **consequent**, which form a statement that says "if this happens, then that happens." The antecedent comprises certain combinations of feature values, while the consequent specifies the class value to assign if the rule's conditions are met. A simple rule might state, "if the hard drive is making a clicking sound, then it is about to fail."

Rule learners are a closely related sibling of decision tree learners and are often used for similar types of tasks. Like decision trees, they can be used for applications that generate knowledge for future action, such as:

- Identifying conditions that lead to hardware failure in mechanical devices
- Describing the key characteristics of groups of people for customer segmentation
- Finding conditions that precede large drops or increases in the prices of shares on the stock market

Rule learners do have some distinct contrasts relative to decision trees. Unlike a tree that must be followed through a series of branching decisions, rules are propositions that can be read much like independent statements of fact. Additionally, for reasons that will be discussed later, the results of a rule learner can be more simple, direct, and easier to understand than a decision tree built on the same data.

 You may have already realized that the branches of decision trees are almost identical to if-else statements of rule learning algorithms, and in fact, rules can be generated from trees. So, why bother with a separate group of rule learning algorithms? Read further to discover the nuances that differentiate the two approaches.

Rule learners are generally applied to problems where the features are primarily or entirely nominal. They do well at identifying rare events, even if the rare event occurs only for a very specific interaction among feature values.

Separate and conquer

Classification rule learning algorithms utilize a heuristic known as **separate and conquer**. The process involves identifying a rule that covers a subset of examples in the training data and then separating this partition from the remaining data. As rules are added, additional subsets of data are separated until the entire dataset has been covered and no more examples remain. Although separate and conquer is in many ways similar to the divide and conquer heuristic covered earlier, it differs in subtle ways that will become clear soon.

One way to imagine the rule learning process of separate and conquer is to imagine drilling down into the data by creating increasingly specific rules to identify class values. Suppose you were tasked with creating rules to identify whether or not an animal is a mammal. You could depict the set of all animals as a large space, as shown in the following diagram:

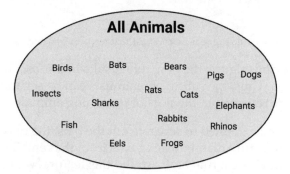

Figure 5.7: A rule learning algorithm may help divide animals into groups of mammals and non-mammals

A rule learner begins by using the available features to find homogeneous groups. For example, using a feature that indicates whether the species travels via land, sea, or air, the first rule might suggest that any land-based animals are mammals:

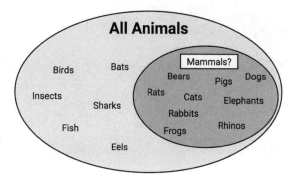

Figure 5.8: A potential rule considers animals that travel on land to be mammals

Do you notice any problems with this rule? If you're an animal lover, you might have realized that frogs are amphibians, not mammals. Therefore, our rule needs to be a bit more specific. Let's drill down further by suggesting that mammals must walk on land and have a tail:

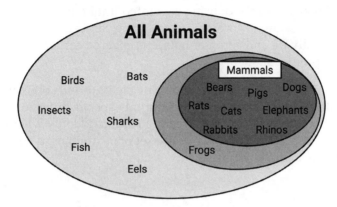

Figure 5.9: A more specific rule suggests that animals that walk on land and have tails are mammals

As shown in the previous figure, the new rule results in a subset of animals that are entirely mammals. Thus, the subset of mammals can be separated from the other data and the frogs are returned to the pool of remaining animals—no pun intended!

An additional rule can be defined to separate out the bats, the only remaining mammal. A potential feature distinguishing bats from the remaining animals would be the presence of fur. Using a rule built around this feature, we have then correctly identified all the animals:

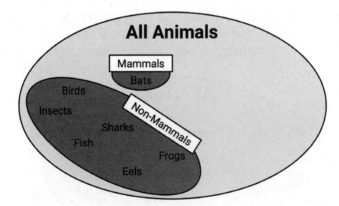

Figure 5.10: A rule stating that animals with fur are mammals perfectly classifies the remaining animals

At this point, since all of the training instances have been classified, the rule learning process would stop. We learned a total of three rules:

- Animals that walk on land and have tails are mammals

- If the animal does not have fur, it is not a mammal

- Otherwise, the animal is a mammal

The previous example illustrates how rules gradually consume larger and larger segments of data to eventually classify all instances. As the rules seem to cover portions of the data, separate and conquer algorithms are also known as **covering algorithms**, and the resulting rules are called covering rules. In the next section, we will learn how covering rules are applied in practice by examining a simple rule learning algorithm. We will then examine a more complex rule learner and apply both algorithms to a real-world problem.

The 1R algorithm

Suppose a television game show has an animal hidden behind a large curtain. You are asked to guess whether or not it is a mammal and if correct, you win a large cash prize. You are not given any clues about the animal's characteristics, but you know that a very small portion of the world's animals are mammals. Consequently, you guess "non-mammal." What do you think about your chances?

Choosing this, of course, maximizes your odds of winning the prize, as it is the most likely outcome under the assumption the animal was chosen at random. Clearly, this game show is a bit ridiculous, but it demonstrates the simplest classifier, **ZeroR**, which is a rule learner that considers no features and literally learns no rules (hence the name). For every unlabeled example, regardless of the values of its features, it predicts the most common class. This algorithm has very little real-world utility, except that it provides a simple baseline for comparison to other, more sophisticated, rule learners.

The **1R algorithm** (**One Rule** or **OneR**), improves over ZeroR by selecting a single rule. Although this may seem overly simplistic, it tends to perform better than you might expect. As demonstrated in empirical studies, the accuracy of this algorithm can approach that of much more sophisticated algorithms for many real-world tasks.

 For an in-depth look at the surprising performance of 1R, see *Very Simple Classification Rules Perform Well on Most Commonly Used Datasets, Holte, RC, Machine Learning, 1993, Vol. 11, pp. 63-91.*

The strengths and weaknesses of the 1R algorithm are shown in the following table:

Strengths	Weaknesses
• Generates a single, easy-to-understand, human-readable rule • Often performs surprisingly well • Can serve as a benchmark for more complex algorithms	• Uses only a single feature • Probably overly simplistic

The way this algorithm works is simple. For each feature, 1R divides the data into groups with similar values of the feature. Then, for each segment, the algorithm predicts the majority class. The error rate for the rule based on each feature is calculated and the rule with the fewest errors is chosen as the one rule.

The following tables show how this would work for the animal data we looked at earlier:

Animal	Travels By	Has Fur	Mammal
Bats	Air	Yes	Yes
Bears	Land	Yes	Yes
Birds	Air	No	No
Cats	Land	Yes	Yes
Dogs	Land	Yes	Yes
Eels	Sea	No	No
Elephants	Land	No	Yes
Fish	Sea	No	No
Frogs	Land	No	No
Insects	Air	No	No
Pigs	Land	No	Yes
Rabbits	Land	Yes	Yes
Rats	Land	Yes	Yes
Rhinos	Land	No	Yes
Sharks	Sea	No	No

Full Dataset

Travels By	Predicted	Mammal	
Air	No	Yes	✖
Air	No	No	
Air	No	No	
Land	Yes	Yes	
Land	Yes	Yes	
Land	Yes	Yes	
Land	Yes	Yes	
Land	Yes	No	✖
Land	Yes	Yes	
Land	Yes	Yes	
Land	Yes	Yes	
Land	Yes	Yes	
Sea	No	No	
Sea	No	No	
Sea	No	No	

Rule for "Travels By"
Error Rate = 2 / 15

Has Fur	Predicted	Mammal	
No	No	No	
No	No	No	
No	No	Yes	✖
No	No	No	
No	No	No	
No	No	No	
No	No	Yes	✖
No	No	Yes	✖
No	No	No	
Yes	Yes	Yes	
Yes	Yes	Yes	
Yes	Yes	Yes	
Yes	Yes	Yes	
Yes	Yes	Yes	
Yes	Yes	Yes	

Rule for "Has Fur"
Error Rate = 3 / 15

Figure 5.11: The 1R algorithm chooses the single rule with the lowest misclassification rate

For the **Travels By** feature, the dataset was divided into three groups: **Air**, **Land**, and **Sea**. Animals in the **Air** and **Sea** groups were predicted to be non-mammal, while animals in the **Land** group were predicted to be mammals. This resulted in two errors: bats and frogs.

The **Has Fur** feature divided animals into two groups. Those with fur were predicted to be mammals, while those without fur were not. Three errors were counted: pigs, elephants, and rhinos. As the **Travels By** feature resulted in fewer errors, the 1R algorithm would return the following:

- If the animal travels by air, it is not a mammal
- If the animal travels by land, it is a mammal
- If the animal travels by sea, it is not a mammal

The algorithm stops here, having found the single most important rule.

Obviously, this rule learning algorithm may be too basic for some tasks. Would you want a medical diagnosis system to consider only a single symptom, or an automated driving system to stop or accelerate your car based on only a single factor? For these types of tasks, a more sophisticated rule learner might be useful. We'll learn about one in the following section.

The RIPPER algorithm

Early rule learning algorithms were plagued by a couple of problems. First, they were notorious for being slow, which made them ineffective for the increasing number of large datasets. Secondly, they were often prone to being inaccurate on noisy data.

A first step toward solving these problems was proposed in 1994 by Johannes Furnkranz and Gerhard Widmer. Their **incremental reduced error pruning (IREP) algorithm** uses a combination of pre-pruning and post-pruning methods that grow very complex rules and prune them before separating the instances from the full dataset. Although this strategy helped the performance of rule learners, decision trees often still performed better.

For more information on IREP, see *Incremental Reduced Error Pruning, Furnkranz, J* and *Widmer, G, Proceedings of the 11th International Conference on Machine Learning, 1994, pp. 70-77.*

Rule learners took another step forward in 1995 when William W. Cohen introduced the **repeated incremental pruning to produce error reduction (RIPPER) algorithm**, which improved upon IREP to generate rules that match or exceed the performance of decision trees.

For more detail on RIPPER, see *Fast Effective Rule Induction, Cohen, WW, Proceedings of the 12th International Conference on Machine Learning, 1995, pp. 115-123.*

As outlined in the following table, the strengths and weaknesses of RIPPER are generally comparable to decision trees. The chief benefit is that they may result in a slightly more parsimonious model.

Strengths	Weaknesses
• Generates easy-to-understand, human-readable rules • Efficient on large and noisy datasets • Generally, produces a simpler model than a comparable decision tree	• May result in rules that seem to defy common sense or expert knowledge • Not ideal for working with numeric data • Might not perform as well as more complex models

Having evolved from several iterations of rule learning algorithms, the RIPPER algorithm is a patchwork of efficient heuristics for rule learning. Due to its complexity, a discussion of the implementation details is beyond the scope of this book. However, it can be understood in general terms as a three-step process:

1. Grow

2. Prune

3. Optimize

The growing phase uses the separate and conquer technique to greedily add conditions to a rule until it perfectly classifies a subset of data or runs out of attributes for splitting. Similar to decision trees, the information gain criterion is used to identify the next splitting attribute. When increasing a rule's specificity no longer reduces entropy, the rule is immediately pruned. Steps one and two are repeated until reaching a stopping criterion, at which point the entire set of rules is optimized using a variety of heuristics.

The RIPPER algorithm can create much more complex rules than the 1R algorithm can, as it can consider more than one feature. This means that it can create rules with multiple antecedents such as "if an animal flies and has fur, then it is a mammal." This improves the algorithm's ability to model complex data, but just like decision trees, it means that the rules can quickly become difficult to comprehend.

 The evolution of classification rule learners didn't stop with RIPPER. New rule learning algorithms are being proposed rapidly. A survey of literature shows algorithms called IREP++, SLIPPER, TRIPPER, among many others.

Rules from decision trees

Classification rules can also be obtained directly from decision trees. Beginning at a leaf node and following the branches back to the root, you will have obtained a series of decisions. These can be combined into a single rule. The following figure shows how rules could be constructed from the decision tree for predicting movie success:

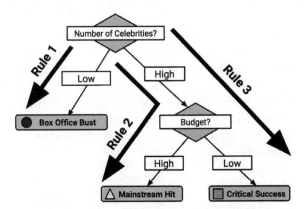

Figure 5.12: Rules can be generated from decision trees by following paths from the root node to each leaf node

Following the paths from the root node down to each leaf, the rules would be:

1. If the number of celebrities is low, then the movie will be a **Box Office Bust**.
2. If the number of celebrities is high and the budget is high, then the movie will be a **Mainstream Hit**.
3. If the number of celebrities is high and the budget is low, then the movie will be a **Critical Success**.

For reasons that will be made clear in the following section, the chief downside to using a decision tree to generate rules is that the resulting rules are often more complex than those learned by a rule learning algorithm. The divide and conquer strategy employed by decision trees biases the results differently to that of a rule learner. On the other hand, it is sometimes more computationally efficient to generate rules from trees.

 The C5.0() function in the C50 package will generate a model using classification rules if you specify rules = TRUE when training the model.

What makes trees and rules greedy?

Decision trees and rule learners are known as **greedy learners** because they use data on a first-come, first-served basis. Both the divide and conquer heuristic used by decision trees and the separate and conquer heuristic used by rule learners attempt to make partitions one at a time, finding the most homogeneous partition first, followed by the next best, and so on, until all examples have been classified.

The downside to the greedy approach is that greedy algorithms are not guaranteed to generate the optimal, most accurate, or smallest number of rules for a particular dataset. By taking the low-hanging fruit early, a greedy learner may quickly find a single rule that is accurate for one subset of data; however, in doing so, the learner may miss the opportunity to develop a more nuanced set of rules with better overall accuracy on the entire set of data. However, without using the greedy approach to rule learning, it is likely that for all but the smallest of datasets, rule learning would be computationally infeasible.

Figure 5.13: Both decision trees and classification rule learners are greedy algorithms

Though both trees and rules employ greedy learning heuristics, there are subtle differences in how they build rules. Perhaps the best way to distinguish them is to note that once divide and conquer splits on a feature, the partitions created by the split may not be re-conquered, only further subdivided. In this way, a tree is permanently limited by its history of past decisions. In contrast, once separate and conquer finds a rule, any examples not covered by all of the rule's conditions may be re-conquered.

To illustrate this contrast, consider the previous case in which we built a rule learner to determine whether an animal was a mammal. The rule learner identified three rules that perfectly classify the example animals:

- Animals that walk on land and have tails are mammals (bears, cats, dogs, elephants, pigs, rabbits, rats, and rhinos)

- If the animal does not have fur, it is not a mammal (birds, eels, fish, frogs, insects, and sharks)

- Otherwise, the animal is a mammal (bats)

In contrast, a decision tree built on the same data might have come up with four rules to achieve the same perfect classification:

- If an animal walks on land and has a tail, then it is a mammal (bears, cats, dogs, elephants, pigs, rabbits, rats, and rhinos)

- If an animal walks on land and does not have a tail, then it is not a mammal (frogs)

- If the animal does not walk on land and has fur, then it is a mammal (bats)

- If the animal does not walk on land and does not have fur, then it is not a mammal (birds, insects, sharks, fish, and eels)

The different result across these two approaches has to do with what happens to the frogs after they are separated by the "walk on land" decision. Where the rule learner allows frogs to be re-conquered by the "do not have fur" decision, the decision tree cannot modify the existing partitions and therefore must place the frog into its own rule.

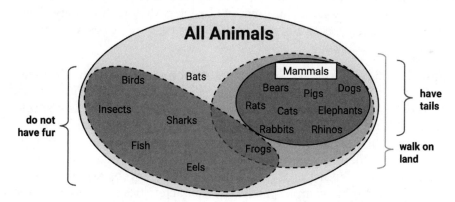

Figure 5.14: The handling of frogs distinguishes divide and the conquer and separate and conquer heuristics. The latter approach allows the frogs to be re-conquered by later rules.

On the one hand, because rule learners can reexamine cases that were considered but ultimately not covered as part of prior rules, rule learners often find a more parsimonious set of rules than those generated from decision trees. On the other hand, this reuse of data means that the computational cost of rule learners may be somewhat higher than for decision trees.

Example – identifying poisonous mushrooms with rule learners

Each year, many people fall ill and sometimes even die from ingesting poisonous wild mushrooms. Since many mushrooms are very similar to each other in appearance, occasionally even experienced mushroom gatherers are poisoned.

Unlike the identification of harmful plants, such as a poison oak or poison ivy, there are no clear rules like "leaves of three, let them be" for identifying whether a wild mushroom is poisonous or edible. Complicating matters, many traditional rules such as "poisonous mushrooms are brightly colored" provide dangerous or misleading information. If simple, clear, and consistent rules were available for identifying poisonous mushrooms, they could save the lives of foragers.

As one of the strengths of rule learning algorithms is the fact that they generate easy-to-understand rules, they seem like an appropriate fit for this classification task. However, the rules will only be as useful as they are accurate.

Step 1 – collecting data

To identify rules for distinguishing poisonous mushrooms, we will utilize the Mushroom dataset by Jeff Schlimmer of Carnegie Mellon University. The raw dataset is available freely at the UCI Machine Learning Repository (http:// archive.ics.uci.edu/ml).

The dataset includes information on 8,124 mushroom samples from 23 species of gilled mushrooms listed in the *Audubon Society Field Guide to North American Mushrooms* (1981). In the field guide, each of the mushroom species is identified as "definitely edible," "definitely poisonous," or "likely poisonous, and not recommended to be eaten." For the purposes of this dataset, the latter group was combined with the "definitely poisonous" group to make two classes: poisonous and non-poisonous. The data dictionary available on the UCI website describes the 22 features of the mushroom samples, including characteristics such as cap shape, cap color, odor, gill size and color, stalk shape, and habitat.

> This chapter uses a slightly modified version of the mushroom data. If you plan on following along with the example, download the mushrooms.csv file from the Packt Publishing website and save it to your R working directory.

Step 2 – exploring and preparing the data

We begin by using `read.csv()` to import the data for our analysis. Since all 22 features and the target class are nominal, we will set `stringsAsFactors =` TRUE to take advantage of the automatic factor conversion:

```
> mushrooms <- read.csv("mushrooms.csv", stringsAsFactors = TRUE)
```

The output of the `str(mushrooms)` command notes that the data contains 8,124 observations of 23 variables as the data dictionary had described. While most of the `str()` output is unremarkable, one feature is worth mentioning. Do you notice anything peculiar about the `veil_type` variable in the following line?

```
$ veil_type : Factor w/ 1 level "partial": 1 1 1 1 1 ...
```

If you think it is odd that a factor has only one level, you are correct. The data dictionary lists two levels for this feature: `partial` and `universal`, however, all examples in our data are classified as `partial`. It is likely that this data element was somehow coded incorrectly. In any case, since the veil type does not vary across samples, it does not provide any useful information for prediction. We will drop this variable from our analysis using the following command:

```
> mushrooms$veil_type <- NULL
```

By assigning NULL to the `veil_type` vector, R eliminates the feature from the `mushrooms` data frame.

Before going much further, we should take a quick look at the distribution of the mushroom `type` variable in our dataset:

```
> table(mushrooms$type)

  edible poisonous
    4208      3916
```

About 52 percent of the mushroom samples (N = 4,208) are edible, while 48 percent (N = 3,916) are poisonous.

For the purposes of this experiment, we will consider the 8,214 samples in the mushroom data to be an exhaustive set of all the possible wild mushrooms. This is an important assumption because it means that we do not need to hold some samples out of the training data for testing purposes. We are not trying to develop rules that cover unforeseen types of mushrooms; we are merely trying to find rules that accurately depict the complete set of known mushroom types. Therefore, we can build and test the model on the same data.

Step 3 – training a model on the data

If we trained a hypothetical ZeroR classifier on this data, what would it predict? Since ZeroR ignores all of the features and simply predicts the target's mode, in plain language, its rule would state that "all mushrooms are edible." Obviously, this is not a very helpful classifier because it would leave a mushroom gatherer sick or dead for nearly half of the mushroom samples! Our rules will need to do much better than this benchmark in order to provide safe advice that can be published. At the same time, we need simple rules that are easy to remember.

Since simple rules can still be useful, let's see how a very simple rule learner performs on the mushroom data. Toward this end, we will apply the 1R classifier, which will identify the single feature that is the most predictive of the target class and use this feature to construct a rule.

We will use the 1R implementation found in the `OneR` package by Holger von Jouanne-Diedrich at the Aschaffenburg University of Applied Sciences. This is a relatively new package, which implements 1R in native R code for speed and ease of use. If you don't already have this package, it can be installed using the command `install.packages("OneR")` and loaded by typing `library(OneR)`.

1R classification rule syntax

using the **OneR()** function in the **OneR** package

Building the classifier:

```
m <- OneR(class ~ predictors, data = mydata)
```

- **class** is the column in the **mydata** data frame to be predicted
- **predictors** is an R formula specifying the features in the **mydata** data frame to use for prediction
- **data** is the data frame in which **class** and **predictors** can be found

The function will return a OneR model object that can be used to make predictions.

Making predictions:

```
p <- predict(m, test)
```

- **m** is a model trained by the **OneR()** function
- **test** is a data frame containing test data with the same features as the training data used to build the classifier.

The function will return a vector of predicted class values.

Example:

```
mushroom_classifier <- OneR(type ~ odor + cap_color,
                            data = mushroom_train)
mushroom_prediction <- predict(mushroom_classifier,
                               mushroom_test)
```

The OneR() function uses the **R formula syntax** to specify the model to be trained. The formula syntax uses the ~ operator (known as the tilde) to express the relationship between a target variable and its predictors. The class variable to be learned goes to the left of the tilde and the predictor features are written on the right, separated by + operators. If you would like to model the relationship between the class y and predictors x1 and x2, you would write the formula as y ~ x1 + x2. To include all of the variables in the model, the period character is used. For example, y ~ . specifies the relationship between y and all other features in the dataset.

> The R formula syntax is used across many R functions and offers some powerful features to describe the relationships among predictor variables. We will explore some of these features in later chapters. However, if you're eager for a sneak peak, feel free to read the documentation using the ?formula command.

Using the formula type ~ . with OneR() allows our first rule learner to consider all possible features in the mushroom data when predicting mushroom type:

```
> mushroom_1R <- OneR(type ~ ., data = mushrooms)
```

To examine the rules it created, we can type the name of the classifier object:

```
> mushroom_1R
```

```
Call:
OneR.formula(formula = type ~ ., data = mushrooms)

Rules:
If odor = almond   then type = edible
If odor = anise    then type = edible
If odor = creosote then type = poisonous
If odor = fishy    then type = poisonous
If odor = foul     then type = poisonous
If odor = musty    then type = poisonous
If odor = none     then type = edible
If odor = pungent  then type = poisonous
If odor = spicy    then type = poisonous

Accuracy:
8004 of 8124 instances classified correctly (98.52%)
```

Examining the output, we see that the `odor` feature was selected for rule generation. The categories of `odor`, such as `almond`, `anise`, and so on, specify rules for whether the mushroom is likely to be `edible` or `poisonous`. For instance, if the mushroom smells `fishy`, `foul`, `musty`, `pungent`, `spicy`, or like `creosote`, the mushroom is likely to be `poisonous`. On the other hand, mushrooms with more pleasant smells, like `almond` and `anise`, and those with no smell at all, are predicted to be `edible`. For the purposes of a field guide for mushroom gathering, these rules could be summarized in a simple rule of thumb: "if the mushroom smells unappetizing, then it is likely to be poisonous."

Step 4 – evaluating model performance

The last line of the output notes that the rules correctly predict the edibility 8,004 of the 8,124 mushroom samples, or nearly 99 percent. Anything short of perfection, however, runs the risk of poisoning someone if the model were to classify a poisonous mushroom as edible.

To determine whether or not this occurred, let's examine a confusion matrix of the predicted versus actual values. This requires us to first generate the 1R model's predictions, then compare the predictions to the actual values:

```
> mushroom_1R_pred <- predict(mushroom_1R, mushrooms)
> table(actual = mushrooms$type, predicted = mushroom_1R_pred)
             predicted
actual       edible poisonous
  edible        4208         0
  poisonous      120      3796
```

Here, we can see where our rules went wrong. The table's columns indicate the predicted edibility of the mushroom while the table's rows divide the 4,208 actually edible mushrooms and the 3,916 actually poisonous mushrooms. Examining the table, we can see that although the 1R classifier did not classify any edible mushrooms as poisonous, it did classify 120 poisonous mushrooms as edible — which makes for an incredibly dangerous mistake!

Considering that the learner utilized only a single feature, it did reasonably well; if you avoid unappetizing smells when foraging for mushrooms, you will almost always avoid a trip to the hospital. That said, close does not cut it when lives are involved, not to mention the field guide publisher might not be happy about the prospect of a lawsuit when its readers fall ill. Let's see if we can add a few more rules and develop an even better classifier.

Step 5 – improving model performance

For a more sophisticated rule learner, we will use JRip(), a Java-based implementation of the RIPPER algorithm. The JRip() function is included in the RWeka package, which you may recall was described in *Chapter 1, Introducing Machine Learning*, during the tutorial on installing and loading packages. If you have not installed this package already, you will need to use the install.packages("RWeka") command after installing Java on your machine according to the system-specific instructions. With these steps complete, load the package using the library(RWeka) command.

RIPPER classification rule syntax

using the **JRip()** function in the **RWeka** package

Building the classifier:

```
m <- JRip(class ~ predictors, data = mydata)
```

- **class** is the column in the **mydata** data frame to be predicted
- **predictors** is an R formula specifying the features in the **mydata** data frame to use for prediction
- **data** is the data frame in which **class** and **predictors** can be found

The function will return a RIPPER model object that can be used to make predictions.

Making predictions:

```
p <- predict(m, test)
```

- **m** is a model trained by the **JRip()** function
- **test** is a data frame containing test data with the same features as the training data used to build the classifier.

The function will return a vector of predicted class values.

Example:

```
mushroom_classifier <- JRip(type ~ odor + cap_color,
                            data = mushroom_train)

mushroom_prediction <- predict(mushroom_classifier,
                               mushroom_test)
```

As shown in the syntax box, the process of training a JRip() model is very similar to the training of a OneR() model. This is one of the pleasant benefits of the R formula interface: the syntax is consistent across algorithms, which makes it simple to compare a variety of models.

Let's train the JRip() rule learner as we had done with OneR(), allowing it to find rules among all of the available features:

```
> mushroom_JRip <- JRip(type ~ ., data = mushrooms)
```

To examine the rules, type the name of the classifier:

```
> mushroom_JRip
```

```
JRIP rules:
===========
(odor = foul) => type=poisonous (2160.0/0.0)
(gill_size = narrow) and (gill_color = buff)
  => type=poisonous (1152.0/0.0)
(gill_size = narrow) and (odor = pungent)
  => type=poisonous (256.0/0.0)
(odor = creosote) => type=poisonous (192.0/0.0)
(spore_print_color = green) => type=poisonous (72.0/0.0)
(stalk_surface_below_ring = scaly)
  and (stalk_surface_above_ring = silky)
    => type=poisonous (68.0/0.0)
(habitat = leaves) and (gill_attachment = free)
  and (population = clustered)
  => type=poisonous (16.0/0.0)
 => type=edible (4208.0/0.0)
```

```
Number of Rules : 8
```

The `JRip()` classifier learned a total of eight rules from the mushroom data. An easy way to read these rules is to think of them as a list of if–else statements, similar to programming logic. The first three rules could be expressed as:

- If the odor is foul, then the mushroom type is poisonous
- If the gill size is narrow and the gill color is buff, then the mushroom type is poisonous
- If the gill size is narrow and the odor is pungent, then the mushroom type is poisonous

Finally, the eighth rule implies that any mushroom sample that was not covered by the preceding seven rules is edible. Following the example of our programming logic, this can be read as:

- Else, the mushroom is edible

The numbers next to each rule indicate the number of instances covered by the rule and a count of misclassified instances. Notably, there were no misclassified mushroom samples using these eight rules. As a result, the number of instances covered by the last rule is exactly equal to the number of edible mushrooms in the data (N = 4,208).

The following figure provides a rough illustration of how the rules are applied to the mushroom data. If you imagine the large oval as containing all mushroom species, the rule learner identified features, or sets of features, which separate homogeneous segments from the larger group. First, the algorithm found a large group of poisonous mushrooms uniquely distinguished by their foul odor. Next, it found smaller and more specific groups of poisonous mushrooms. By identifying covering rules for each of the varieties of poisonous mushrooms, all of the remaining mushrooms were edible. Thanks to Mother Nature, each variety of mushrooms was unique enough that the classifier was able to achieve 100 percent accuracy.

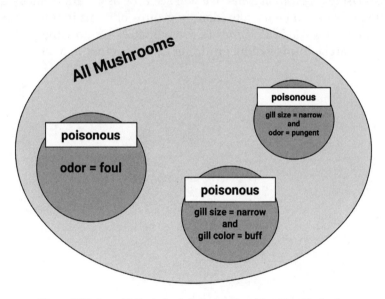

Figure 5.15: A sophisticated rule learning algorithm identified rules to perfectly cover all types of poisonous mushrooms

Summary

This chapter covered two classification methods that use so-called "greedy" algorithms to partition the data according to feature values. Decision trees use a divide and conquer strategy to create flowchart-like structures, while rule learners separate and conquer data to identify logical if-else rules. Both methods produce models that can be interpreted without a statistical background.

One popular and highly configurable decision tree algorithm is C5.0. We used the C5.0 algorithm to create a tree to predict whether a loan applicant will default. Using options for boosting and cost-sensitive errors, we were able to improve our accuracy and avoid risky loans that could cost the bank more money.

We also used two rule learners, 1R and RIPPER, to develop rules for identifying poisonous mushrooms. The 1R algorithm used a single feature to achieve 99 percent accuracy in identifying potentially fatal mushroom samples. On the other hand, the set of eight rules generated by the more sophisticated RIPPER algorithm correctly identified the edibility of every mushroom.

This chapter merely scratched the surface of how trees and rules can be used. The next chapter, *Chapter 6, Forecasting Numeric Data – Regression Methods*, describes techniques known as regression trees and model trees, which use decision trees for numeric prediction rather than classification. In *Chapter 8, Finding Patterns – Market Basket Analysis Using Association Rules*, we will see how association rules—a relative of classification rules—can be used to identify groups of items in transactional data. In *Chapter 11, Improving Model Performance*, we will discover how the performance of decision trees can be improved by grouping them together in a model known as a random forest.

6
Forecasting Numeric Data – Regression Methods

Mathematical relationships help us to make sense of many aspects of everyday life. For example, body weight is a function of one's calorie intake; income is often related to education and job experience; and poll numbers help to estimate a presidential candidate's odds of being re-elected.

When such patterns are formulated with numbers, we gain additional clarity. For example, an additional 250 kilocalories consumed daily may result in nearly a kilogram of weight gain per month; each year of job experience may be worth an additional $1,000 in yearly salary; and a president is more likely to be re-elected when the economy is strong. Obviously, these equations do not perfectly fit every situation, but we expect that they are reasonably correct most of the time.

This chapter extends our machine learning toolkit by going beyond the classification methods covered previously and introducing techniques for estimating relationships among numeric data. While examining several real-world numeric prediction tasks, you will learn:

- The basic statistical principles used in regression, a technique that models the size and strength of numeric relationships
- How to prepare data for regression analysis, and estimate and interpret a regression model
- A pair of hybrid techniques known as regression trees and model trees, which adapt decision tree classifiers for numeric prediction tasks

Based on a large body of work in the field of statistics, the methods used in this chapter are a bit heavier on math than those covered previously, but don't worry! Even if your algebra skills are a bit rusty, R takes care of the heavy lifting.

Understanding regression

Regression involves specifying the relationship between a single numeric **dependent variable** (the value to be predicted) and one or more numeric **independent variables** (the predictors). As the name implies, the dependent variable depends upon the value of the independent variable or variables. The simplest forms of regression assume that the relationship between the independent and dependent variables follows a straight line.

> The origin of the term "regression" to describe the process of fitting lines to data is rooted in a study of genetics by Sir Francis Galton in the late 19th century. He discovered that fathers who were extremely short or tall tended to have sons whose heights were closer to the average height. He called this phenomenon "regression to the mean."

You might recall from basic algebra that lines can be defined in a **slope-intercept form** similar to $y = a + bx$. In this form, the letter y indicates the dependent variable and x indicates the independent variable. The **slope** term b specifies how much the line rises for each increase in x. Positive values define lines that slope upward while negative values define lines that slope downward. The term a is known as the **intercept** because it specifies the point where the line crosses, or intercepts, the vertical y axis. It indicates the value of y when $x = 0$.

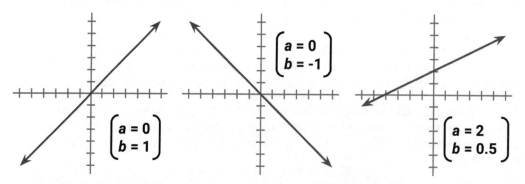

Figure 6.1: Examples of lines with various slopes and intercepts

Regression equations model data using a similar slope-intercept format. The machine's job is to identify values of a and b such that the specified line is best able to relate the supplied x values to the values of y. There may not always be a single function that perfectly relates the values, so the machine must also have some way to quantify the margin of error. We'll discuss this in depth shortly.

Regression analysis is used for a huge variety of tasks—it is almost surely the most widely used machine learning method. It can be used both for explaining the past and extrapolating into the future, and can be applied to nearly any task. Some specific use cases include:

- Examining how populations and individuals vary by their measured characteristics, in scientific studies in the fields of economics, sociology, psychology, physics, and ecology

- Quantifying the causal relationship between an event and its response, in cases such as clinical drug trials, engineering safety tests, or marketing research

- Identifying patterns that can be used to forecast future behavior given known criteria, such as for predicting insurance claims, natural disaster damage, election results, and crime rates

Regression methods are also used for **statistical hypothesis testing**, which determines whether a premise is likely to be true or false in light of observed data. The regression model's estimates of the strength and consistency of a relationship provide information that can be used to assess whether the observations are due to chance alone.

 Hypothesis testing is extremely nuanced and falls outside the scope of machine learning. If you are interested in this topic, an introductory statistics textbook is a good place to get started, for instance *Intuitive Introductory Statistics, Wolfe, DA* and *Schneider, G, Springer, 2017.*

Regression analysis is not synonymous with a single algorithm. Rather, it is an umbrella term for a large number of methods that can be adapted to nearly any machine learning task. If you were limited to choosing only a single machine learning method to study, regression would be a good choice. One could devote an entire career to nothing else and perhaps still have much to learn.

In this chapter, we'll focus only on the most basic **linear regression** models—those that use straight lines. The case when there is only a single independent variable is known as **simple linear regression**. In the case of two or more independent variables, it is known as **multiple linear regression**, or simply **multiple regression**. Both of these techniques assume a single dependent variable that is measured on a continuous scale.

Regression can also be used for other types of dependent variables and even for some classification tasks. For instance, **logistic regression** is used to model a binary categorical outcome, while **Poisson regression** — named after the French mathematician Siméon Poisson — models integer count data. The method known as **multinomial logistic regression** models a categorical outcome and can therefore be used for classification. Because the same statistical principles apply across all regression methods, after understanding the linear case, learning the other variants is straightforward.

Many of the specialized regression methods fall in a class of **generalized linear models** (**GLM**). Using a GLM, linear models can be generalized to other patterns via the use of a **link function**, which specifies more complex forms for the relationship between x and y. This allows regression to be applied to almost any type of data.

We'll begin with the basic case of simple linear regression. Despite the name, this method is not too simple to address complex problems. In the next section, we'll see how the use of a simple linear regression model might have averted a tragic engineering disaster.

Simple linear regression

On January 28, 1986, seven crew members of the United States space shuttle *Challenger* were killed when a rocket booster failed, causing a catastrophic disintegration. In the aftermath, experts quickly focused on the launch temperature as a potential culprit. The rubber O-rings responsible for sealing the rocket joints had never been tested below 40° F (4° C), and the weather on launch day was unusually cold and below freezing.

With the benefit of hindsight, the accident has been a case study for the importance of data analysis and visualization. Although it is unclear what information was available to the rocket engineers and decision makers leading up to the launch, it is undeniable that better data, utilized carefully, might very well have averted this disaster.

This section's analysis is based on data presented in *Risk Analysis of the Space Shuttle: Pre-Challenger Prediction of Failure*, Dalal SR, Fowlkes EB, and Hoadley B, *Journal of the American Statistical Association, 1989, Vol. 84, pp. 945-957*. For one perspective on how data may have changed the result, see *Visual Explanations: Images And Quantities, Evidence And Narrative*, Tufte, ER, Cheshire, CT: Graphics Press, 1997. For a counterpoint, see *Representation and misrepresentation: Tufte and the Morton Thiokol engineers on the Challenger*, Robison, W, Boisjoly, R, Hoeker, D, and Young, S, *Science and Engineering Ethics, 2002, Vol. 8, pp. 59-81.*

The rocket engineers almost certainly knew that cold temperatures could make the components more brittle and less able to seal properly, which would result in a higher chance of a dangerous fuel leak. However, given the political pressure to continue with the launch, they needed data to support this hypothesis. A regression model that demonstrated a link between temperature and O-ring failures, and could forecast the chance of failure given the expected temperature at launch, might have been very helpful.

To build the regression model, the scientists might have used the data on launch temperature and component distresses recorded during the 23 previous successful shuttle launches. A component distress indicates one of two types of problems. The first problem, called erosion, occurs when excessive heat burns up the O-ring. The second problem, called blowby, occurs when hot gasses leak through or "blow by" a poorly sealed O-ring. Since the shuttle had a total of six primary O-rings, up to six distresses could occur per flight. Though the rocket could survive one or more distress events or be destroyed with as few as one, each additional distress increased the probability of a catastrophic failure. The following scatterplot shows a plot of primary O-ring distresses detected for the previous 23 launches, as compared to the temperature at launch:

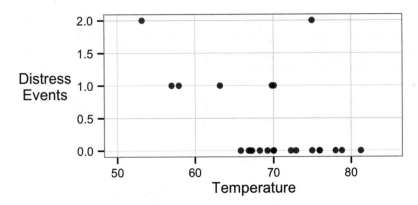

Figure 6.2: A visualization of space shuttle O-ring distresses versus launch temperature

Examining the plot, there is an apparent trend: launches occurring at higher temperatures tend to have fewer O-ring distress events. Additionally, the coldest launch (53° F) had two distress events, a level which had only been reached in one other launch. With this information in mind, the fact that the Challenger was scheduled to launch in conditions more than 20 degrees colder seems concerning. But exactly how concerned should they have been? To answer this question, we can turn to simple linear regression.

A simple linear regression model defines the relationship between a dependent variable and a single independent predictor variable using a line defined by an equation in the following form:

$$y = \alpha + \beta x$$

Aside from the Greek characters, this equation is virtually identical to the slope-intercept form described previously. The intercept, *a* (alpha), describes where the line crosses the *y* axis, while the slope, *β* (beta), describes the change in *y* given an increase of *x*. For the shuttle launch data, the slope would tell us the expected change in O-ring failures for each degree the launch temperature increases.

Greek characters are often used in the field of statistics to indicate variables that are parameters of a statistical function. Therefore, performing a regression analysis involves finding **parameter estimates** for *a* and *β*. The parameter estimates for alpha and beta are typically denoted using *a* and *b*, although you may find that some of this terminology and notation is used interchangeably.

Suppose we know that the estimated regression parameters in the equation for the shuttle launch data are *a* = 3.70 and *b* = -0.048. Consequently, the full linear equation is *y* = 3.70 – 0.048*x*. Ignoring for a moment how these numbers were obtained, we can plot the line on the scatterplot like this:

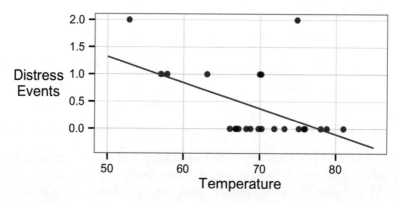

Figure 6.3: A regression line modeling the relationship between distress events and launch temperature

As the line shows, at 60 degrees Fahrenheit, we predict less than one O-ring distress. At 50 degrees Fahrenheit, we expect around 1.3 failures. If we use the model to extrapolate all the way out to 31 degrees — the forecasted temperature for the Challenger launch — we would expect about 3.70 - 0.048 * 31 = 2.21 O-ring distress events.

Assuming that each O-ring failure is equally likely to cause a catastrophic fuel leak, this means that the Challenger launch at 31 degrees was nearly three times riskier than the typical launch at 60 degrees, and over eight times riskier than a launch at 70 degrees.

Notice that the line doesn't pass through each data point exactly. Instead, it cuts through the data somewhat evenly, with some predictions lower or higher than the line. In the next section, we will learn about why this particular line was chosen.

Ordinary least squares estimation

In order to determine the optimal estimates of a and β, an estimation method known as **ordinary least squares (OLS)** was used. In OLS regression, the slope and intercept are chosen such that they minimize the **sum of the squared errors (SSE)**. The errors, also known as **residuals**, are the vertical distance between the predicted y value and the actual y value. Because the errors can be over-estimates or under-estimates, they can be positive or negative values. These are illustrated for several points in the preceding diagram:

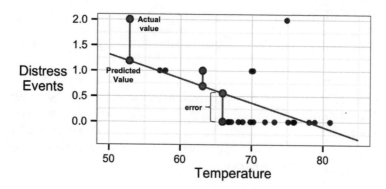

Figure 6.4: The regression line predictions differ from the actual values by a residual

In mathematical terms, the goal of OLS regression can be expressed as the task of minimizing the following equation:

$$\sum \left(y_i - \hat{y}_i \right)^2 = \sum e_i^2$$

In plain language, this equation defines e (the error) as the difference between the actual y value and the predicted y value. The error values are squared to eliminate the negative values and summed across all points in the data.

The caret character (^) above the y term is a commonly used feature of statistical notation. It indicates that the term is an estimate for the true y value. This is referred to as the y *hat*, and is pronounced exactly like the hat you'd wear on your head.

The solution for a depends on the value of b. It can be obtained using the following formula:

$$a = \bar{y} - b\bar{x}$$

To understand these equations, you'll need to know another bit of statistical notation. The horizontal bar appearing over the x and y terms indicates the mean value of x or y. This is referred to as the x *bar* or y *bar*, and is pronounced just like the establishment you'd go to for an alcoholic drink.

Though the proof is beyond the scope of this book, it can be shown using calculus that the value of b that results in the minimum squared error is:

$$b = \frac{\sum(x_i - \bar{x})(y_i - \bar{y})}{\sum(x_i - \bar{x})^2}$$

If we break this equation apart into its component pieces, we can simplify it somewhat. The denominator for b should look familiar; it is very similar to the variance of x, which is denoted as $Var(x)$. As we learned in *Chapter 2, Managing and Understanding Data*, the variance involves finding the average squared deviation from the mean of x. This can be expressed as:

$$Var(x) = \frac{\sum(x_i - \bar{x})^2}{n}$$

The numerator involves taking the sum of each data point's deviation from the mean x value multiplied by that point's deviation away from the mean y value. This is similar to the **covariance** function for x and y, denoted as $Cov(x, y)$. The covariance formula is:

$$Cov(x, y) = \frac{\sum(x_i - \bar{x})(y_i - \bar{y})}{n}$$

If we divide the covariance function by the variance function, the *n* terms in the numerator and denominator cancel each other and we can rewrite the formula for *b* as:

$$b = \frac{\text{Cov}(x, y)}{\text{Var}(x)}$$

Given this restatement, it is easy to calculate the value of *b* using built-in R functions. Let's apply them to the shuttle launch data to estimate the regression line.

 If you would like to follow along with these examples, download the `challenger.csv` file from the Packt Publishing website and load it to a data frame using the `launch <- read.csv("challenger.csv")` command.

If the shuttle launch data is stored in a data frame named `launch`, the independent variable *x* is named `temperature`, and the dependent variable *y* is named `distress_ct`, we can then use the R functions `cov()` and `var()` to estimate b:

```
> b <- cov(launch$temperature, launch$distress_ct) /
        var(launch$temperature)
> b
[1] -0.04753968
```

We can then estimate a using the computed b value and applying the `mean()` function:

```
> a <- mean(launch$distress_ct) - b * mean(launch$temperature)
> a
[1] 3.698413
```

Estimating the regression equation by hand is obviously less than ideal, so R predictably provides a function for fitting regression models automatically. We will use this function shortly. Before then, it is important to expand your understanding of the regression model's fit by first learning a method for measuring the strength of a linear relationship. Additionally, you will soon learn how to apply multiple linear regression to problems with more than one independent variable.

Correlations

The **correlation** between two variables is a number that indicates how closely their relationship follows a straight line. Without additional qualification, correlation typically refers to the **Pearson correlation coefficient**, which was developed by the 20th century mathematician Karl Pearson. A correlation ranges between -1 and +1. The maximum and minimum values indicate a perfectly linear relationship, while a correlation close to zero indicates the absence of a linear relationship.

The following formula defines Pearson's correlation:

$$\rho_{x,y} = \text{Corr}(x, y) = \frac{\text{Cov}(x, y)}{\sigma_x \sigma_y}$$

More Greek notation has been introduced here: the first symbol (which looks like a lowercase p) is *rho*, and it is used to denote the Pearson correlation statistic. The symbols that look like q characters rotated counter-clockwise are the Greek letter *sigma*, and they indicate the standard deviation of x or y.

Using this formula, we can calculate the correlation between the launch temperature and the number of O-ring distress events. Recall that the covariance function is `cov()` and the standard deviation function is `sd()`. We'll store the result in `r`, a letter that is commonly used to indicate the estimated correlation:

```
> r <- cov(launch$temperature, launch$distress_ct) /
        (sd(launch$temperature) * sd(launch$distress_ct))
> r
[1] -0.5111264
```

Alternatively, we can obtain the same result with the `cor()` correlation function:

```
> cor(launch$temperature, launch$distress_ct)
[1] -0.5111264
```

The correlation between the temperature and the number of distressed O-rings is -0.51. The negative correlation implies that increases in temperature are related to decreases in the number of distressed O-rings. To the NASA engineers studying the O-ring data, this would have been a very clear indicator that a low temperature launch could be problematic. The correlation also tells us about the relative strength of the relationship between temperature and O-ring distress. Because -0.51 is halfway to the maximum negative correlation of -1, this implies that there is a moderately strong negative linear association.

There are various rules of thumb used to interpret correlation strength. One method assigns a status of "weak" to values between 0.1 and 0.3; "moderate" to the range of 0.3 to 0.5; and "strong" to values above 0.5 (these also apply to similar ranges of negative correlations). However, these thresholds may be too strict or too lax for certain purposes. Often, the correlation must be interpreted in context. For data involving human beings, a correlation of 0.5 may be considered very high; for data generated by mechanical processes, a correlation of 0.5 may be very weak.

You have probably heard the expression "correlation does not imply causation." This is rooted in the fact that a correlation only describes the association between a pair of variables, yet there could be other unmeasured explanations. For example, there may be a strong association between life expectancy and time per day spent watching movies, but before doctors start recommending that we all watch more movies, we need to rule out another explanation: younger people watch more movies and are less likely to die.

Measuring the correlation between two variables gives us a way to quickly check for linear relationships among independent variables and the dependent variable. This will be increasingly important as we start defining regression models with a larger number of predictors.

Multiple linear regression

Most real-world analyses have more than one independent variable. Therefore, it is likely that you will be using multiple linear regression for most numeric prediction tasks. The strengths and weaknesses of multiple linear regression are shown in the following table:

Strengths	Weaknesses
• By far the most common approach for modeling numeric data • Can be adapted to model almost any modeling task • Provides estimates of both the size and strength of the relationships among features and the outcome	• Makes strong assumptions about the data • The model's form must be specified by the user in advance • Does not handle missing data • Only works with numeric features, so categorical data requires additional preparation • Requires some knowledge of statistics to understand the model

We can understand multiple regression as an extension of simple linear regression. The goal in both cases is similar—to find values of slope coefficients that minimize the prediction error of a linear equation. The key difference is that there are additional terms for the additional independent variables.

Multiple regression models are in the form of the following equation. The dependent variable y is specified as the sum of an intercept term a plus, for each of i features, the product of the estimated β value and the x variable. An error term ε (denoted by the Greek letter *epsilon*) has been added here as a reminder that the predictions are not perfect. This represents the residual term noted previously.

$$y = \alpha + \beta_1 x_1 + \beta_2 x_2 + \ldots + \beta_i x_i + \varepsilon$$

Let's consider for a moment the interpretation of the estimated regression parameters. You will note that in the preceding equation, a coefficient is provided for each feature. This allows each feature to have a separate estimated effect on the value of y. In other words, y changes by the amount β_i for each unit increase in feature x_i. The intercept a is then the expected value of y when the independent variables are all zero.

Since the intercept term a is really no different than any other regression parameter, it is also sometimes denoted as β_0 (pronounced *beta naught*) as shown in the following equation:

$$y = \beta_0 + \beta_1 x_1 + \beta_2 x_2 + \ldots + \beta_i x_i + \varepsilon$$

Just like before, the intercept is unrelated to any of the independent x variables. However, for reasons that will become clear shortly, it helps to imagine β_0 as if it were being multiplied by a term x_0. We assign x_0 to be a constant with the value of 1.

$$y = \beta_0 x_0 + \beta_1 x_1 + \beta_2 x_2 + \ldots + \beta_i x_i + \varepsilon$$

In order to estimate the regression parameters, each observed value of the dependent variable y must be related to observed values of the independent x variables using the regression equation in the previous form. The following figure is a graphical representation of the setup of a multiple regression task:

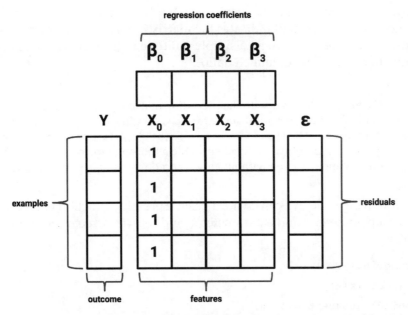

Figure 6.5: Multiple regression seeks to find the β values that relate the X values to Y while minimizing ε

The many rows and columns of data illustrated in the preceding figure can be described in a condensed formulation using **matrix notation** in bold font to indicate that each of the terms represents multiple values. Simplified in this way, the formula is as follows:

$$Y = \beta X + \varepsilon$$

In matrix notation, the dependent variable is a vector, **Y**, with a row for every example. The independent variables have been combined into a matrix, **X**, with a column for each feature plus an additional column of "1" values for the intercept. Each column has a row for every example. The regression coefficients **β** and residual errors **ε** are also now vectors.

The goal now is to solve for **β**, the vector of regression coefficients that minimizes the sum of the squared errors between the predicted and actual **Y** values. Finding the optimal solution requires the use of matrix algebra; therefore, the derivation deserves more careful attention than can be provided in this text. However, if you're willing to trust the work of others, the best estimate of the vector **β** can be computed as:

$$\hat{\beta} = \left(X^T X \right)^{-1} X^T Y$$

This solution uses a pair of matrix operations: the **T** indicates the **transpose** of matrix **X**, while the negative exponent indicates the **matrix inverse**. Using R's built-in matrix operations, we can thus implement a simple multiple regression learner. Let's apply this formula to the Challenger launch data.

If you are unfamiliar with the preceding matrix operations, the Wolfram MathWorld pages for transpose (http://mathworld.wolfram.com/Transpose.html) and matrix inverse (http://mathworld.wolfram.com/MatrixInverse.html) provide a thorough introduction and are understandable even without an advanced mathematics background.

Using the following code, we can create a basic regression function named `reg()`, which takes a parameter y and a parameter x, and returns a vector of estimated beta coefficients:

```
reg <- function(y, x) {
  x <- as.matrix(x)
  x <- cbind(Intercept = 1, x)
  b <- solve(t(x) %*% x) %*% t(x) %*% y
  colnames(b) <- "estimate"
  print(b)
}
```

The `reg()` function created here uses several R commands that we have not used previously. First, since we will be using the function with sets of columns from a data frame, the `as.matrix()` function converts the data frame into matrix form. Next, the `cbind()` function binds an additional column onto the x matrix; the command `Intercept = 1` instructs R to name the new column `Intercept` and to fill the column with repeating 1 values. Then, a series of matrix operations are performed on the x and y objects:

- `solve()` takes the inverse of a matrix
- `t()` is used to transpose a matrix
- `%*%` multiplies two matrices

By combining these as shown, our function will return a vector b, which contains estimated parameters for the linear model relating x to y. The final two lines in the function give the b vector a name and print the result on screen.

Let's apply this function to the shuttle launch data. As shown in the following code, the dataset includes three features and the distress count (`distress_ct`), which is the outcome of interest:

```
> str(launch)
'data.frame':    23 obs. of  4 variables:
 $ distress_ct       : int  0 1 0 0 0 0 0 0 1 1 ...
 $ temperature       : int  66 70 69 68 67 72 73 70 57 63 ...
 $ field_check_pressure: int  50 50 50 50 50 50 100 100 200 ...
 $ flight_num        : int  1 2 3 4 5 6 7 8 9 10 ...
```

We can confirm that our function is working correctly by comparing its result for the simple linear regression model of O-ring failures versus temperature, which we found earlier to have parameters $a = 3.70$ and $b = -0.048$. Since temperature is in the second column of the launch data, we can run the `reg()` function as follows:

```
> reg(y = launch$distress_ct, x = launch[2])
               estimate
Intercept    3.69841270
temperature -0.04753968
```

These values exactly match our prior result, so let's use the function to build a multiple regression model. We'll apply it just as before, but this time we will specify columns two through four for the x parameter to add two additional predictors:

```
> reg(y = launch$distress_ct, x = launch[2:4])
                       estimate
Intercept            3.527093383
temperature         -0.051385940
field_check_pressure  0.001757009
flight_num            0.014292843
```

This model predicts the number of O-ring distress events using temperature, field check pressure, and the launch ID number. Notably, the inclusion of the two new predictors did not change our finding from the simple linear regression model. Just as before, the coefficient for the temperature variable is negative, which suggests that as temperature increases, the number of expected O-ring events decreases. The magnitude of the effect is also approximately the same: roughly 0.05 fewer distress events are expected for each degree increase in launch temperature.

The two new predictors also contribute to the predicted distress events. The field check pressure refers to the amount of pressure applied to the O-ring during pre-launch testing. Although the check pressure had originally been 50 psi, it was raised to 100 and 200 psi for some launches, which led some to believe that it may be responsible for O-ring erosion. The coefficient is positive, but small, providing at least a little evidence for this hypothesis. The flight number accounts for the shuttle's age. With each flight, it gets older, and parts may be more brittle or prone to fail. The small positive association between flight number and distress count may reflect this fact.

Overall, our retrospective analysis of the space shuttle data suggests that there was reason to believe that the Challenger launch was highly risky given the weather conditions. Perhaps if the engineers had applied linear regression beforehand, a disaster could have been averted. Of course, the reality of the situation, and all of the political implications involved, were surely not as simple as they appear in hindsight.

This study only scratched the surface of what is possible with linear regression modeling. Although the work was helpful for understanding exactly how regression models are built, there is more involved in modeling complex phenomena. R's built-in regression functions include additional functionality necessary to fit these more sophisticated models, and provide additional diagnostic output to aid model interpretation and assess fit. Let's apply these functions and expand our knowledge of regression by attempting a more challenging learning task.

Example – predicting medical expenses using linear regression

In order for a health insurance company to make money, it needs to collect more in yearly premiums than it spends on medical care to its beneficiaries. Consequently, insurers invest a great deal of time and money to develop models that accurately forecast medical expenses for the insured population.

Medical expenses are difficult to estimate because the costliest conditions are rare and seemingly random. Still, some conditions are more prevalent for certain segments of the population. For instance, lung cancer is more likely among smokers than non-smokers, and heart disease may be more likely among the obese.

The goal of this analysis is to use patient data to forecast the average medical care expenses for such population segments. These estimates could be used to create actuarial tables that set the price of yearly premiums higher or lower according to the expected treatment costs.

Step 1 – collecting data

For this analysis, we will use a simulated dataset containing hypothetical medical expenses for patients in the United States. This data was created for this book using demographic statistics from the US Census Bureau, and thus, approximately reflect real-world conditions.

If you would like to follow along interactively, download the insurance.csv file from the Packt Publishing website and save it to your R working folder.

The insurance.csv file includes 1,338 examples of beneficiaries currently enrolled in the insurance plan, with features indicating characteristics of the patient as well as the total medical expenses charged to the plan for the calendar year. The features are:

- age: An integer indicating the age of the primary beneficiary (excluding those above 64 years, as they are generally covered by the government).

- sex: The policy holder's gender: either male or female.

- bmi: The body mass index (BMI), which provides a sense of how over or underweight a person is relative to their height. BMI is equal to weight (in kilograms) divided by height (in meters) squared. An ideal BMI is within the range of 18.5 to 24.9.

- children: An integer indicating the number of children/dependents covered by the insurance plan.

- smoker: A yes or no categorical variable that indicates whether the insured regularly smokes tobacco.

- region: The beneficiary's place of residence in the US, divided into four geographic regions: northeast, southeast, southwest, or northwest.

It is important to give some thought to how these variables may relate to billed medical expenses. For instance, we might expect that older people and smokers are at higher risk of large medical expenses. Unlike many other machine learning methods, in regression analysis, the relationships among the features are typically specified by the user rather than detected automatically. We'll explore some of these potential relationships in the next section.

Step 2 – exploring and preparing the data

As we have done before, we will use the read.csv() function to load the data for analysis. We can safely use stringsAsFactors = TRUE because it is appropriate to convert the three nominal variables to factors:

```
> insurance <- read.csv("insurance.csv", stringsAsFactors = TRUE)
```

The str() function confirms that the data is formatted as we had expected:

```
> str(insurance)
'data.frame':    1338 obs. of  7 variables:
```

```
$ age      : int  19 18 28 33 32 31 46 37 37 60 ...
$ sex      : Factor w/ 2 levels "female","male": 1 2 2 2 2 1 ...
$ bmi      : num  27.9 33.8 33 22.7 28.9 25.7 33.4 27.7 ...
$ children : int  0 1 3 0 0 0 1 3 2 0 ...
$ smoker   : Factor w/ 2 levels "no","yes": 2 1 1 1 1 1 1 1 ...
$ region   : Factor w/ 4 levels "northeast","northwest",..: ...
$ expenses : num  16885 1726 4449 21984 3867 ...
```

Our model's dependent variable is `expenses`, which measures the medical costs each person charged to the insurance plan for the year. Prior to building a regression model, it is often helpful to check for normality. Although linear regression does not strictly require a normally distributed dependent variable, the model often fits better when this is true. Let's take a look at the summary statistics:

```
> summary(insurance$expenses)
   Min. 1st Qu.  Median    Mean 3rd Qu.    Max.
   1122    4740    9382   13270   16640   63770
```

Because the mean value is greater than the median, this implies that the distribution of insurance expenses is right-skewed. We can confirm this visually using a histogram:

```
> hist(insurance$expenses)
```

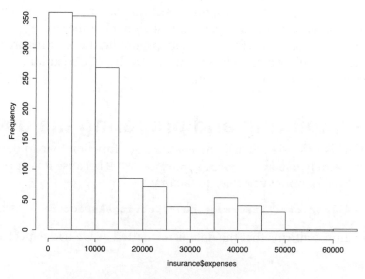

Figure 6.6: The distribution of annual insurance expenses

As expected, the figure shows a right-skewed distribution. It also shows that the majority of people in our data have yearly medical expenses between zero and $15,000, in spite of the fact that the tail of the distribution extends far past these peaks. Although this distribution is not ideal for a linear regression, knowing this weakness ahead of time may help us design a better-fitting model later on.

Before we address that issue, another problem is at hand. Regression models require that every feature is numeric, yet we have three factor-type features in our data frame. For instance, the `sex` variable is divided into `male` and `female` levels, while `smoker` has categories for `yes` and `no`. From the `summary()` output, we know that `region` has four levels, but we need to take a closer look to see how they are distributed:

```
> table(insurance$region)

northeast northwest southeast southwest
    324       325       364       325
```

Here, we see that the data has been divided nearly evenly among four geographic regions. We will see how R's linear regression function handles these factor variables shortly.

Exploring relationships among features – the correlation matrix

Before fitting a regression model to data, it can be useful to determine how the independent variables are related to the dependent variable and each other. A **correlation matrix** provides a quick overview of these relationships. Given a set of variables, it provides a correlation for each pairwise relationship.

To create a correlation matrix for the four numeric variables in the `insurance` data frame, use the `cor()` command:

```
> cor(insurance[c("age", "bmi", "children", "expenses")])
                 age        bmi   children   expenses
age       1.0000000 0.10934101 0.04246900 0.29900819
bmi       0.1093410 1.00000000 0.01264471 0.19857626
children  0.0424690 0.01264471 1.00000000 0.06799823
expenses  0.2990082 0.19857626 0.06799823 1.00000000
```

At the intersection of each row and column pair, the correlation is listed for the variables indicated by that row and column. The diagonal is always `1.0000000` since there is always a perfect correlation between a variable and itself. The values above and below the diagonal are identical since correlations are symmetrical. In other words, `cor(x, y)` is equal to `cor(y, x)`.

None of the correlations in the matrix are very strong, but there are some notable associations. For instance, age and bmi appear to have a weak positive correlation, meaning that as someone ages, their body mass tends to increase. There are also positive correlations between age and expenses, bmi and expenses, and children and expenses. These associations imply that as age, body mass, and number of children increase, the expected cost of insurance goes up. We'll try to tease out these relationships more clearly when we build our final regression model.

Visualizing relationships among features – the scatterplot matrix

It can also be helpful to visualize the relationships among numeric features with scatterplots. Although we could create a scatterplot for each possible relationship, doing so for a large number of features quickly becomes tedious.

An alternative is to create a **scatterplot matrix** (sometimes abbreviated as **SPLOM**), which is simply a collection of scatterplots arranged in a grid. It is used to detect patterns among three or more variables. The scatterplot matrix is not a true multi-dimensional visualization because only two features are examined at a time. Still, it provides a general sense of how the data may be interrelated.

We can use R's graphical capabilities to create a scatterplot matrix for the four numeric features: age, bmi, children, and expenses. The pairs() function is provided in the default R installation and provides basic functionality for producing scatterplot matrices. To invoke the function, simply provide it the data frame to plot. Here, we'll limit the insurance data frame to the four numeric variables of interest:

```
> pairs(insurance[c("age", "bmi", "children", "expenses")])
```

This produces the following diagram:

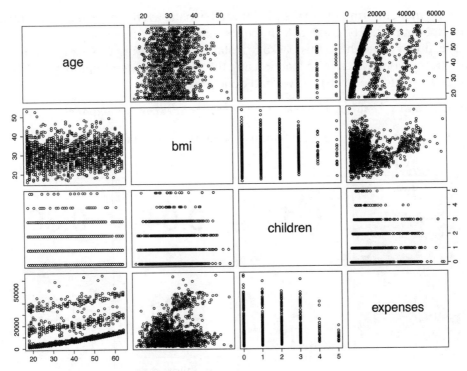

Figure 6.7: A scatterplot matrix of the numeric features in the insurance dataset

In the scatterplot matrix, the intersection of each row and column holds the scatterplot of the variables indicated by the row and column pair. The diagrams above and below the diagonal are transpositions because the *x* axis and *y* axis have been swapped.

Do you notice any patterns in these plots? Although some look like random clouds of points, a few seem to display some trends. The relationship between age and expenses displays several relatively straight lines, while the bmi versus expenses plot has two distinct groups of points. It is difficult to detect trends in any of the other plots.

By adding more information to the plot, it can be made even more useful. An enhanced scatterplot matrix can be created with the `pairs.panels()` function in the `psych` package. If you do not have this package installed, type `install.packages("psych")` to install it on your system and load it using the `library(psych)` command. Then, we can create a scatterplot matrix as we had done previously:

```
> pairs.panels(insurance[c("age", "bmi", "children", "expenses")])
```

This produces a slightly more informative scatterplot matrix, as follows:

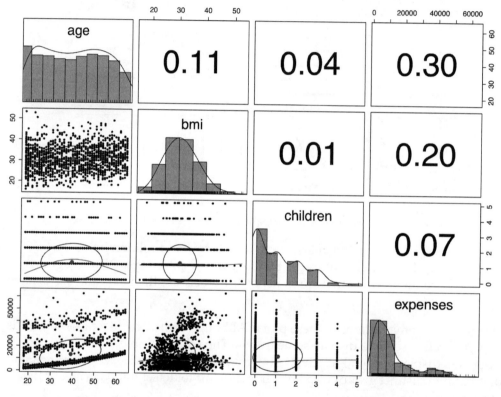

Figure 6.8: The pairs.panels() function adds detail to the scatterplot matrix

In the `pairs.panels()` output, the scatterplots above the diagonal are replaced with a correlation matrix. The diagonal now contains histograms depicting the distribution of values for each feature. Finally, the scatterplots below the diagonal are presented with additional visual information.

The oval-shaped object on each scatterplot is a **correlation ellipse**. It provides a visualization of correlation strength. The more the ellipse is stretched, the stronger the correlation. An almost perfectly round oval, as with `bmi` and `children`, indicates a very weak correlation (in this case 0.01).

The ellipse for age and expenses in much more stretched, reflecting its stronger correlation (0.30). The dot at the center of the ellipse is a point reflecting the means of the *x* axis and *y* axis variables.

The curve drawn on the scatterplot is called a **loess curve**. It indicates the general relationship between the *x* axis and *y* axis variables. It is best understood by example. The curve for age and children is an upside-down U, peaking around middle age. This means that the oldest and youngest people in the sample have fewer children on the insurance plan than those around middle age. Because this trend is nonlinear, this finding could not have been inferred from the correlations alone. On the other hand, the loess curve for age and bmi is a line sloping up gradually, implying that body mass increases with age, but we had already inferred this from the correlation matrix.

Step 3 – training a model on the data

To fit a linear regression model to data with R, the lm() function can be used. This is part of the stats package, which should be included and loaded by default with your R installation. The lm() syntax is as follows:

Multiple regression modeling syntax

using the `lm()` function in the `stats` package

Building the model:

```
m <- lm(dv ~ iv, data = mydata)
```

- **dv** is the dependent variable in the **mydata** data frame to be modeled
- **iv** is an R formula specifying the independent variables in the **mydata** data frame to use in the model
- **data** specifies the data frame in which the **dv** and **iv** variables can be found

The function will return a regression model object that can be used to make predictions. Interactions between independent variables can be specified using the * operator.

Making predictions:

```
p <- predict(m, test)
```

- **m** is a model trained by the `lm()` function
- **test** is a data frame containing test data with the same features as the training data used to build the model.

The function will return a vector of predicted values.

Example:

```
ins_model <- lm(charges ~ age + sex + smoker,
                data = insurance)
ins_pred <- predict(ins_model, insurance_test)
```

The following command fits a linear regression model that relates the six independent variables to the total medical expenses. The R formula syntax uses the tilde character (~) to describe the model; the dependent variable expenses goes to the left of the tilde while the independent variables go to the right, separated by + signs. There is no need to specify the regression model's intercept term, as it is included by default:

```
> ins_model <- lm(expenses ~ age + children + bmi + sex +
    smoker + region, data = insurance)
```

Because the period character (.) can be used to specify all features (excluding those already specified in the formula), the following command is equivalent to the prior command:

```
> ins_model <- lm(expenses ~ ., data = insurance)
```

After building the model, simply type the name of the model object to see the estimated beta coefficients:

```
> ins_model

Call:
lm(formula = expenses ~ ., data = insurance)


Coefficients:
    (Intercept)              age           sexmale
       -11941.6            256.8            -131.4
            bmi         children         smokeryes
          339.3            475.7           23847.5
regionnorthwest  regionsoutheast   regionsouthwest
         -352.8          -1035.6            -959.3
```

Understanding the regression coefficients is fairly straightforward. The intercept is the predicted value of expenses when the independent variables are equal to zero. However, in many cases, the intercept is of little explanatory value by itself, as it is often impossible to have values of zero for all of the features. This is the case here, where no person can exist with age zero and BMI zero, and consequently, the intercept has no real-world interpretation. For this reason, in practice, the intercept is often ignored.

The beta coefficients indicate the estimated increase in expenses for an increase of one unit in each of the features, assuming all other values are held constant. For instance, for each additional year of age, we would expect $256.80 higher medical expenses on average, assuming everything else is held equal.

Similarly, each additional child results in an average of $475.70 in additional medical expenses each year, and each unit increase in BMI is associated with an average increase of $339.30 in yearly medical expenses, all else equal.

You might notice that although we only specified six features in our model formula, there are eight coefficients reported in addition to the intercept. This happened because the `lm()` function automatically applies dummy coding to each of the factor type variables included in the model.

As explained in *Chapter 2, Managing and Understanding Data*, dummy coding allows a nominal feature to be treated as numeric by creating a binary variable for each category of the feature. The dummy variable is set to 1 if the observation falls into the specified category or 0 otherwise. For instance, the `sex` feature has two categories: `male` and `female`. This will be split into two binary variables, which R names `sexmale` and `sexfemale`. For observations where `sex` = `male`, then `sexmale` = 1 and `sexfemale` = 0; conversely, if `sex` = `female`, then `sexmale` = 0 and `sexfemale` = 1. The same coding applies to variables with three or more categories. For example, R split the four-category feature `region` into four dummy variables: `regionnorthwest`, `regionsoutheast`, `regionsouthwest`, and `regionnortheast`.

When adding a dummy variable to a regression model, one category is always left out to serve as the reference category. The estimates are then interpreted relative to the reference. In our model, R automatically held out the `sexfemale`, `smokerno`, and `regionnortheast` variables, making female non-smokers in the northeast region the reference group. Thus, males have $131.40 less medical expenses each year relative to females and smokers cost an average of $23,847.50 more than non-smokers per year. The coefficient for each of the three regions in the model is negative, which implies that the reference group, the northeast region, tends to have the highest average expenses.

By default, R uses the first level of the factor variable as the reference. If you would prefer to use another level, the `relevel()` function can be used to specify the reference group manually. Use the `?relevel` command in R for more information.

The results of the linear regression model make logical sense: old age, smoking, and obesity tend to be linked to additional health issues, while additional family member dependents may result in an increase in physician visits and preventive care such as vaccinations and yearly physical exams. However, we currently have no sense of how well the model is fitting the data. We'll answer this question in the next section.

Step 4 – evaluating model performance

The parameter estimates we obtained by typing `ins_model` tell us about how the independent variables are related to the dependent variable, but they tell us nothing about how well the model fits our data. To evaluate the model performance, we can use the `summary()` command on the stored model:

```
> summary(ins_model)
```

This produces the following output, which has been annotated for illustrative purposes:

```
Call:
lm(formula = expenses ~ ., data = insurance)

Residuals:
     Min       1Q   Median       3Q      Max
-11302.7  -2850.9   -979.6   1383.9  29981.7    (1)

Coefficients:
                  Estimate Std. Error t value Pr(>|t|)
(Intercept)       -11941.6      987.8 -12.089  < 2e-16 ***
age                  256.8       11.9  21.586  < 2e-16 ***    (2)
sexmale             -131.3      332.9  -0.395 0.693255
bmi                  339.3       28.6  11.864  < 2e-16 ***
children             475.7      137.8   3.452 0.000574 ***
smokeryes          23847.5      413.1  57.723  < 2e-16 ***
regionnorthwest     -352.8      476.3  -0.741 0.458976
regionsoutheast    -1035.6      478.7  -2.163 0.030685 *
regionsouthwest     -959.3      477.9  -2.007 0.044921 *
---
Signif. codes:  0 '***' 0.001 '**' 0.01 '*' 0.05 '.' 0.1 ' ' 1

Residual standard error: 6062 on 1329 degrees of freedom
Multiple R-squared:  0.7509,  Adjusted R-squared:  0.7494    (3)
F-statistic: 500.9 on 8 and 1329 DF,  p-value: < 2.2e-16
```

The `summary()` output may seem overwhelming at first, but the basics are easy to pick up. As indicated by the numbered labels in the preceding output, there are three key ways to evaluate the performance, or fit, of our model:

1. The **Residuals** section provides summary statistics for the errors in our predictions, some of which are apparently quite substantial. Since a residual is equal to the true value minus the predicted value, the maximum error of 29981.7 suggests that the model under-predicted expenses by nearly $30,000 for at least one observation. On the other hand, 50 percent of errors fall within the 1Q and 3Q values (the first and third quartile), so the majority of predictions were between $2,850.90 over the true value and $1,383.90 under the true value.

2. For each estimated regression coefficient, the **p-value**, denoted `Pr(>|t|)`, provides an estimate of the probability that the true coefficient is zero given the value of the estimate. Small p-values suggest that the true coefficient is very unlikely to be zero, which means that the feature is extremely unlikely to have no relationship with the dependent variable. Note that some of the p-values have stars (`***`), which correspond to the footnotes that specify the **significance level** met by the estimate. This level is a threshold, chosen prior to building the model, which will be used to indicate "real" findings, as opposed to those due to chance alone; p-values less than the significance level are considered **statistically significant**. If the model had few such terms, it may be a cause for concern, since this would indicate that the features used are not very predictive of the outcome. Here, our model has several highly significant variables, and they seem to be related to the outcome in logical ways.

3. The **Multiple R-squared** value (also called the coefficient of determination) provides a measure of how well our model as a whole explains the values of the dependent variable. It is similar to the correlation coefficient in that the closer the value is to 1.0, the better the model perfectly explains the data. Since the R-squared value is 0.7494, we know that the model explains nearly 75 percent of the variation in the dependent variable. Because models with more features always explain more variation, the **Adjusted R-squared** value corrects R-squared by penalizing models with a large number of independent variables. It is useful for comparing the performance of models with different numbers of explanatory variables.

Given the preceding three performance indicators, our model is performing fairly well. It is not uncommon for regression models of real-world data to have fairly low R-squared values; a value of 0.75 is actually quite good. The size of some of the errors is a bit concerning, but not surprising given the nature of medical expense data. However, as shown in the next section, we may be able to improve the model's performance by specifying the model in a slightly different way.

Step 5 – improving model performance

As mentioned previously, a key difference between regression modeling and other machine learning approaches is that regression typically leaves feature selection and model specification to the user. Consequently, if we have subject matter knowledge about how a feature is related to the outcome, we can use this information to inform the model specification and potentially improve the model's performance.

Model specification – adding nonlinear relationships

In linear regression, the relationship between an independent variable and the dependent variable is assumed to be linear, yet this may not necessarily be true. For example, the effect of age on medical expenditures may not be constant across all age values; the treatment may become disproportionately expensive for the oldest populations.

If you recall, a typical regression equation follows a form similar to this:

$$y = \alpha + \beta_1 x$$

To account for a nonlinear relationship, we can add a higher order term to the regression equation, treating the model as a polynomial. In effect, we will be modeling a relationship like this:

$$y = \alpha + \beta_1 x + \beta_2 x^2$$

The difference between these two models is that an additional beta will be estimated, which is intended to capture the effect of the x^2 term. This allows the impact of age to be measured as a function of age squared.

To add the nonlinear age to the model, we simply need to create a new variable:

```
> insurance$age2 <- insurance$age^2
```

Then, when we produce our improved model, we'll add both age and age2 to the lm() formula using the form expenses ~ age + age2. This will allow the model to separate the linear and nonlinear impact of age on medical expenses.

Transformation – converting a numeric variable to a binary indicator

Suppose we have a hunch that the effect of a feature is not cumulative, but rather that it has an effect only after a specific threshold has been reached. For instance, BMI may have zero impact on medical expenditures for individuals in the normal weight range, but it may be strongly related to higher costs for the obese (that is, BMI of 30 or more).

We can model this relationship by creating a binary obesity indicator variable that is 1 if the BMI is at least 30 and 0 if it is less than 30. The estimated beta for this binary feature would then indicate the average net impact on medical expenses for individuals with BMI of 30 or above, relative to those with BMI less than 30.

To create the feature, we can use the `ifelse()` function that, for each element in a vector, tests a specified condition and returns a value depending on whether the condition is true or false. For BMI greater than or equal to 30, we will return `1`, otherwise we'll return `0`:

```
> insurance$bmi30 <- ifelse(insurance$bmi >= 30, 1, 0)
```

We can then include the `bmi30` variable in our improved model, either replacing the original `bmi` variable or in addition to it, depending on whether or not we think the effect of obesity occurs in addition to a separate linear BMI effect. Without good reason to do otherwise, we'll include both in our final model.

If you have trouble deciding whether or not to include a variable, a common practice is to include it and examine the p-value. If the variable is not statistically significant, you have evidence to support excluding it in the future.

Model specification – adding interaction effects

So far, we have only considered each feature's individual contribution to the outcome. What if certain features have a combined impact on the dependent variable? For instance, smoking and obesity may have harmful effects separately, but it is reasonable to assume that their combined effect may be worse than the sum of each one alone.

When two features have a combined effect, this is known as an **interaction**. If we suspect that two variables interact, we can test this hypothesis by adding their interaction to the model. Interaction effects are specified using the R formula syntax. To interact the obesity indicator (`bmi30`) with the smoking indicator (`smoker`), we would write a formula in the form `expenses ~ bmi30*smoker`.

The `*` operator is shorthand that instructs R to model `expenses ~ bmi30 + smokeryes + bmi30:smokeryes`. The colon operator (`:`) in the expanded form indicates that `bmi30:smokeryes` is the interaction between the two variables. Note that the expanded form automatically also included the individual `bmi30` and `smokeryes` variables as well as their interaction.

Interactions should never be included in a model without also adding each of the interacting variables. If you always create interactions using the `*` operator, this will not be a problem since R will add the required components for you automatically.

Putting it all together – an improved regression model

Based on a bit of subject matter knowledge of how medical costs may be related to patient characteristics, we developed what we think is a more accurately specified regression formula. To summarize the improvements, we:

- Added a nonlinear term for age
- Created an indicator for obesity
- Specified an interaction between obesity and smoking

We'll train the model using the `lm()` function as before, but this time we'll add the newly constructed variables and the interaction term:

```
> ins_model2 <- lm(expenses ~ age + age2 + children + bmi + sex +
                    bmi30*smoker + region, data = insurance)
```

Next, we summarize the results:

```
> summary(ins_model2)
```

The output is as follows:

```
Call:
lm(formula = expenses ~ age + age2 + children + bmi + sex + bmi30 *
    smoker + region, data = insurance)

Residuals:
     Min       1Q   Median       3Q      Max
-17297.1  -1656.0  -1262.7   -727.8  24161.6

Coefficients:
                    Estimate Std. Error t value Pr(>|t|)
(Intercept)        139.0053  1363.1359    0.102 0.918792
age                -32.6181    59.8250   -0.545 0.585690
age2                 3.7307     0.7463    4.999 6.54e-07 ***
children           678.6017   105.8855    6.409 2.03e-10 ***
bmi                119.7715    34.2796    3.494 0.000492 ***
sexmale           -496.7690   244.3713   -2.033 0.042267 *
bmi30             -997.9355   422.9607   -2.359 0.018449 *
smokeryes        13404.5952   439.9591   30.468  < 2e-16 ***
regionnorthwest   -279.1661   349.2826   -0.799 0.424285
regionsoutheast   -828.0345   351.6484   -2.355 0.018682 *
regionsouthwest  -1222.1619   350.5314   -3.487 0.000505 ***
bmi30:smokeryes  19810.1534   604.6769   32.762  < 2e-16 ***
---
Signif. codes:  0 '***' 0.001 '**' 0.01 '*' 0.05 '.' 0.1 ' ' 1

Residual standard error: 4445 on 1326 degrees of freedom
Multiple R-squared:  0.8664,  Adjusted R-squared:  0.8653
F-statistic: 781.7 on 11 and 1326 DF,  p-value: < 2.2e-16
```

The model fit statistics help to determine whether our changes improved the performance of the regression model. Relative to our first model, the R-squared value has improved from 0.75 to about 0.87.

Similarly, the adjusted R-squared value, which takes into account the fact that the model grew in complexity, improved from 0.75 to 0.87. Our model is now explaining 87 percent of the variation in medical treatment costs. Additionally, our theories about the model's functional form seem to be validated. The higher-order age2 term is statistically significant, as is the obesity indicator, bmi30. The interaction between obesity and smoking suggests a massive effect; in addition to the increased costs of over \$13,404 for smoking alone, obese smokers spend another \$19,810 per year. This may suggest that smoking exacerbates diseases associated with obesity.

> Strictly speaking, regression modeling makes some strong assumptions about the data. These assumptions are not as important for numeric forecasting, as the model's worth is not based upon whether it truly captures the underlying process—we simply care about the accuracy of its predictions. However, if you would like to make firm inferences from the regression model coefficients, it is necessary to run diagnostic tests to ensure that the regression assumptions have not been violated. For an excellent introduction to this topic, see: *Multiple Regression: A Primer, Allison, PD, Pine Forge Press, 1998.*

Making predictions with a regression model

After examining the estimated regression coefficients and fit statistics, we can also use the model to predict the expenses of future enrollees on the health insurance plan. To illustrate the process of making predictions, let's first apply the model to the original training data using the predict() function as follows:

```
> insurance$pred <- predict(ins_model2, insurance)
```

This saves the predictions as a new vector named pred in the insurance data frame. We can then compute the correlation between the predicted and actual costs of insurance:

```
> cor(insurance$pred, insurance$expenses)
[1] 0.9307999
```

The correlation of 0.93 suggests a very strong linear relationship between the predicted and actual values. This is a good sign—it suggests that the model is highly accurate! It can also be useful to examine this finding as a scatterplot. The following R commands plot the relationship and then add an identity line with an intercept equal to zero and slope equal to one. The col, lwd, and lty parameters affect the line color, width, and type, respectively:

```
> plot(insurance$pred, insurance$expenses)
> abline(a = 0, b = 1, col = "red", lwd = 3, lty = 2)
```

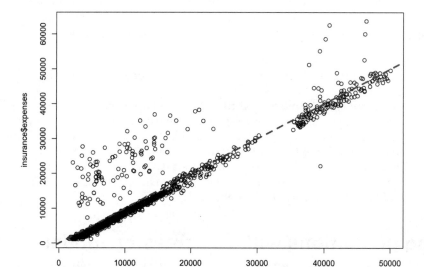

Figure 6.9: In this scatterplot, points falling on or near the diagonal dashed line where $y = x$ indicate the predictions that were very close to the actual values

The off-diagonal points falling above the line are cases where the actual expenses were greater than expected, while cases falling below the line are those less than expected. We can see here that a small number of patients with much larger-than-expected medical expenses are balanced by a larger number of patients with slightly smaller-than-expected expenses.

Now, suppose you would like to forecast the expenses for potential new enrollees on the insurance plan. To do this, you must provide the predict() function a data frame with the prospective patients' data. In the case of many patients, you may consider creating a CSV spreadsheet file to load in R, or for a smaller number, you may simply create a data frame within the predict() function itself. For example, to estimate the insurance expenses for a 30 year old, overweight, male non-smoker with two children in the Northeast, type:

```
> predict(ins_model2,
          data.frame(age = 30, age2 = 30^2, children = 2,
                     bmi = 30, sex = "male", bmi30 = 1,
                     smoker = "no", region = "northeast"))
       1
5973.774
```

Using this value, the insurance company might need to set its prices to about $6,000 per year, or $500 per month in order to break even for this demographic group. To compare the rate for a female who is otherwise similar, use the predict() function in much the same way:

```
> predict(ins_model2,
          data.frame(age = 30, age2 = 30^2, children = 2,
                     bmi = 30, sex = "female", bmi30 = 1,
                     smoker = "no", region = "northeast"))
       1
6470.543
```

Note that the difference between these two values, *5,973.774 - 6,470.543 = -496.769*, is the same as the estimated regression model coefficient for sexmale. On average, males are estimated to have about $496 less in expenses for the plan per year, all else being equal.

This illustrates the more general fact that the predicted expenses are a sum of each of the regression coefficients times their corresponding value in the prediction data frame. For instance, using the model's regression coefficient of *678.6017* for the number of children we can predict that reducing the children from two to zero will result in a drop in expenses of *2 * 678.6017 = 1,357.203* as follows:

```
> predict(ins_model2,
          data.frame(age = 30, age2 = 30^2, children = 0,
                     bmi = 30, sex = "female", bmi30 = 1,
                     smoker = "no", region = "northeast"))
       1
5113.34
```

```
> 6470.543 - 5113.34
[1] 1357.203
```

Following similar steps for a number of additional customer segments, the insurance company would be able to develop a profitable pricing structure for various demographics.

Exporting the model's regression coefficients allows you to build your own forecasting function. One potential use case for doing so would be to implement the regression model in a customer database for real-time prediction.

Understanding regression trees and model trees

If you recall from *Chapter 5, Divide and Conquer – Classification Using Decision Trees and Rules*, a decision tree builds a model, much like a flowchart, in which decision nodes, leaf nodes, and branches define a series of decisions that are used to classify examples. Such trees can also be used for numeric prediction by making only small adjustments to the tree growing algorithm. In this section, we will consider the ways in which trees for numeric prediction differ from trees used for classification.

Trees for numeric prediction fall into two categories. The first, known as **regression trees**, were introduced in the 1980s as part of the seminal **classification and regression tree (CART)** algorithm. Despite the name, regression trees do not use linear regression methods as described earlier in this chapter; rather, they make predictions based on the average value of examples that reach a leaf.

The CART algorithm is described in detail in *Classification and Regression Trees, Breiman L, Friedman, JH, Stone, CJ, Olshen, RA, Chapman and Hall, 1984.*

The second type of trees for numeric prediction are known as **model trees**. Introduced several years later than regression trees, they are less widely known, but perhaps more powerful. Model trees are grown in much the same way as regression trees, but at each leaf, a multiple linear regression model is built from the examples reaching that node. Depending on the number of leaf nodes, a model tree may build tens or even hundreds of such models. This makes model trees more difficult to understand than the equivalent regression tree, with the benefit that they may result in a more accurate model.

The earliest model tree algorithm, **M5**, is described in *Learning with continuous classes*, Quinlan, JR, *Proceedings of the 5th Australian Joint Conference on Artificial Intelligence*, 1992, pp. 343-348.

Adding regression to trees

Trees that can perform numeric prediction offer a compelling, yet often overlooked, alternative to regression modeling. The strengths and weaknesses of regression trees and model trees relative to the more common regression methods are listed in the following table:

Strengths	Weaknesses
• Combines the strengths of decision trees with the ability to model numeric data • Does not require the user to specify the model in advance • Uses automatic feature selection, which allows the approach to be used with a very large number of features • May fit some types of data much better than linear regression • Does not require knowledge of statistics to interpret the model	• Not as well-known as linear regression • Requires a large amount of training data • Difficult to determine the overall net effect of individual features on the outcome • Large trees can become more difficult to interpret than a regression model

Though traditional regression methods are typically the first choice for numeric prediction tasks, in some cases, numeric decision trees offer distinct advantages. For instance, decision trees may be better suited for tasks with many features or many complex, nonlinear relationships among features and the outcome; these situations present challenges for regression. Regression modeling also makes assumptions about the data that are often violated in real-world data; this is not the case for trees.

Trees for numeric prediction are built in much the same way as they are for classification. Beginning at the root node, the data is partitioned using a divide and conquer strategy according to the feature that will result in the greatest increase in homogeneity in the outcome after a split is performed. In classification trees, you will recall that homogeneity is measured by entropy. This is undefined for numeric data. Instead, for numeric decision trees, homogeneity is measured by statistics such as variance, standard deviation, or absolute deviation from the mean.

One common splitting criterion is called the **standard deviation reduction (SDR)**. It is defined by the following formula:

$$\text{SDR} = sd\left(T\right) - \sum_{i} \frac{|T_i|}{|T|} \times sd\left(T_i\right)$$

In this formula, the *sd(T)* function refers to the standard deviation of the values in set *T*, while T_1, T_2, ..., T_n are sets of values resulting from a split on a feature. The $|T|$ term refers to the number of observations in set *T*. Essentially, the formula measures the reduction in standard deviation by comparing the standard deviation pre-split to the weighted standard deviation post-split.

For example, consider the following case in which a tree is deciding whether to perform a split on binary feature A or a split on binary feature B:

original data	1	1	1	2	2	3	4	5	5	6	6	7	7	7	7
split on feature A	1	1	1	2	2	3	4	5	5	6	6	7	7	7	7
split on feature B	1	1	1	2	2	3	4	5	5	6	6	7	7	7	7

$$T_1 \qquad\qquad\qquad\qquad T_2$$

Figure 6.10: The algorithm considers splits on features A and B, which creates different T_1 and T_2 groups

Using the groups that would result from the proposed splits, we can compute the SDR for A and B as follows. The `length()` function used here returns the number of elements in a vector. Note that the overall group T is named `tee` to avoid overwriting R's built-in `T()` and `t()` functions.

```
> tee <- c(1, 1, 1, 2, 2, 3, 4, 5, 5, 6, 6, 7, 7, 7, 7)
> at1 <- c(1, 1, 1, 2, 2, 3, 4, 5, 5)
> at2 <- c(6, 6, 7, 7, 7, 7)
> bt1 <- c(1, 1, 1, 2, 2, 3, 4)
> bt2 <- c(5, 5, 6, 6, 7, 7, 7, 7)
> sdr_a <- sd(tee) - (length(at1) / length(tee) * sd(at1) +
            length(at2) / length(tee) * sd(at2))
> sdr_b <- sd(tee) - (length(bt1) / length(tee) * sd(bt1) +
            length(bt2) / length(tee) * sd(bt2))
```

Let's compare the SDR of A against the SDR of B:

```
> sdr_a
[1] 1.202815
> sdr_b
[1] 1.392751
```

The SDR for the split on feature A was about 1.2 versus 1.4 for the split on feature B. Since the standard deviation was reduced more for the split on B, the decision tree would use B first. It results in slightly more homogeneous sets than does A.

Suppose that the tree stopped growing here using this one and only split. A regression tree's work is done. It can make predictions for new examples depending on whether the example's value on feature B places the example into group T_1 or T_2. If the example ends up in T_1, the model would predict *mean(bt1)* = 2, otherwise it would predict *mean(bt2)* = 6.25.

In contrast, a model tree would go one step further. Using the seven training examples falling in group T_1 and the eight in T_2, the model tree could build a linear regression model of the outcome versus feature A. Note that feature B is of no help in building the regression model because all examples at the leaf have the same value of B—they were placed into T_1 or T_2 according to their value of B. The model tree can then make predictions for new examples using either of the two linear models.

To further illustrate the differences between these two approaches, let's work through a real-world example.

Example – estimating the quality of wines with regression trees and model trees

Winemaking is a challenging and competitive business that offers the potential for great profit. However, there are numerous factors that contribute to the profitability of a winery. As an agricultural product, variables as diverse as the weather and the growing environment impact the quality of a varietal. The bottling and manufacturing can also affect the flavor for better or worse. Even the way the product is marketed, from the bottle design to the price point, can affect the customer's perception of taste.

As a consequence, the winemaking industry has invested heavily in data collection and machine learning methods that may assist with the decision science of winemaking. For example, machine learning has been used to discover key differences in the chemical composition of wines from different regions, and to identify the chemical factors that lead a wine to taste sweeter.

More recently, machine learning has been employed to assist with rating the quality of wine—a notoriously difficult task. A review written by a renowned wine critic often determines whether the product ends up on the top or bottom shelf, in spite of the fact that even expert judges are inconsistent when rating a wine in a blinded test.

In this case study, we will use regression trees and model trees to create a system capable of mimicking expert ratings of wine. Because trees result in a model that is readily understood, this could allow winemakers to identify key factors that contribute to better-rated wines. Perhaps more importantly, the system does not suffer from the human elements of tasting, such as the rater's mood or palate fatigue. Computer-aided wine testing may therefore result in a better product as well as more objective, consistent, and fair ratings.

Step 1 – collecting data

To develop the wine rating model, we will use data donated to the UCI Machine Learning Repository (http://archive.ics.uci.edu/ml) by P. Cortez, A. Cerdeira, F. Almeida, T. Matos, and J. Reis. Their dataset includes examples of red and white Vinho Verde wines from Portugal — one of the world's leading wine-producing countries. Because the factors that contribute to a highly rated wine may differ between the red and white varieties, for this analysis we will examine only the more popular white wines.

 To follow along with this example, download the whitewines.csv file from the Packt Publishing website and save it to your R working directory. The redwines.csv file is also available in case you would like to explore this data on your own.

The white wine data includes information on 11 chemical properties of 4,898 wine samples. For each wine, a laboratory analysis measured characteristics such as the acidity, sugar content, chlorides, sulfur, alcohol, pH, and density. The samples were then rated in a blind tasting by panels of no less than three judges on a quality scale ranging from zero (very bad) to 10 (excellent). In the case that the judges disagreed on the rating, the median value was used.

The study by Cortez evaluated the ability of three machine learning approaches to model the wine data: multiple regression, artificial neural networks, and support vector machines. We covered multiple regression earlier in this chapter, and we will learn about neural networks and support vector machines in *Chapter 7, Black Box Methods – Neural Networks and Support Vector Machines*. The study found that the support vector machine offered significantly better results than the linear regression model. However, unlike regression, the support vector machine model is difficult to interpret. Using regression trees and model trees, we may be able to improve the regression results while still having a model that is easy to understand.

To read more about the wine study described here, please refer to *Modeling wine preferences by data mining from physicochemical properties*, Cortez, P, Cerdeira, A, Almeida, F, Matos, T, and Reis, J, Decision Support Systems, 2009, Vol. 47, pp. 547-553.

Step 2 – exploring and preparing the data

As usual, we will use the `read.csv()` function to load the data into R. Since all of the features are numeric, we can safely ignore the `stringsAsFactors` parameter:

```
> wine <- read.csv("whitewines.csv")
```

The wine data includes 11 features and the quality outcome, as follows:

```
> str(wine)
'data.frame':    4898 obs. of  12 variables:
 $ fixed.acidity       : num  6.7 5.7 5.9 5.3 6.4 7 7.9 ...
 $ volatile.acidity    : num  0.62 0.22 0.19 0.47 0.29 0.12 ...
 $ citric.acid         : num  0.24 0.2 0.26 0.1 0.21 0.41 ...
 $ residual.sugar      : num  1.1 16 7.4 1.3 9.65 0.9 ...
 $ chlorides           : num  0.039 0.044 0.034 0.036 0.041 ...
 $ free.sulfur.dioxide : num  6 41 33 11 36 22 33 17 34 40 ...
 $ total.sulfur.dioxide: num  62 113 123 74 119 95 152 ...
 $ density             : num  0.993 0.999 0.995 0.991 0.993 ...
 $ pH                  : num  3.41 3.22 3.49 3.48 2.99 3.25 ...
 $ sulphates           : num  0.32 0.46 0.42 0.54 0.34 0.43 ...
 $ alcohol             : num  10.4 8.9 10.1 11.2 10.9 ...
 $ quality             : int  5 6 6 4 6 6 6 6 6 7 ...
```

Compared with other types of machine learning models, one of the advantages of trees is that they can handle many types of data without preprocessing. This means we do not need to normalize or standardize the features.

However, a bit of effort to examine the distribution of the outcome variable is needed to inform our evaluation of the model's performance. For instance, suppose that there was very little variation in quality from wine to wine, or that wines fell into a bimodal distribution: either very good or very bad. This may impact the way we design the model. To check for such extremes, we can examine the distribution of wine quality using a histogram:

```
> hist(wine$quality)
```

This produces the following figure:

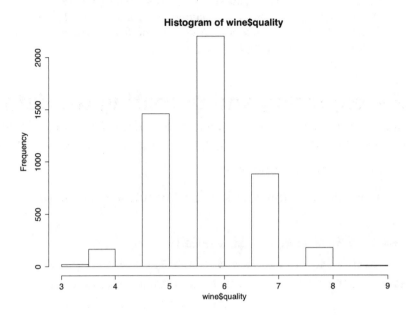

Figure 6.11: The distribution of the quality ratings of white wines

The wine quality values appear to follow a fairly normal, bell-shaped distribution, centered around a value of six. This makes sense intuitively, because most wines are of average quality; few are particularly bad or good. Although the results are not shown here, it is also useful to examine the `summary(wine)` output for outliers or other potential data problems. Even though trees are fairly robust to messy data, it is always prudent to check for severe problems. For now, we'll assume that the data is reliable.

Our last step, then, is to divide the dataset into training and testing sets. Since the wine dataset was already sorted randomly, we can partition into two sets of contiguous rows as follows:

```
> wine_train <- wine[1:3750, ]
> wine_test <- wine[3751:4898, ]
```

In order to mirror the conditions used by Cortez, we used sets of 75 percent and 25 percent for training and testing, respectively. We'll evaluate the performance of our tree-based models on the testing data to see if we can obtain results comparable to the prior research study.

Step 3 – training a model on the data

We will begin by training a regression tree model. Although almost any implementation of decision trees can be used to perform regression tree modeling, the rpart (recursive partitioning) package offers the most faithful implementation of regression trees as they were described by the CART team. As the classic R implementation of CART, the rpart package is also well-documented and supported with functions for visualizing and evaluating the rpart models.

Install the rpart package using the install.packages("rpart") command. It can then be loaded into your R session using the library(rpart) statement. The following syntax will train a tree using the default settings, which typically work fairly well. If you need more finely-tuned settings, refer to the documentation for the control parameters using the ?rpart.control command.

Regression trees syntax

using the **rpart()** function in the **rpart** package

Building the model:

```
m <- rpart(dv ~ iv, data = mydata)
```

- **dv** is the dependent variable in the **mydata** data frame to be modeled
- **iv** is an R formula specifying the independent variables in the **mydata** data frame to use in the model
- **data** specifies the data frame in which the **dv** and **iv** variables can be found

The function will return a regression tree model object that can be used to make predictions.

Making predictions:

```
p <- predict(m, test, type = "vector")
```

- **m** is a model trained by the **rpart()** function
- **test** is a data frame containing test data with the same features as the training data used to build the model
- **type** specifies the type of prediction to return, either **"vector"** (for predicted numeric values), **"class"** for predicted classes, or **"prob"** (for predicted class probabilities)

The function will return a vector of predictions depending on the **type** parameter.

Example:

```
wine_model <- rpart(quality ~ alcohol + sulfates,
                    data = wine_train)
wine_predictions <- predict(wine_model, wine_test)
```

Using the R formula interface, we can specify quality as the outcome variable and use the dot notation to allow all other columns in the wine_train data frame to be used as predictors. The resulting regression tree model object is named m.rpart to distinguish it from the model tree we will train later:

```
> m.rpart <- rpart(quality ~ ., data = wine_train)
```

For basic information about the tree, simply type the name of the model object:

```
> m.rpart
n= 3750

node), split, n, deviance, yval
     * denotes terminal node

 1) root 3750 2945.53200 5.870933
   2) alcohol< 10.85 2372 1418.86100 5.604975
     4) volatile.acidity>=0.2275 1611  821.30730 5.432030
       8) volatile.acidity>=0.3025 688  278.97670 5.255814 *
       9) volatile.acidity< 0.3025 923  505.04230 5.563380 *
     5) volatile.acidity< 0.2275 761  447.36400 5.971091 *
   3) alcohol>=10.85 1378 1070.08200 6.328737
     6) free.sulfur.dioxide< 10.5 84   95.55952 5.369048 *
     7) free.sulfur.dioxide>=10.5 1294  892.13600 6.391036
      14) alcohol< 11.76667 629  430.11130 6.173291
        28) volatile.acidity>=0.465 11   10.72727 4.545455 *
        29) volatile.acidity< 0.465 618  389.71680 6.202265 *
      15) alcohol>=11.76667 665  403.99400 6.596992 *
```

For each node in the tree, the number of examples reaching the decision point is listed. For instance, all 3,750 examples begin at the root node, of which 2,372 have alcohol < 10.85 and 1,378 have alcohol >= 10.85. Because alcohol was used first in the tree, it is the single most important predictor of wine quality.

Nodes indicated by * are terminal or leaf nodes, which means that they result in a prediction (listed here as yval). For example, node 5 has a yval of 5.971091. When the tree is used for predictions, any wine samples with alcohol < 10.85 and volatile.acidity < 0.2275 would therefore be predicted to have a quality value of 5.97.

A more detailed summary of the tree's fit, including the mean squared error for each of the nodes and an overall measure of feature importance, can be obtained using the command summary(m.rpart).

Visualizing decision trees

Although the tree can be understood using only the preceding output, it is often more readily understood using visualization. The rpart.plot package by Stephen Milborrow provides an easy-to-use function that produces publication-quality decision trees.

For more information on `rpart.plot`, including additional examples of the types of decision tree diagrams the function can produce, refer to the author's website at `http://www.milbo.org/rpart-plot/`.

After installing the package using the `install.packages("rpart.plot")` command, the `rpart.plot()` function produces a tree diagram from any `rpart` model object. The following commands plot the regression tree we built earlier:

```
> library(rpart.plot)
> rpart.plot(m.rpart, digits = 3)
```

The resulting tree diagram is as follows:

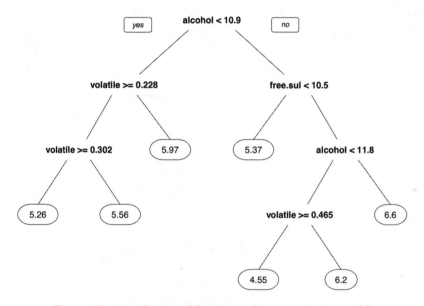

Figure 6.12: A visualization of the wine quality regression tree model

In addition to the `digits` parameter that controls the number of numeric digits to include in the diagram, many other aspects of the visualization can be adjusted. The following command shows just a few of the useful options:

```
> rpart.plot(m.rpart, digits = 4, fallen.leaves = TRUE,
             type = 3, extra = 101)
```

The `fallen.leaves` parameter forces the leaf nodes to be aligned at the bottom of the plot, while the `type` and `extra` parameters affect the way the decisions and nodes are labeled. The numbers 3 and 101 refer to specific style formats, which can be found in the command's documentation, or via experimentation with various numbers.

The result of these changes is a very different looking tree diagram:

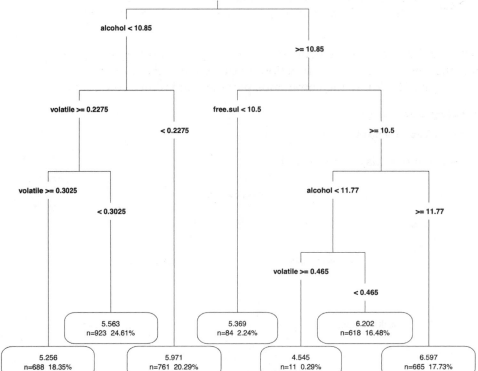

Figure 6.13: Changing the plot function parameters allows customization of the tree visualization

Visualizations like these may assist with the dissemination of regression tree results, as they are readily understood even without a mathematics background. In both cases, the numbers shown in the leaf nodes are the predicted values for the examples reaching that node. Showing the diagram to the wine producers may thus help to identify the key factors that predict the higher-rated wines.

Step 4 – evaluating model performance

To use the regression tree model to make predictions on the test data, we use the predict() function. By default, this returns the estimated numeric value for the outcome variable, which we'll save in a vector named p.rpart:

```
> p.rpart <- predict(m.rpart, wine_test)
```

A quick look at the summary statistics of our predictions suggests a potential problem: the predictions fall on a much narrower range than the true values:

```
> summary(p.rpart)
   Min. 1st Qu.  Median    Mean 3rd Qu.    Max.
  4.545   5.563   5.971   5.893   6.202   6.597
> summary(wine_test$quality)
   Min. 1st Qu.  Median    Mean 3rd Qu.    Max.
  3.000   5.000   6.000   5.901   6.000   9.000
```

This finding suggests that the model is not correctly identifying the extreme cases, in particular, the best and worst wines. On the other hand, between the first and third quartile, we may be doing well.

The correlation between the predicted and actual quality values provides a simple way to gauge the model's performance. Recall that the cor() function can be used to measure the relationship between two equal-length vectors. We'll use this to compare how well the predicted values correspond to the true values:

```
> cor(p.rpart, wine_test$quality)
[1] 0.5369525
```

A correlation of 0.54 is certainly acceptable. However, the correlation only measures how strongly the predictions are related to the true value; it is not a measure of how far off the predictions were from the true values.

Measuring performance with the mean absolute error

Another way to think about the model's performance is to consider how far, on average, its prediction was from the true value. This measurement is called the **mean absolute error** (**MAE**). The equation for MAE is as follows, where n indicates the number of predictions and e_i indicates the error for prediction i:

$$\text{MAE} = \frac{1}{n}\sum_{i=1}^{n}|e_i|$$

As the name implies, this equation takes the mean of the absolute value of the errors. Since the error is just the difference between the predicted and actual values, we can create a simple MAE() function as follows:

```
> MAE <- function(actual, predicted) {
    mean(abs(actual - predicted))
}
```

The MAE for our predictions is then:

```
> MAE(p.rpart, wine_test$quality)
[1] 0.5872652
```

This implies that, on average, the difference between our model's predictions and the true quality score was about 0.59. On a quality scale from zero to 10, this seems to suggest that our model is doing fairly well.

On the other hand, recall that most wines were neither very good nor very bad; the typical quality score was around five to six. Therefore, a classifier that did nothing but predict the mean value may still do fairly well according to this metric.

The mean quality rating in the training data is as follows:

```
> mean(wine_train$quality)
[1] 5.870933
```

If we predicted the value 5.87 for every wine sample, we would have a mean absolute error of only about 0.67:

```
> MAE(5.87, wine_test$quality)
[1] 0.6722474
```

Our regression tree (*MAE = 0.59*) comes closer on average to the true quality score than the imputed mean (*MAE = 0.67*), but not by much. In comparison, Cortez reported an MAE of 0.58 for the neural network model and an MAE of 0.45 for the support vector machine. This suggests that there is room for improvement.

Step 5 – improving model performance

To improve the performance of our learner, let's apply a model tree algorithm, which is a more complex application of trees to numeric prediction. Recall that a model tree extends regression trees by replacing the leaf nodes with regression models. This often results in more accurate results than regression trees, which use only a single numeric value for the prediction at the leaf nodes.

The current state-of-the-art in model trees is the **Cubist** algorithm, which itself is an enhancement of the M5 model tree algorithm—both of which were published by J.R. Quinlan in the early 1990s. Though the implementation details are beyond the scope of this book, the Cubist algorithm involves building a decision tree, creating decision rules based on the branches of the tree, and building a regression model at each of the leaf nodes. Additional heuristics, such as pruning and boosting, are used to improve the quality of the predictions and smoothness across the range of predicted values.

For more background on the Cubist and M5 algorithms, see *Learning With Continuous Classes, Quinlan, JR, Proceedings of the 5th Australian Joint Conference on Artificial Intelligence, 1992; pp. 343-348*. Additionally, see *Combining Instance-Based and Model-Based Learning, Quinlan, JR, Proceedings of the Tenth International Conference on Machine Learning, 1993, pp. 236-243*.

The Cubist algorithm is available in R via the `Cubist` package and the associated `cubist()` function. The syntax of this function is shown in the following table:

Model trees syntax
using the **cubist()** function in the **Cubist** package
Building the model: `m <- cubist(train, class)` • **train** is a data frame or matrix containing training data • **class** is a factor vector with the class for each row in the training data The function will return a cubist model tree object that can be used to make predictions.
Making predictions: `p <- predict(m, test)` • **m** is a model trained by the **cubist()** function • **test** is a data frame containing test data with the same features as the training data used to build the model The function will return a vector of predicted numeric values.

Example:
 	
`wine_model <- cubist(wine_train, wine_quality)`
`wine_predictions <- predict(wine_model, wine_test)`

We'll fit the Cubist model tree using a slightly different syntax from what was used for the regression tree, as the `cubist()` function does not accept the R formula syntax. Instead, we must specify the data frame columns used for the x independent variables and the y dependent variable. With the wine quality to be predicted residing in column 12, and using all other columns as predictors, the full command is as follows:

```
> library(Cubist)
> m.cubist <- cubist(x = wine_train[-12], y = wine_train$quality)
```

Basic information about the model tree can be examined by typing its name:

```
> m.cubist
```

```
Call:
cubist.default(x = wine_train[-12], y = wine_train$quality)


Number of samples: 3750
Number of predictors: 11


Number of committees: 1
Number of rules: 25
```

In this output, we see that the algorithm generated 25 rules to model the wine quality. To examine some of these rules, we can apply the summary() function to the model object. Since the complete tree is very large, only the first few lines of output depicting the first decision rule are included here:

```
> summary(m.cubist)


  Rule 1: [21 cases, mean 5.0, range 4 to 6, est err 0.5]

    if
          free.sulfur.dioxide > 30
          total.sulfur.dioxide > 195
          total.sulfur.dioxide <= 235
          sulphates > 0.64
          alcohol > 9.1
    then
          outcome = 573.6 + 0.0478 total.sulfur.dioxide
                      - 573 density - 0.788 alcohol
                      + 0.186 residual.sugar - 4.73 volatile.acidity
```

You will note that the if portion of the output is somewhat similar to the regression tree we built earlier. A series of decisions based on the wine properties of sulfur dioxide, sulphates, and alcohol creates a rule culminating in the final prediction. A key difference between this model tree output and the earlier regression tree output, however, is that the nodes here terminate not in a numeric prediction, but rather a linear model.

The linear model for this rule is shown in the then output following the outcome = statement. The numbers can be interpreted exactly the same as the multiple regression models we built earlier in this chapter. Each value is the estimated beta of the associated feature, that is, the net effect of that feature on the predicted wine quality. For example, the coefficient of 0.186 for residual sugar implies that for an increase of 1 unit of residual sugar, the wine quality rating is expected to increase by 0.186.

It is important to note that the regression effects estimated by this model apply only to wine samples reaching this node; an examination of the entirety of the Cubist output reveals that a total of 25 linear models were built in this model tree, one for each decision rule, and each with different parameter estimates of the impact of residual sugar and the 10 other features.

To examine the performance of this model, we'll look at how well it performs on the unseen test data. The `predict()` function gets us a vector of predicted values:

```
> p.cubist <- predict(m.cubist, wine_test)
```

The model tree appears to be predicting a wider range of values than the regression tree:

```
> summary(p.cubist)
   Min. 1st Qu.  Median    Mean 3rd Qu.    Max.
  3.677   5.416   5.906   5.848   6.238   7.393
```

The correlation also seems to be substantially higher:

```
> cor(p.cubist, wine_test$quality)
[1] 0.6201015
```

Furthermore, the model slightly reduced the mean absolute error:

```
> MAE(wine_test$quality, p.cubist)
[1] 0.5339725
```

Although we did not improve a great deal beyond the regression tree, we surpassed the performance of the neural network model published by Cortez, and we are getting closer to the published mean absolute error value of 0.45 for the support vector machine model, all while using a much simpler learning method.

Not surprisingly, we have confirmed that predicting the quality of wines is a difficult problem; wine tasting, after all, is inherently subjective. If you would like additional practice, you may try revisiting this problem after reading *Chapter 11, Improving Model Performance*, which covers additional techniques that may lead to better results.

Summary

In this chapter, we studied two methods for modeling numeric data. The first method, linear regression, involves fitting straight lines to data. The second method uses decision trees for numeric prediction. The latter comes in two forms: regression trees, which use the average value of examples at leaf nodes to make numeric predictions, and model trees, which build a regression model at each leaf node in a hybrid approach that is, in some ways, the best of both worlds.

We began to understand the utility of regression modeling by using it to investigate the causes of the Challenger space shuttle disaster. We then used linear regression modeling to calculate the expected medical costs for various segments of the population. Because the relationship between the features and the target variable are well-described by the estimated regression model, we were able to identify certain demographics, such as smokers and the obese, who may need to be charged higher insurance rates to cover the higher-than-average medical expenses.

Regression trees and model trees were used to model the subjective quality of wines from measurable characteristics. In doing so, we learned how regression trees offer a simple way to explain the relationship between features and a numeric outcome, but the more complex model trees may be more accurate. Along the way, we learned new methods for evaluating the performance of numeric models.

In stark contrast to this chapter, which covered machine learning methods that result in a clear understanding of the relationships between the input and the output, the next chapter covers methods that result in nearly-incomprehensible models. The upside is that they are extremely powerful techniques — among the most powerful stock classifiers — that can be applied to both classification and numeric prediction problems.

7

Black Box Methods – Neural Networks and Support Vector Machines

The late science fiction author Arthur C. Clarke wrote, "Any sufficiently advanced technology is indistinguishable from magic." This chapter covers a pair of machine learning methods that may appear at first glance to be magic. Though they are extremely powerful, their inner workings can be difficult to understand.

In engineering, these are referred to as **black box** processes because the mechanism that transforms the input into the output is obfuscated by an imaginary box. For instance, the black box of closed-source software intentionally conceals proprietary algorithms, the black box of political lawmaking is rooted in bureaucratic processes, and the black box of sausage making involves a bit of purposeful (but tasty) ignorance. In the case of machine learning, the black box is due to the complex mathematics allowing them to function.

Although they may not be easy to understand, it is dangerous to apply black box models blindly. Thus, in this chapter, we'll peek inside the box and investigate the statistical sausage making involved in fitting such models. You'll discover how:

- Neural networks mimic living brains to model mathematic functions
- Support vector machines use multidimensional surfaces to define the relationship between features and outcomes
- Despite their complexity, these can be applied easily to real-world problems

With any luck, you'll realize that you don't need a black belt in statistics to tackle black box machine learning methods — there's no need to be intimidated!

Understanding neural networks

An **artificial neural network (ANN)** models the relationship between a set of input signals and an output signal using a model derived from our understanding of how a biological brain responds to stimuli from sensory inputs. Just like a brain uses a network of interconnected cells called **neurons** to provide vast learning capability, the ANN uses a network of artificial neurons or **nodes** to solve challenging learning problems.

The human brain is made up of about 85 billion neurons, resulting in a network capable of representing a tremendous amount of knowledge. As you might expect, this dwarfs the brains of other living creatures. For instance, a cat has roughly a billion neurons, a mouse has about 75 million neurons, and a cockroach has only about a million neurons. In contrast, many ANNs contain far fewer neurons, typically only several hundred, so we're in no danger of creating an artificial brain in the near future—even a fruit fly with 100,000 neurons far exceeds a state-of-the-art ANN.

Though it may be infeasible to completely model a cockroach's brain, a neural network may still provide an adequate heuristic model of its behavior. Suppose that we develop an algorithm that can mimic how a roach flees when discovered. If the behavior of the robot roach is convincing, does it matter whether its brain is as sophisticated as the living creature? This question is the basis of the controversial **Turing test**, proposed in 1950 by the pioneering computer scientist Alan Turing, which grades a machine as intelligent if a human being cannot distinguish its behavior from a living creature's.

[

For more about the intrigue and controversy that surrounds the Turing test, refer to the *Stanford Encyclopedia of Philosophy*: https://plato.stanford.edu/entries/turing-test/.
]

Rudimentary ANNs have been used for over 50 years to simulate the brain's approach to problem solving. At first, this involved learning simple functions like the logical AND function or the logical OR function. These early exercises were used primarily to help scientists understand how biological brains might operate. However, as computers have become increasingly powerful in recent years, the complexity of ANNs has likewise increased so much that they are now frequently applied to more practical problems, including:

- Speech, handwriting, and image recognition programs like those used by smartphone applications, mail sorting machines, and search engines

- The automation of smart devices, such as an office building's environmental controls, or the control of self-driving cars and self-piloting drones

- Sophisticated models of weather and climate patterns, tensile strength, fluid dynamics, and many other scientific, social, or economic phenomena

Broadly speaking, ANNs are versatile learners that can be applied to nearly any learning task: classification, numeric prediction, and even unsupervised pattern recognition.

> Whether deserving or not, ANN learners are often reported in the media with great fanfare. For instance, an "artificial brain" developed by Google was touted for its ability to identify cat videos on YouTube. Such hype may have less to do with anything unique to ANNs and more to do with the fact that ANNs are captivating because of their similarities to living minds.

ANNs are often applied to problems where the input data and output data are well-defined, yet the process that relates the input to the output is extremely complex and hard to define. As a black box method, ANNs work well for these types of black box problems.

From biological to artificial neurons

Because ANNs were intentionally designed as conceptual models of human brain activity, it is helpful to first understand how biological neurons function. As illustrated in the following figure, incoming signals are received by the cell's **dendrites** through a biochemical process. The process allows the impulse to be weighted according to its relative importance or frequency. As the **cell body** begins to accumulate the incoming signals, a threshold is reached at which the cell fires and the output signal is transmitted via an electrochemical process down the **axon**. At the axon's terminals, the electric signal is again processed as a chemical signal to be passed to the neighboring neurons across a tiny gap known as a **synapse**.

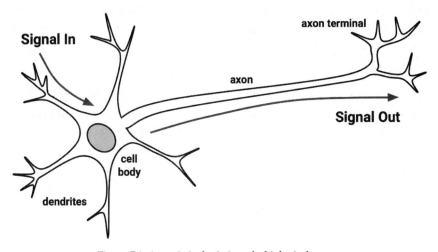

Figure 7.1: An artistic depiction of a biological neuron

The model of a single artificial neuron can be understood in terms very similar to the biological model. As depicted in the following figure, a directed network diagram defines a relationship between the input signals received by the dendrites (*x* variables) and the output signal (*y* variable). Just as with the biological neuron, each dendrite's signal is weighted (*w* values) according to its importance—ignore, for now, how these weights are determined. The input signals are summed by the cell body and the signal is passed on according to an **activation function** denoted by *f*.

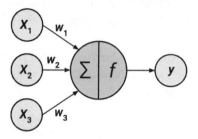

Figure 7.2: An artificial neuron is designed to mimic the structure and function of a biological neuron

A typical artificial neuron with *n* input dendrites can be represented by the formula that follows. The *w* weights allow each of the *n* inputs (denoted by x_i) to contribute a greater or lesser amount to the sum of input signals. The net total is used by the activation function *f(x)*, and the resulting signal, *y(x)*, is the output axon:

$$y(x) = f\left(\sum_{i=1}^{n} w_i x_i \right)$$

Neural networks use neurons defined in this way as building blocks to construct complex models of data. Although there are numerous variants of neural networks, each can be defined in terms of the following characteristics:

- An **activation function**, which transforms a neuron's net input signal into a single output signal to be broadcasted further in the network

- A **network topology** (or architecture), which describes the number of neurons in the model as well as the number of layers and manner in which they are connected

- The **training algorithm**, which specifies how connection weights are set in order to inhibit or excite neurons in proportion to the input signal

Let's take a look at some of the variations within each of these categories to see how they can be used to construct typical neural network models.

Activation functions

The activation function is the mechanism by which the artificial neuron processes incoming information and passes it throughout the network. Just as the artificial neuron is modeled after the biological version, so too is the activation function modeled after nature's design.

In the biological case, the activation function could be imagined as a process that involves summing the total input signal and determining whether it meets the firing threshold. If so, the neuron passes on the signal; otherwise, it does nothing. In ANN terms, this is known as a **threshold activation function**, as it results in an output signal only once a specified input threshold has been attained.

The following figure depicts a typical threshold function; in this case, the neuron fires when the sum of input signals is at least zero. Because its shape resembles a stair, it is sometimes called a **unit step activation function**.

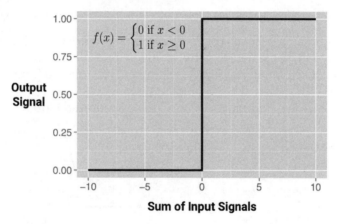

$$f(x) = \begin{cases} 0 \text{ if } x < 0 \\ 1 \text{ if } x \geq 0 \end{cases}$$

Figure 7.3: The threshold activation function is "on" only after the input signals meet a threshold

Although the threshold activation function is interesting due to its parallels with biology, it is rarely used in ANNs. Freed from the limitations of biochemistry, ANN activation functions can be chosen based on their ability to demonstrate desirable mathematical characteristics and their ability to accurately model relationships among data.

Perhaps the most commonly used alternative is the **sigmoid activation function** (more specifically the *logistic* sigmoid) shown in the following figure. Note that in the formula shown, *e* is the base of the natural logarithm (approximately 2.72). Although it shares a similar step or "S" shape with the threshold activation function, the output signal is no longer binary; output values can fall anywhere in the range from zero to one.

Additionally, the sigmoid is **differentiable**, which means that it is possible to calculate the derivative across the entire range of inputs. As you will learn later, this feature is crucial for creating efficient ANN optimization algorithms.

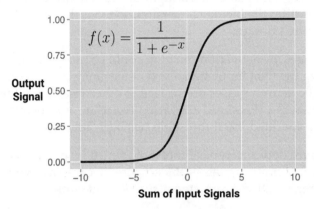

Figure 7.4: The sigmoid activation function mimics the biological activation function with a smooth curve

Although the sigmoid is perhaps the most commonly used activation function and is often used by default, some neural network algorithms allow a choice of alternatives. A selection of such activation functions is shown in the following figure:

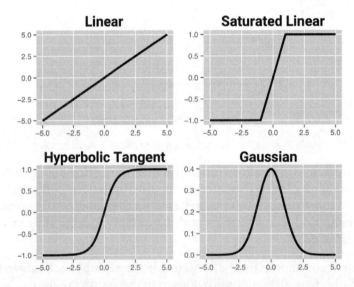

Figure 7.5: Several common neural network activation functions

The primary detail that differentiates these activation functions is the output signal range. Typically, this is one of (0, 1), (-1, +1), or (-inf, +inf). The choice of activation function biases the neural network such that it may fit certain types of data more appropriately, allowing the construction of specialized neural networks.

For instance, a linear activation function results in a neural network very similar to a linear regression model, while a Gaussian activation function is the basis of a **radial basis function (RBF) network**. Each of these has strengths better suited for certain learning tasks and not others.

It's important to recognize that for many of the activation functions, the range of input values that affect the output signal is relatively narrow. For example, in the case of the sigmoid, the output signal is very near zero for an input signal below negative five and very near one for an input signal above positive five. The compression of the signal in this way results in a saturated signal at the high and low ends of very dynamic inputs, just as turning a guitar amplifier up too high results in a distorted sound due to clipping of the peaks of sound waves. Because this essentially squeezes the input values into a smaller range of outputs, activation functions like the sigmoid are sometimes called **squashing functions**.

One solution to the squashing problem is to transform all neural network inputs such that the feature values fall within a small range around zero. This may involve standardizing or normalizing the features. By restricting the range of input values, the activation function will have action across the entire range. A side benefit is that the model may also be faster to train, since the algorithm can iterate more quickly through the actionable range of input values.

 Although theoretically a neural network can adapt to a very dynamic feature by adjusting its weight over many iterations, in extreme cases many algorithms will stop iterating long before this occurs. If your model is failing to converge, double-check that you've correctly standardized the input data. Choosing a different activation function may also be appropriate.

Network topology

The capacity of a neural network to learn is rooted in its **topology**, or the patterns and structures of interconnected neurons. Although there are countless forms of network architecture, they can be differentiated by three key characteristics:

- The number of layers
- Whether information in the network is allowed to travel backward
- The number of nodes within each layer of the network

The topology determines the complexity of tasks that can be learned by the network. Generally, larger and more complex networks are capable of identifying more subtle patterns and more complex decision boundaries. However, the power of a network is not only a function of the network size, but also the way units are arranged.

The number of layers

To define topology, we need terminology that distinguishes artificial neurons based on their position in the network. The figure that follows illustrates the topology of a very simple network. A set of neurons called **input nodes** receives unprocessed signals directly from the input data. Each input node is responsible for processing a single feature in the dataset; the feature's value will be transformed by the corresponding node's activation function. The signals sent by the input nodes are received by the **output node**, which uses its own activation function to generate a final prediction (denoted here as p).

The input and output nodes are arranged in groups known as **layers**. Because the input nodes process the incoming data exactly as received, the network has only one set of connection weights (labeled here as w_1, w_2, and w_3). It is therefore termed a **single-layer network**. Single-layer networks can be used for basic pattern classification, particularly for patterns that are linearly separable, but more sophisticated networks are required for most learning tasks.

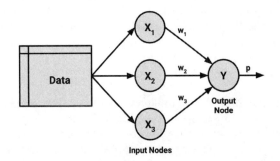

Figure 7.6: A simple single-layer ANN with three input nodes

As you might expect, an obvious way to create more complex networks is by adding additional layers. As depicted here, a **multilayer network** adds one or more **hidden layers** that process the signals from the input nodes prior to reaching the output node. Most multilayer networks are **fully connected**, which means that every node in one layer is connected to every node in the next layer, but this is not required.

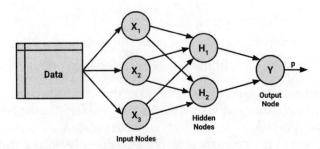

Figure 7.7: A multilayer network with a single two-node hidden layer

The direction of information travel

You may have noticed that in the prior examples, arrowheads were used to indicate signals traveling in only one direction. Networks in which the input signal is fed continuously in one direction from the input layer to the output layer are called **feedforward networks**.

In spite of the restriction on information flow, feedforward networks offer a surprising amount of flexibility. For instance, the number of levels and nodes at each level can be varied, multiple outcomes can be modeled simultaneously, or multiple hidden layers can be applied. A neural network with multiple hidden layers is called a **deep neural network (DNN)**, and the practice of training such networks is referred to as **deep learning**. Deep neural networks trained on large datasets are capable of human-like performance on complex tasks like image recognition and text processing.

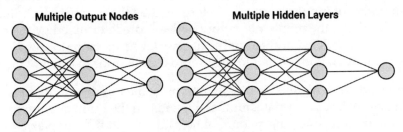

Figure 7.8: Complex ANNs can have multiple output nodes or multiple hidden layers

In contrast to feedforward networks, a **recurrent network** (or **feedback network**) allows signals to travel backward using loops. This property, which more closely mirrors how a biological neural network works, allows extremely complex patterns to be learned. The addition of a short-term memory, or **delay**, increases the power of recurrent networks immensely. Notably, this includes the capability to understand sequences of events over a period of time. This could be used for stock market prediction, speech comprehension, or weather forecasting. A simple recurrent network is depicted as follows:

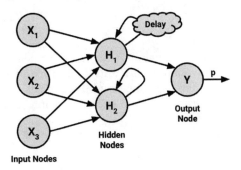

Figure 7.9: Allowing information to travel backward in the network can model a time delay

DNNs and recurrent networks are increasingly being used for a variety of high-profile applications and consequently have become highly popular. However, building such networks uses techniques and software outside the scope of this book, and often requires access to specialized computing hardware or cloud servers. On the other hand, simpler feedforward networks are also very capable of modeling many real-world tasks. In fact, the multilayer feedforward network, also known as the **multilayer perceptron** (**MLP**), is the de facto standard ANN topology. If you are interested in deep learning, understanding the MLP topology provides a strong theoretical basis for building more complex DNN models later on.

The number of nodes in each layer

In addition to variations in the number of layers and the direction of information travel, neural networks can also vary in complexity by the number of nodes in each layer. The number of input nodes is predetermined by the number of features in the input data. Similarly, the number of output nodes is predetermined by the number of outcomes to be modeled or the number of class levels in the outcome. However, the number of hidden nodes is left to the user to decide prior to training the model.

Unfortunately, there is no reliable rule to determine the number of neurons in the hidden layer. The appropriate number depends on the number of input nodes, the amount of training data, the amount of noisy data, and the complexity of the learning task among many other factors.

In general, more complex network topologies with a greater number of network connections allow the learning of more complex problems. A greater number of neurons will result in a model that more closely mirrors the training data, but this runs a risk of overfitting; it may generalize poorly to future data. Large neural networks can also be computationally expensive and slow to train.

The best practice is to use the fewest nodes that result in adequate performance on a validation dataset. In most cases, even with only a small number of hidden nodes — often as few as a handful — the neural network can offer a tremendous amount of learning ability.

 It has been proven that a neural network with at least one hidden layer of sufficiently many neurons is a **universal function approximator**. This means that neural networks can be used to approximate any continuous function to an arbitrary precision over a finite interval.

Training neural networks with backpropagation

The network topology is a blank slate that by itself has not learned anything. Like a newborn child, it must be trained with experience. As the neural network processes the input data, connections between the neurons are strengthened or weakened, similar to how a baby's brain develops as he or she experiences the environment. The network's connection weights are adjusted to reflect the patterns observed over time.

Training a neural network by adjusting connection weights is very computationally intensive. Consequently, though they had been studied for decades prior, ANNs were rarely applied to real-world learning tasks until the mid-to-late 1980s, when an efficient method of training an ANN was discovered. The algorithm, which used a strategy of back-propagating errors, is now known simply as **backpropagation**.

Coincidentally, several research teams independently discovered and published the backpropagation algorithm around the same time. Among them, perhaps the most often cited work is *Learning representations by back-propagating errors, Rumelhart, DE, Hinton, GE, Williams, RJ, Nature, 1986, Vol. 323, pp. 533-566.*

Although still somewhat computationally expensive relative to many other machine learning algorithms, the backpropagation method led to a resurgence of interest in ANNs. As a result, multilayer feedforward networks that use the backpropagation algorithm are now common in the field of data mining. Such models offer the following strengths and weaknesses:

Strengths	Weaknesses
• Can be adapted to classification or numeric prediction problems • Capable of modeling more complex patterns than nearly any algorithm • Makes few assumptions about the data's underlying relationships	• Extremely computationally intensive and slow to train, particularly if the network topology is complex • Very prone to overfitting training data • Results in a complex black box model that is difficult, if not impossible, to interpret

In its most general form, the backpropagation algorithm iterates through many cycles of two processes. Each cycle is known as an **epoch**. Because the network contains no *a priori* (existing) knowledge, the starting weights are typically set at random. Then, the algorithm iterates through the processes until a stopping criterion is reached. Each epoch in the backpropagation algorithm includes:

- A **forward phase**, in which the neurons are activated in sequence from the input layer to the output layer, applying each neuron's weights and activation function along the way. Upon reaching the final layer, an output signal is produced.

- A **backward phase**, in which the network's output signal resulting from the forward phase is compared to the true target value in the training data. The difference between the network's output signal and the true value results in an error that is propagated backwards in the network to modify the connection weights between neurons and reduce future errors.

Over time, the algorithm uses the information sent backward to reduce the total error of the network. Yet one question remains: because the relationship between each neuron's inputs and outputs is complex, how does the algorithm determine how much a weight should be changed? The answer to this question involves a technique called **gradient descent**. Conceptually, this works similarly to how an explorer trapped in the jungle might find a path to water. By examining the terrain and continually walking in the direction with the greatest downward slope, the explorer will eventually reach the lowest valley, which is likely to be a riverbed.

In a similar process, the backpropagation algorithm uses the derivative of each neuron's activation function to identify the gradient in the direction of each of the incoming weights—hence the importance of having a differentiable activation function. The gradient suggests how steeply the error will be reduced or increased for a change in the weight. The algorithm will attempt to change the weights that result in the greatest reduction in error by an amount known as the **learning rate**. The greater the learning rate, the faster the algorithm will attempt to descend down the gradients, which could reduce training time at the risk of overshooting the valley.

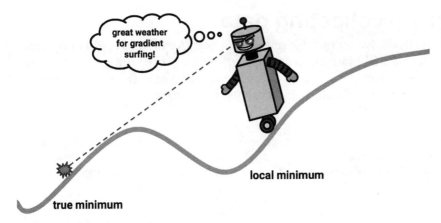

Figure 7.10: The gradient decent algorithm seeks the minimum error but may also find a local minimum

Although this process seems complex, it is easy to apply in practice. Let's apply our understanding of multilayer feedforward networks to a real-world problem.

Example – modeling the strength of concrete with ANNs

In the field of engineering, it is crucial to have accurate estimates of the performance of building materials. These estimates are required in order to develop safety guidelines governing the materials used in the construction of buildings, bridges, and roadways.

Estimating the strength of concrete is a challenge of particular interest. Although it is used in nearly every construction project, concrete performance varies greatly due to a wide variety of ingredients that interact in complex ways. As a result, it is difficult to accurately predict the strength of the final product. A model that could reliably predict concrete strength given a listing of the composition of the input materials could result in safer construction practices.

Step 1 – collecting data

For this analysis, we will utilize data on the compressive strength of concrete donated to the UCI Machine Learning Repository (http://archive.ics.uci.edu/ml) by I-Cheng Yeh. As he found success using neural networks to model these data, we will attempt to replicate Yeh's work using a simple neural network model in R.

> For more information on Yeh's approach to this learning task, refer to *Modeling of Strength of High-Performance Concrete Using Artificial Neural Networks, Yeh, IC, Cement and Concrete Research, 1998, Vol. 28, pp. 1797-1808.*

According to the website, the dataset contains 1,030 examples of concrete, with eight features describing the components used in the mixture. These features are thought to be related to the final compressive strength, and include the amount (in kilograms per cubic meter) of cement, slag, ash, water, superplasticizer, coarse aggregate, and fine aggregate used in the product, in addition to the aging time (measured in days).

> To follow along with this example, download the concrete. csv file from the Packt Publishing website and save it to your R working directory.

Step 2 – exploring and preparing the data

As usual, we'll begin our analysis by loading the data into an R object using the read.csv() function and confirming that it matches the expected structure:

```
> concrete <- read.csv("concrete.csv")
> str(concrete)
'data.frame':    1030 obs. of  9 variables:
 $ cement      : num  141 169 250 266 155 ...
 $ slag        : num  212 42.2 0 114 183.4 ...
 $ ash         : num  0 124.3 95.7 0 0 ...
 $ water       : num  204 158 187 228 193 ...
 $ superplastic: num  0 10.8 5.5 0 9.1 0 0 6.4 0 9 ...
 $ coarseagg   : num  972 1081 957 932 1047 ...
 $ fineagg     : num  748 796 861 670 697 ...
 $ age         : int  28 14 28 28 28 90 7 56 28 28 ...
 $ strength    : num  29.9 23.5 29.2 45.9 18.3 ...
```

The nine variables in the data frame correspond to the eight features and one outcome we expected, although a problem has become apparent. Neural networks work best when the input data are scaled to a narrow range around zero, and here we see values ranging anywhere from zero to over a thousand.

Typically, the solution to this problem is to rescale the data with a normalizing or standardization function. If the data follow a bell-shaped curve (a normal distribution as described in *Chapter 2, Managing and Understanding Data*), then it may make sense to use standardization via R's built-in `scale()` function. On the other hand, if the data follow a uniform distribution or are severely non-normal, then normalization to a zero to one range may be more appropriate. In this case, we'll use the latter.

In *Chapter 3, Lazy Learning – Classification Using Nearest Neighbors*, we defined our own `normalize()` function as:

```
> normalize <- function(x) {
    return((x - min(x)) / (max(x) - min(x)))
  }
```

After executing this code, our `normalize()` function can be applied to every column in the concrete data frame using the `lapply()` function as follows:

```
> concrete_norm <- as.data.frame(lapply(concrete, normalize))
```

To confirm that the normalization worked, we can see that the minimum and maximum strength are now zero and one, respectively:

```
> summary(concrete_norm$strength)
   Min. 1st Qu.  Median    Mean 3rd Qu.     Max.
 0.0000  0.2664  0.4001  0.4172  0.5457  1.0000
```

In comparison, the original minimum and maximum values were 2.33 and 82.60:

```
> summary(concrete$strength)
   Min. 1st Qu.  Median    Mean 3rd Qu.     Max.
   2.33   23.71   34.44   35.82   46.14   82.60
```

Any transformation applied to the data prior to training the model will have to be applied in reverse later on in order to convert back to the original units of measurement. To facilitate the rescaling, it is wise to save the original data, or at least the summary statistics of the original data.

Following Yeh's precedent in the original publication, we will partition the data into a training set with 75 percent of the examples and a testing set with 25 percent. The CSV file we used was already sorted in random order, so we simply need to divide it into two portions:

```
> concrete_train <- concrete_norm[1:773, ]
> concrete_test <- concrete_norm[774:1030, ]
```

We'll use the training dataset to build the neural network and the testing dataset to evaluate how well the model generalizes to future results. As it is easy to overfit a neural network, this step is very important.

Step 3 – training a model on the data

To model the relationship between the ingredients used in concrete and the strength of the finished product, we will use a multilayer feedforward neural network. The neuralnet package by Stefan Fritsch and Frauke Guenther provides a standard and easy-to-use implementation of such networks. It also offers a function to plot the network topology. For these reasons, the neuralnet implementation is a strong choice for learning more about neural networks, though that's not to say that it cannot be used to accomplish real work as well—it's quite a powerful tool, as you will soon see.

There are several other commonly used packages to train ANN models in R, each with unique strengths and weaknesses. Because it ships as part of the standard R installation, the nnet package is perhaps the most frequently cited ANN implementation. It uses a slightly more sophisticated algorithm than standard backpropagation. Another option is the RSNNS package, which offers a complete suite of neural network functionality, with the downside being that it is more difficult to learn.

As neuralnet is not included in base R, you will need to install it by typing install.packages("neuralnet") and load it with the library(neuralnet) command. The included neuralnet() function can be used for training neural networks for numeric prediction using the following syntax:

Neural network syntax

using the **neuralnet()** function in the **neuralnet** package

Building the model:

```
m <- neuralnet(target ~ predictors, data = mydata,
                hidden = 1, act.fct = "logistic")
```

- **target** is the outcome in the **mydata** data frame to be modeled
- **predictors** is an R formula specifying the features in the **mydata** data frame to use for prediction
- **data** specifies the data frame where the **target** and **predictors** are found
- **hidden** specifies the number of neurons in the hidden layer (by default, 1)
  - *note:* use an integer vector to specify multiple hidden layers, e.g., **c(2, 2)**
- **act.fct** specifies the activation function, either **"logistic"** or **"tanh"**
  - note: a *differentiable* custom activation function can also be supplied

The function will return a neural network object that can be used to make predictions.

Making predictions:

```
p <- compute(m, test)
```

- **m** is a model trained by the **neuralnet()** function
- **test** is a data frame containing test data with the same features as the training data used to build the classifier

The function will return a list with two components: **$neurons**, which stores the neurons for each layer in the network, and **$net.result**, which stores the model's predicted values.

Example:

```
concrete_model <- neuralnet(strength ~ cement + slag + ash,
    data = concrete, hidden = c(5, 5), act.fct = "tanh")
model_results <- compute(concrete_model, concrete_data)
strength_predictions <- model_results$net.result
```

We'll begin by training the simplest multilayer feedforward network with the default settings using only a single hidden node:

```
> concrete_model <- neuralnet(strength ~ cement + slag
  + ash + water + superplastic + coarseagg + fineagg + age,
  data = concrete_train)
```

We can then visualize the network topology using the `plot()` function on the resulting model object:

```
> plot(concrete_model)
```

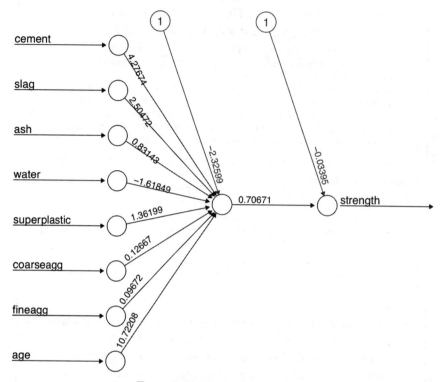

Error: 5.077438   Steps: 4882

Figure 7.11: Topology visualization of our simple multilayer feedforward network

In this simple model, there is one input node for each of the eight features, followed by a single hidden node and a single output node that predicts the concrete strength. The weights for each of the connections are also depicted, as are the **bias terms** indicated by the nodes labeled with the number 1. The bias terms are numeric constants that allow the value at the indicated nodes to be shifted upward or downward, much like the intercept in a linear equation.

A neural network with a single hidden node can be thought of as a cousin of the linear regression models we studied in *Chapter 6, Forecasting Numeric Data – Regression Methods*. The weight between each input node and the hidden node is similar to the beta coefficients, and the weight for the bias term is similar to the intercept.

At the bottom of the figure, R reports the number of training steps and an error measure called the **sum of squared errors** (SSE), which, as you might expect, is the sum of the squared differences between the predicted and actual values. The lower the SSE, the more closely the model conforms to the training data, which tells us about performance on the training data but little about how it will perform on unseen data.

Step 4 – evaluating model performance

The network topology diagram gives us a peek into the black box of the ANN, but it doesn't provide much information about how well the model fits future data. To generate predictions on the test dataset, we can use the `compute()` function as follows:

```
> model_results <- compute(concrete_model, concrete_test[1:8])
```

The `compute()` function works a bit differently from the `predict()` functions we've used so far. It returns a list with two components: `$neurons`, which stores the neurons for each layer in the network, and `$net.result`, which stores the predicted values. We'll want the latter:

```
> predicted_strength <- model_results$net.result
```

Because this is a numeric prediction problem rather than a classification problem, we cannot use a confusion matrix to examine model accuracy. Instead, we'll measure the correlation between our predicted concrete strength and the true value. If the predicted and actual values are highly correlated, the model is likely to be a useful gauge of concrete strength.

Recall that the `cor()` function is used to obtain a correlation between two numeric vectors:

```
> cor(predicted_strength, concrete_test$strength)
              [,1]
[1,]  0.8064655576
```

> Don't be alarmed if your result differs. Because the neural network begins with random weights, the predictions can vary from model to model. If you'd like to match these results exactly, try using `set.seed(12345)` before building the neural network.

Correlations close to one indicate strong linear relationships between two variables. Therefore, the correlation here of about 0.806 indicates a fairly strong relationship. This implies that our model is doing a fairly good job, even with only a single hidden node.

Given that we only used one hidden node, it is likely that we can improve the performance of our model. Let's try to do a bit better.

Step 5 – improving model performance

As networks with more complex topologies are capable of learning more difficult concepts, let's see what happens when we increase the number of hidden nodes to five. We use the `neuralnet()` function as before, but add the parameter `hidden = 5`:

```
> concrete_model2 <- neuralnet(strength ~ cement + slag +
                        ash + water + superplastic +
                        coarseagg + fineagg + age,
                        data = concrete_train, hidden = 5)
```

Plotting the network again, we see a drastic increase in the number of connections. How did this impact performance?

```
> plot(concrete_model2)
```

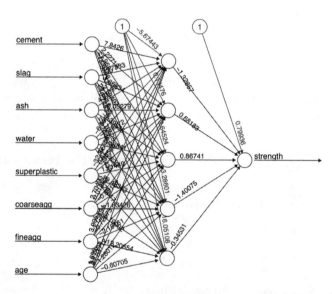

Error: 1.626684   Steps: 86849

Figure 7.12: Topology visualization of an increased number of hidden nodes

Notice that the reported error (measured again by SSE) has been reduced from 5.08 in the previous model to 1.63 here. Additionally, the number of training steps rose from 4,882 to 86,849, which should come as no surprise given how much more complex the model has become. More complex networks take many more iterations to find the optimal weights.

Applying the same steps to compare the predicted values to the true values, we now obtain a correlation around 0.92, which is a considerable improvement over the previous result of 0.80 with a single hidden node:

```
> model_results2 <- compute(concrete_model2, concrete_test[1:8])
> predicted_strength2 <- model_results2$net.result
> cor(predicted_strength2, concrete_test$strength)
            [,1]
[1,]  0.9244533426
```

Despite these substantial improvements, there is still more we can do to attempt to improve the model performance. In particular, we have the ability to add additional hidden layers and to change the network's activation function. In making these changes, we create the foundations of a very simple deep neural network.

The choice of activation function is usually very important for deep learning. The best function for a particular learning task is typically identified through experimentation, then shared more widely within the machine learning research community.

Recently, an activation function known as a **rectifier** has become extremely popular due to its success on complex tasks such as image recognition. A node in a neural network that uses the rectifier activation function is known as a **rectified linear unit (ReLU)**. As depicted in the following figure, the rectifier activation function is defined such that it returns x if x is at least zero, and zero otherwise. The significance of this function is due to the fact that it is nonlinear yet has simple mathematical properties that make it both computationally inexpensive and highly efficient for gradient descent. Unfortunately, its derivative is undefined at $x = 0$ and therefore cannot be used with the `neuralnet()` function.

Instead, we can use a smooth approximation of the ReLU known as **softplus** or **SmoothReLU**, an activation function defined as $log(1 + e^x)$. As shown in the following figure, the softplus function is nearly zero for x less than zero and approximately x when x is greater than zero:

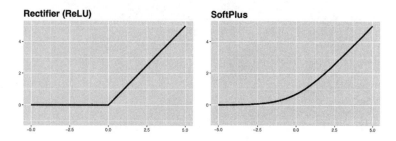

Figure 7.13: The softplus activation function provides a smooth, differentiable approximation of ReLU

To define a `softplus()` function in R, use the following code:

```
> softplus <- function(x) { log(1 + exp(x)) }
```

This activation function can be provided to `neuralnet()` using the `act.fct` parameter. Additionally, we will add a second hidden layer of five nodes by supplying the `hidden` parameter the integer vector `c(5, 5)`. This creates a two-layer network, each having five nodes, all using the softplus activation function:

```
> set.seed(12345)
> concrete_model3 <- neuralnet(strength ~ cement + slag +
                               ash + water + superplastic +
                               coarseagg + fineagg + age,
                               data = concrete_train,
                               hidden = c(5, 5),
                               act.fct = softplus)
```

As before, the network can be visualized:

```
> plot(concrete_model3)
```

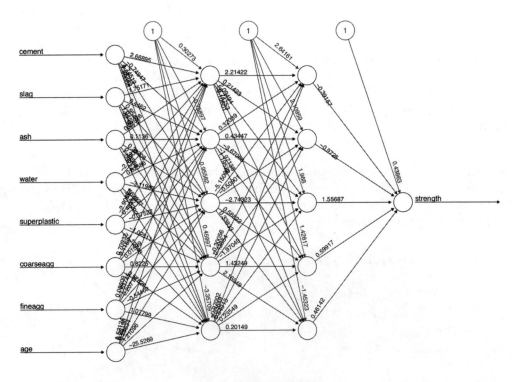

Error: 1.666068   Steps: 88240

Figure 7.14: Visualizing our network with two layers of hidden nodes using the softplus activation function

And the correlation between the predicted and actual concrete strength can be computed:

```
> model_results3 <- compute(concrete_model3, concrete_test[1:8])
> predicted_strength3 <- model_results3$net.result
> cor(predicted_strength3, concrete_test$strength)
              [,1]
[1,]  0.9348395359
```

The correlation between the predicted and actual strength was 0.935, which is our best performance yet. Interestingly, in the original publication, Yeh reported a correlation of 0.885. This means that with relatively little effort, we were able to match and even exceed the performance of a subject matter expert. Of course, Yeh's results were published in 1998, giving us the benefit of over 20 years of additional neural network research!

One important thing to be aware of is that, because we had normalized the data prior to training the model, the predictions are also on a normalized scale from zero to one. For example, the following code shows a data frame comparing the original dataset's concrete strength values to their corresponding predictions side-by-side:

```
> strengths <- data.frame(
    actual = concrete$strength[774:1030],
    pred = predicted_strength3
  )
> head(strengths, n = 3)
    actual           pred
774  30.14  0.2860639091
775  44.40  0.4777304648
776  24.50  0.2840964250
```

Examining the correlation, we see that the choice of normalized or unnormalized data does not affect our computed performance statistic — the correlation of 0.935 is exactly the same as before:

```
> cor(strengths$pred, strengths$actual)
[1]  0.9348395359
```

However, if we were to compute a different performance metric, such as the absolute difference between the predicted and actual values, the choice of scale would matter quite a bit.

With this in mind, we can create an `unnormalize()` function that reverses the min-max normalization procedure and allow us to convert the normalized predictions to the original scale:

```
> unnormalize <- function(x) {
    return((x * (max(concrete$strength)) -
            min(concrete$strength)) + min(concrete$strength))
  }
```

After applying the custom `unnormalize()` function to the predictions, we can see that the new predictions are on a similar scale to the original concrete strength values. This allows us to compute a meaningful absolute error value. Additionally, the correlation between the unnormalized and original strength values remains the same:

```
> strengths$pred_new <- unnormalize(strengths$pred)
> strengths$error <- strengths$pred_new - strengths$actual

> head(strengths, n = 3)
    actual          pred    pred_new           error
774  30.14 0.2860639091 23.62887889  -6.511121108
775  44.40 0.4777304648 39.46053639  -4.939463608
776  24.50 0.2840964250 23.46636470  -1.033635298

> cor(strengths$pred_new, strengths$actual)
[1] 0.9348395359
```

When applying neural networks to your own projects, you will need to perform a similar series of steps to return the data to its original scale.

You may also find that neural networks quickly become much more complicated as they are applied to more challenging learning tasks. For example, you may find that you run into the so-called **vanishing gradient problem** and closely-related **exploding gradient problem**, where the backpropagation algorithm fails to find a useful solution due to an inability to converge in a reasonable time. As a remedy to these problems, one may perhaps try varying the number of hidden nodes, applying different activation functions such as the ReLU, adjusting the learning rate, and so on. The `?neuralnet` help page provides more information on the various parameters that can be adjusted. This leads to another problem, however, in which testing a large number of parameters becomes a bottleneck to building a strong-performing model. This is the tradeoff of ANNs and even more so, DNNs: harnessing their great potential requires a great investment of time and computing power.

Just as is often the case in life more generally, it is possible to trade time and money in machine learning. Using paid cloud computing resources such as Amazon Web Services (AWS) and Microsoft Azure allows one to build more complex models or test many models more quickly. More on this subject can be found in *Chapter 12, Specialized Machine Learning Topics*.

Understanding support vector machines

A **support vector machine** (**SVM**) can be imagined as a surface that creates a boundary between points of data plotted in a multidimensional space representing examples and their feature values. The goal of an SVM is to create a flat boundary called a **hyperplane**, which divides the space to create fairly homogeneous partitions on either side. In this way, SVM learning combines aspects of both the instance-based nearest neighbor learning presented in *Chapter 3, Lazy Learning – Classification Using Nearest Neighbors*, and the linear regression modeling described in *Chapter 6, Forecasting Numeric Data – Regression Methods*. The combination is extremely powerful, allowing SVMs to model highly complex relationships.

Although the basic mathematics that drive SVMs have been around for decades, interest in them grew greatly after they were adopted by the machine learning community. Their popularity exploded after high-profile success stories on difficult learning problems, as well as the development of award-winning SVM algorithms that were implemented in well-supported libraries across many programming languages, including R. SVMs have thus been adopted by a wide audience, which might have otherwise been unable to apply the somewhat complex mathematics needed to implement an SVM. The good news is that although the mathematics may be difficult, the basic concepts are understandable.

SVMs can be adapted for use with nearly any type of learning task, including both classification and numeric prediction. Many of the algorithm's key successes have come in pattern recognition. Notable applications include:

- Classification of microarray gene expression data in the field of bioinformatics to identify cancer or other genetic diseases
- Text categorization, such as identification of the language used in a document or classification of documents by subject matter
- The detection of rare yet important events like combustion engine failure, security breaches, or earthquakes

SVMs are most easily understood when used for binary classification, which is how the method has been traditionally applied. Therefore, in the remaining sections we will focus only on SVM classifiers. Principles similar to those presented here also apply when adapting SVMs for numeric prediction.

Classification with hyperplanes

As noted previously, SVMs use a boundary called a hyperplane to partition data into groups of similar class values. For example, the following figure depicts hyperplanes that separate groups of circles and squares in two and three dimensions. Because the circles and squares can be separated perfectly by a straight line or flat surface, they are said to be **linearly separable**. At first, we'll consider only the simple case where this is true, but SVMs can also be extended to problems were the points are not linearly separable.

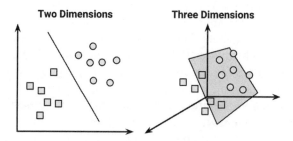

Figure 7.15: The squares and circles are linearly separable in both two and three dimensions

 For convenience, the hyperplane is traditionally depicted as a line in 2D space, but this is simply because it is difficult to illustrate space in greater than two dimensions. In reality, the hyperplane is a flat surface in a high-dimensional space—a concept that can be difficult to get your mind around.

In two dimensions, the task of the SVM algorithm is to identify a line that separates the two classes. As shown in the following figure, there is more than one choice of dividing line between the groups of circles and squares. Three such possibilities are labeled **a**, **b**, and **c**. How does the algorithm choose?

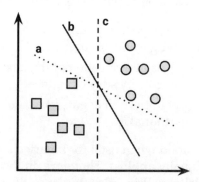

Figure 7.16: Three of many potential lines dividing the squares and circles

The answer to that question involves a search for the **maximum margin hyperplane (MMH)** that creates the greatest separation between the two classes. Although any of the three lines separating the circles and squares would correctly classify all of the data points, the line that leads to the greatest separation will generalize the best to future data. The maximum margin will improve the chance that even if random noise is added, each class will remain on its own side of the boundary.

The **support vectors** (indicated by arrows in the figure that follows) are the points from each class that are the closest to the MMH. Each class must have at least one support vector, but it is possible to have more than one. The support vectors alone define the MMH. This is a key feature of SVMs; the support vectors provide a very compact way to store a classification model, even if the number of features is extremely large.

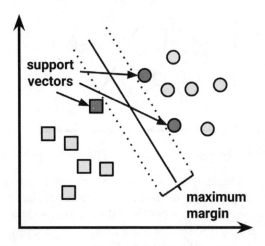

Figure 7.17: The maximum margin hyperplane is defined by the support vectors

The algorithm to identify the support vectors relies on vector geometry and involves some fairly tricky mathematics that is outside the scope of this book. However, the basic principles of the process are fairly straightforward.

> More information on the mathematics of SVMs can be found in the classic paper *Support-Vector Networks, Cortes, C* and *Vapnik, V, Machine Learning, 1995, Vol. 20, pp. 273-297.* A beginner-level discussion can be found in *Support Vector Machines: Hype or Hallelujah?, Bennett, KP* and *Campbell, C, SIGKDD Explorations, 2000, Vol. 2, pp. 1-13.* A more in-depth look can be found in *Support Vector Machines, Steinwart, I* and *Christmann, A, New York: Springer, 2008.*

The case of linearly separable data

Finding the maximum margin is easiest under the assumption that the classes are linearly separable. In this case, the MMH is as far away as possible from the outer boundaries of the two groups of data points. These outer boundaries are known as the **convex hull**. The MMH is then the perpendicular bisector of the shortest line between the two convex hulls. Sophisticated computer algorithms that use a technique known as **quadratic optimization** are capable of finding the maximum margin in this way.

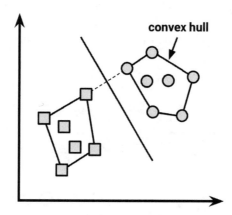

Figure 7.18: The MMH is the perpendicular bisector of the shortest path between convex hulls

An alternative (but equivalent) approach involves a search through the space of every possible hyperplane in order to find a set of two parallel planes that divide the points into homogeneous groups yet themselves are as far apart as possible. To use a metaphor, one can imagine this process as similar to trying to find the thickest mattress that can fit up a stairwell to your bedroom.

To understand this search process, we'll need to define exactly what we mean by a hyperplane. In n-dimensional space, the following equation is used:

$$\vec{w} \cdot \vec{x} + b = 0$$

If you aren't familiar with this notation, the arrows above the letters indicate that they are vectors rather than single numbers. In particular, w is a vector of n weights, that is, $\{w_1, w_2, ..., w_n\}$, and b is a single number known as the bias. The bias is conceptually equivalent to the intercept term in the slope-intercept form discussed in *Chapter 6, Forecasting Numeric Data – Regression Methods*.

If you're having trouble imagining the plane in multidimensional space, don't worry about the details. Simply think of the equation as a way to specify a surface, much like the slope-intercept form ($y = mx + b$) is used to specify lines in 2D space.

Using this formula, the goal of the process is to find a set of weights that specify two hyperplanes, as follows:

$$\vec{w} \cdot \vec{x} + b \geq +1$$
$$\vec{w} \cdot \vec{x} + b \leq -1$$

We will also require that these hyperplanes are specified such that all the points of one class fall above the first hyperplane and all the points of the other class fall beneath the second hyperplane. This is possible so long as the data are linearly separable.

Vector geometry defines the distance between these two planes as:

$$\frac{2}{\|\vec{w}\|}$$

Here, ‖w‖ indicates the **Euclidean norm** (the distance from the origin to vector w). Because ‖w‖ is the denominator, to maximize distance we need to minimize ‖w‖. The task is typically re-expressed as a set of constraints, as follows:

$$\min \frac{1}{2} \|\vec{w}\|^2$$
$$s.t. \ y_i \left(\vec{w} \cdot \vec{x}_i - b \right) \geq 1, \forall \vec{x}_i$$

Although this looks messy, it's really not too complicated to understand conceptually. Basically, the first line implies that we need to minimize the Euclidean norm (squared and divided by two to make the calculation easier). The second line notes that this is subject to (*s.t.*) the condition that each of the y_i data points is correctly classified. Note that y indicates the class value (transformed to either +1 or -1) and the upside-down "A" is shorthand for "for all."

As with the other method for finding the maximum margin, finding a solution to this problem is a task best left for quadratic optimization software. Although it can be processor-intensive, specialized algorithms are capable of solving these problems quickly even on fairly large datasets.

The case of nonlinearly separable data

As we've worked through the theory behind SVMs, you may be wondering about the elephant in the room: what happens in the case that the data are not linearly separable? The solution to this problem is the use of a **slack variable**, which creates a soft margin that allows some points to fall on the incorrect side of the margin. The figure that follows illustrates two points falling on the wrong side of the line with the corresponding slack terms (denoted with the Greek letter Xi):

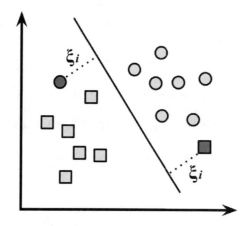

Figure 7.19: Points falling on the wrong side of the boundary come with a cost penalty

A cost value (denoted as *C*) is applied to all points that violate the constraints and rather than finding the maximum margin, the algorithm attempts to minimize the total cost. We can therefore revise the optimization problem to:

$$\min \frac{1}{2}\|\vec{w}\|^2 + C\sum_{i=1}^{n}\xi_i$$
$$s.t. \ \ y_i(\vec{w}\cdot\vec{x}_i - b) \geq 1 - \xi_i, \forall \vec{x}_i, \xi_i \geq 0$$

If you're confused by now, don't worry, you're not alone. Luckily, SVM packages will happily optimize this for you without you having to understand the technical details. The important piece to understand is the addition of the cost parameter, *C*. Modifying this value will adjust the penalty for points that fall on the wrong side of the hyperplane. The greater the cost parameter, the harder the optimization will try to achieve 100 percent separation. On the other hand, a lower cost parameter will place the emphasis on a wider overall margin. It is important to strike a balance between these two in order to create a model that generalizes well to future data.

Using kernels for nonlinear spaces

In many real-world datasets, the relationships between variables are nonlinear. As we just discovered, an SVM can still be trained on such data through the addition of a slack variable, which allows some examples to be misclassified. However, this is not the only way to approach the problem of nonlinearity. A key feature of SVMs is their ability to map the problem into a higher dimension space using a process known as the **kernel trick**. In doing so, a nonlinear relationship may suddenly appear to be quite linear.

Though this seems like nonsense, it is actually quite easy to illustrate by example. In the following figure, the scatterplot on the left depicts a nonlinear relationship between a weather class (Sunny or Snowy) and two features: Latitude and Longitude. The points at the center of the plot are members of the Snowy class, while the points at the margins are all Sunny. Such data could have been generated from a set of weather reports, some of which were obtained from stations near the top of a mountain, while others were obtained from stations around the base of the mountain.

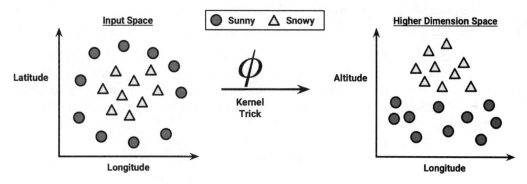

Figure 7.20: The kernel trick can help transform a nonlinear problem into a linear one

On the right side of the figure, after the kernel trick has been applied, we look at the data through the lens of a new dimension: Altitude. With the addition of this feature, the classes are now perfectly linearly separable. This is possible because we have obtained a new perspective on the data. In the left figure, we are viewing the mountain from a bird's-eye view, while on the right, we are viewing the mountain from a distance at ground level. Here, the trend is obvious: snowy weather is found at higher altitudes.

In this way, SVMs with nonlinear kernels add additional dimensions to the data in order to create separation. Essentially, the kernel trick involves a process of constructing new features that express mathematical relationships between measured characteristics.

For instance, the Altitude feature can be expressed mathematically as an interaction between Latitude and Longitude — the closer the point is to the center of each of these scales, the greater the Altitude. This allows the SVM to learn concepts that were not explicitly measured in the original data.

SVMs with nonlinear kernels are extremely powerful classifiers, although they do have some downsides, as shown in the following table:

Strengths	Weaknesses
• Can be used for classification or numeric prediction problems • Not overly influenced by noisy data and not very prone to overfitting • May be easier to use than neural networks, particularly due to the existence of several well-supported SVM algorithms • Gained popularity due to their high accuracy and high-profile wins in data mining competitions	• Finding the best model requires the testing of various combinations of kernels and model parameters • Can be slow to train, particularly if the input dataset has a large number of features or examples • Results in a complex black box model that is difficult, if not impossible, to interpret

Kernel functions, in general, are of the following form. The function denoted by the Greek letter phi, that is, $\phi(x)$, is a mapping of the data into another space. Therefore, the general kernel function applies some transformation to the feature vectors x_i and x_j and combines them using the **dot product**, which takes two vectors and returns a single number.

$$K\left(\vec{x}_i, \vec{x}_j\right) = \phi\left(\vec{x}_i\right) \cdot \phi\left(\vec{x}_j\right)$$

Using this form, kernel functions have been developed for many different domains. A few of the most commonly used kernel functions are listed as follows. Nearly all SVM software packages will include these kernels, among many others.

The **linear kernel** does not transform the data at all. Therefore, it can be expressed simply as the dot product of the features:

$$K\left(\vec{x}_i, \vec{x}_j\right) = \vec{x}_i \cdot \vec{x}_j$$

The **polynomial kernel** of degree d adds a simple nonlinear transformation of the data:

$$K\left(\vec{x}_i, \vec{x}_j\right) = \left(\vec{x}_i \cdot \vec{x}_j + 1\right)^d$$

The **sigmoid kernel** results in an SVM model somewhat analogous to a neural network using a sigmoid activation function. The Greek letters kappa and delta are used as kernel parameters:

$$K\left(\vec{x}_i, \vec{x}_j\right) = \tanh\left(\kappa \vec{x}_i \cdot \vec{x}_j - \delta\right)$$

The **Gaussian RBF kernel** is similar to an RBF neural network. The RBF kernel performs well on many types of data and is thought to be a reasonable starting point for many learning tasks:

$$K\left(\vec{x}_i, \vec{x}_j\right) = e^{\frac{-\left\|\vec{x}_i - \vec{x}_j\right\|^2}{2\sigma^2}}$$

There is no reliable rule for matching a kernel to a particular learning task. The fit depends heavily on the concept to be learned as well as the amount of training data and the relationships among the features. Often, a bit of trial and error is required by training and evaluating several SVMs on a validation dataset. That said, in many cases, the choice of kernel is arbitrary, as the performance may vary only slightly. To see how this works in practice, let's apply our understanding of SVM classification to a real-world problem.

Example – performing OCR with SVMs

Image processing is a difficult task for many types of machine learning algorithms. The relationships linking patterns of pixels to higher concepts are extremely complex and hard to define. For instance, it's easy for a human being to recognize a face, a cat, or the letter "A", but defining these patterns in strict rules is difficult. Furthermore, image data is often noisy. There can be many slight variations in how the image was captured depending on the lighting, orientation, and positioning of the subject.

SVMs are well suited to tackle the challenges of image data. Capable of learning complex patterns without being overly sensitive to noise, they are able to recognize visual patterns with a high degree of accuracy. Moreover, the key weakness of SVMs—the black box model representation—is less critical for image processing. If an SVM can differentiate a cat from a dog, it does not matter much how it is doing so.

In this section, we will develop a model similar to those used at the core of the **optical character recognition (OCR)** software often bundled with desktop document scanners or in smartphone applications. The purpose of such software is to process paper-based documents by converting printed or handwritten text into an electronic form to be saved in a database.

Of course, this is a difficult problem due to the many variants in handwriting style and printed fonts. Even so, software users expect perfection, as errors or typos can result in embarrassing or costly mistakes in a business environment. Let's see whether our SVM is up to the task.

Step 1 – collecting data

When OCR software first processes a document, it divides the paper into a matrix such that each cell in the grid contains a single **glyph**, which is a term referring to a letter, symbol, or number. Next, for each cell, the software will attempt to match the glyph to a set of all characters it recognizes. Finally, the individual characters can be combined into words, which optionally could be spell-checked against a dictionary in the document's language.

In this exercise, we'll assume that we have already developed the algorithm to partition the document into rectangular regions each consisting of a single glyph. We will also assume the document contains only alphabetic characters in English. Therefore, we'll simulate a process that involves matching glyphs to one of the 26 letters, A to Z.

To this end, we'll use a dataset donated to the UCI Machine Learning Repository (`http://archive.ics.uci.edu/ml`) by W. Frey and D. J. Slate. The dataset contains 20,000 examples of 26 English alphabet capital letters as printed using 20 different randomly reshaped and distorted black-and-white fonts.

 For more information about these data, refer to *Letter Recognition Using Holland-Style Adaptive Classifiers, Slate, DJ and Frey, PW, Machine Learning, 1991, Vol. 6, pp. 161-182.*

The following figure, published by Frey and Slate, provides an example of some of the printed glyphs. Distorted in this way, the letters are challenging for a computer to identify, yet are easily recognized by a human being:

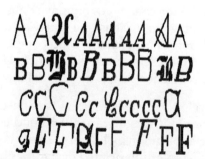

Figure 7.21: Examples of glyphs the SVM algorithm will attempt to identify

Step 2 – exploring and preparing the data

According to the documentation provided by Frey and Slate, when the glyphs are scanned into the computer, they are converted into pixels and 16 statistical attributes are recorded.

The attributes measure such characteristics as the horizontal and vertical dimensions of the glyph; the proportion of black (versus white) pixels; and the average horizontal and vertical position of the pixels. Presumably, differences in the concentration of black pixels across various areas of the box should provide a way to differentiate among the 26 letters of the alphabet.

 To follow along with this example, download the `letterdata.csv` file from the Packt Publishing website and save it to your R working directory.

Reading the data into R, we confirm that we have received the data with the 16 features that define each example of the `letter` class. As expected, it has 26 levels:

```
> letters <- read.csv("letterdata.csv")
> str(letters)
'data.frame':    20000 obs. of 17 variables:
 $ letter: Factor w/ 26 levels "A","B","C","D",..
 $ xbox  : int  2 5 4 7 2 4 4 1 2 11 ...
 $ ybox  : int  8 12 11 11 1 11 2 1 2 15 ...
 $ width : int  3 3 6 6 3 5 5 3 4 13 ...
 $ height: int  5 7 8 6 1 8 4 2 4 9 ...
 $ onpix : int  1 2 6 3 1 3 4 1 2 7 ...
 $ xbar  : int  8 10 10 5 8 8 8 8 10 13 ...
 $ ybar  : int  13 5 6 9 6 8 7 2 6 2 ...
 $ x2bar : int  0 5 2 4 6 6 6 2 2 6 ...
 $ y2bar : int  6 4 6 6 6 9 6 2 6 2 ...
 $ xybar : int  6 13 10 4 6 5 7 8 12 12 ...
 $ x2ybar: int  10 3 3 4 5 6 6 2 4 1 ...
 $ xy2bar: int  8 9 7 10 9 6 6 8 8 9 ...
 $ xedge : int  0 2 3 6 1 0 2 1 1 8 ...
 $ xedgey: int  8 8 7 10 7 8 8 6 6 1 ...
 $ yedge : int  0 4 3 2 5 9 7 2 1 1 ...
 $ yedgex: int  8 10 9 8 10 7 10 7 7 8 ...
```

SVM learners require all features to be numeric, and moreover, that each feature is scaled to a fairly small interval. In this case, every feature is an integer, so we do not need to convert any factors into numbers. On the other hand, some of the ranges for these integer variables appear fairly wide. This indicates that we need to normalize or standardize the data. However, we can skip this step for now, because the R package that we will use for fitting the SVM model will perform the rescaling automatically.

Given that there is no data preparation left to perform, we can move directly to the training and testing phases of the machine learning process. In previous analyses, we randomly divided the data between the training and testing sets. Although we could do so here, Frey and Slate have already randomized the data and therefore suggest using the first 16,000 records (80 percent) for building the model and the next 4,000 records (20 percent) for testing. Following their advice, we can create training and testing data frames as follows:

```
> letters_train <- letters[1:16000, ]
> letters_test  <- letters[16001:20000, ]
```

With our data ready to go, let's start building our classifier.

Step 3 – training a model on the data

When it comes to fitting an SVM model in R, there are several outstanding packages to choose from. The e1071 package from the Department of Statistics at the Vienna University of Technology (TU Wien) provides an R interface to the award-winning LIBSVM library, a widely used open-source SVM program written in C++. If you are already familiar with LIBSVM, you may want to start here.

 For more information on LIBSVM, refer to the authors' website at http://www.csie.ntu.edu.tw/~cjlin/libsvm/.

Similarly, if you're already invested in the SVMlight algorithm, the klaR package from the Department of Statistics at the Dortmund University of Technology (TU Dortmund) provides functions to work with this SVM implementation directly from R.

 For information on SVMlight, see http://svmlight.joachims.org/.

Finally, if you are starting from scratch, it is perhaps best to begin with the SVM functions in the `kernlab` package. An interesting advantage of this package is that it was developed natively in R rather than C or C++, which allows it to be easily customized; none of the internals are hidden behind the scenes. Perhaps even more importantly, unlike the other options, `kernlab` can be used with the `caret` package, which allows SVM models to be trained and evaluated using a variety of automated methods (covered in *Chapter 11, Improving Model Performance*).

 For a more thorough introduction to `kernlab`, please refer to the authors' paper at `http://www.jstatsoft.org/v11/i09/`.

The syntax for training SVM classifiers with `kernlab` is as follows. If you do happen to be using one of the other packages, the commands are largely similar. By default, the `ksvm()` function uses the Gaussian RBF kernel, but a number of other options are provided:

Support vector machine syntax

using the `ksvm()` function in the **kernlab** package

Building the model:

```
m <- ksvm(target ~ predictors, data = mydata,
          kernel = "rbfdot", C = 1)
```

- `target` is the outcome in the **mydata** data frame to be modeled
- `predictors` is an R formula specifying the features in the **mydata** data frame to use for prediction
- `data` specifies the data frame in which the `target` and `predictors` variables can be found
- `kernel` specifies a nonlinear mapping such as `"rbfdot"` (radial basis), `"polydot"` (polynomial), `"tanhdot"` (hyperbolic tangent sigmoid), or `"vanilladot"` (linear)
- `C` is a number that specifies the cost of violating the constraints, i.e., how big of a penalty there is for the "soft margin." Larger values will result in narrower margins

The function will return a SVM object that can be used to make predictions.

Making predictions:

```
p <- predict(m, test, type = "response")
```

- `m` is a model trained by the `ksvm()` function
- `test` is a data frame containing test data with the same features as the training data used to build the classifier
- `type` specifies whether the predictions should be `"response"` (the predicted class) or `"probabilities"` (the predicted probability, one column per class level).

The function will return a vector (or matrix) of predicted classes (or probabilities) depending on the value of the type parameter.

Example:

```
letter_classifier <- ksvm(letter ~ ., data =
  letters_train, kernel = "vanilladot")
letter_prediction <- predict(letter_classifier,
  letters_test)
```

To provide a baseline measure of SVM performance, let's begin by training a simple linear SVM classifier. If you haven't already, install the `kernlab` package to your library using the `install.packages("kernlab")` command. Then, we can call the `ksvm()` function on the training data and specify the linear (that is, "vanilla") kernel using the `vanilladot` option, as follows:

```
> library(kernlab)
> letter_classifier <- ksvm(letter ~ ., data = letters_train,
                            kernel = "vanilladot")
```

Depending on the performance of your computer, this operation may take some time to complete. When it finishes, type the name of the stored model to see some basic information about the training parameters and the fit of the model:

```
> letter_classifier
Support Vector Machine object of class "ksvm"

SV type: C-svc  (classification)
 parameter : cost C = 1

Linear (vanilla) kernel function.

Number of Support Vectors : 7037

Objective Function Value : -14.1746 -20.0072 -23.5628 -6.2009 -7.5524
-32.7694 -49.9786 -18.1824 -62.1111 -32.7284 -16.2209...

Training error : 0.130062
```

This information tells us very little about how well the model will perform in the real world. We'll need to examine its performance on the testing dataset to know whether it generalizes well to unseen data.

Step 4 – evaluating model performance

The `predict()` function allows us to use the letter classification model to make predictions on the testing dataset:

```
> letter_predictions <- predict(letter_classifier, letters_test)
```

Because we didn't specify the `type` parameter, the default `type = "response"` was used. This returns a vector containing a predicted letter for each row of values in the testing data. Using the `head()` function, we can see that the first six predicted letters were U, N, V, X, N, and H:

```
> head(letter_predictions)
[1] U N V X N H
Levels: A B C D E F G H I J K L M N O P Q R S T U V W X Y Z
```

To examine how well our classifier performed, we need to compare the predicted letter to the true letter in the testing dataset. We'll use the `table()` function for this purpose (only a portion of the full table is shown here):

```
> table(letter_predictions, letters_test$letter)
letter_predictions    A    B    C    D    E
                 A  144    0    0    0    0
                 B    0  121    0    5    2
                 C    0    0  120    0    4
                 D    2    2    0  156    0
                 E    0    0    5    0  127
```

The diagonal values of `144`, `121`, `120`, `156`, and `127` indicate the total number of records where the predicted letter matches the true value. Similarly, the number of mistakes is also listed. For example, the value of `5` in row B and column D indicates that there were five cases where the letter D was misidentified as a B.

Looking at each type of mistake individually may reveal some interesting patterns about the specific types of letters the model has trouble with, but this is time consuming. We can simplify our evaluation by instead calculating the overall accuracy. This considers only whether the prediction was correct or incorrect, and ignores the type of error.

The following command returns a vector of TRUE or FALSE values indicating whether the model's predicted letter agrees with (that is, matches) the actual letter in the test dataset:

```
> agreement <- letter_predictions == letters_test$letter
```

Using the `table()` function, we see that the classifier correctly identified the letter in 3,357 out of the 4,000 test records:

```
> table(agreement)
agreement
FALSE   TRUE
 643   3357
```

In percentage terms, the accuracy is about 84 percent:

```
> prop.table(table(agreement))
agreement
   FALSE     TRUE
 0.16075 0.83925
```

Note that when Frey and Slate published the dataset in 1991, they reported a recognition accuracy of about 80 percent. Using just a few lines of R code, we were able to surpass their result, although we also have the benefit of decades of additional machine learning research. With that in mind, it is likely that we are able to do even better.

Step 5 – improving model performance

Let's take a moment to contextualize the performance of the SVM model we trained to identify letters of the alphabet from image data. With one line of R code, the model was able to achieve an accuracy of nearly 84 percent, which slightly surpassed the benchmark percent published by academic researchers in 1991. Although an accuracy of 84 percent is not nearly high enough to be useful for OCR software, the fact that a relatively simple model can reach this level is a remarkable accomplishment in itself. Keep in mind that the probability the model's prediction would match the actual value by dumb luck alone is quite small at under four percent. This implies that our model performs over 20 times better than random chance. As remarkable as this is, perhaps by adjusting the SVM function parameters to train a slightly more complex model, we can also find that the model is useful in the real world.

To calculate the probability of the SVM model's predictions matching the actual values by chance alone, apply the joint probability rule for independent events covered in *Chapter 4, Probabilistic Learning – Classification Using Naive Bayes*. Because there are 26 letters, each appearing at approximately the same rate in the test set, the chance that any one letter is predicted correctly is $(1/26) * (1/26)$. Since there are 26 different letters, the total probability of agreement is $26 * (1/26) * (1/26) = 0.0384$, or 3.84 percent.

Changing the SVM kernel function

Our previous SVM model used the simple linear kernel function. By using a more complex kernel function, we can map the data into a higher dimensional space, and potentially obtain a better model fit.

It can be challenging, however, to choose from the many different kernel functions. A popular convention is to begin with the Gaussian RBF kernel, which has been shown to perform well for many types of data. We can train an RBF-based SVM using the `ksvm()` function as shown here:

```
> letter_classifier_rbf <- ksvm(letter ~ ., data = letters_train,
                                kernel = "rbfdot")
```

Next, we make predictions as before:

```
> letter_predictions_rbf <- predict(letter_classifier_rbf,
                                letters_test)
```

Finally, we'll compare the accuracy to our linear SVM:

```
> agreement_rbf <- letter_predictions_rbf == letters_test$letter
> table(agreement_rbf)
agreement_rbf
FALSE  TRUE
  275  3725
> prop.table(table(agreement_rbf))
agreement_rbf
  FALSE      TRUE
0.06875 0.93125
```

 Your results may differ from those shown here due to randomness in the ksvm RBF kernel. If you'd like them to match exactly, use `RNGversion` `("3.5.2")` and `set.seed(12345)` prior to running the `ksvm()` function.

Simply by changing the kernel function, we were able to increase the accuracy of our character recognition model from 84 percent to 93 percent.

Identifying the best SVM cost parameter

If this level of performance is still unsatisfactory for the OCR program, it is certainly possible to test additional kernels. However, another fruitful approach is to vary the cost parameter, which modifies the width of the SVM decision boundary. This governs the model's balance between overfitting and underfitting the training data—the larger the cost value, the harder the learner will try to perfectly classify every training instance, as there is a higher penalty for each mistake. On the one hand, a high cost can lead the learner to overfit the training data. On the other hand, a cost parameter set too small can cause the learner to miss important, subtle patterns in the training data and underfit the true pattern.

There is no rule of thumb to know the ideal value beforehand, so instead we will examine how the model performs for various values of C, the cost parameter. Rather than repeating the training and evaluation process repeatedly, we can use the sapply() function to apply a custom function to a vector of potential cost values. We begin by using the seq() function to generate this vector as a sequence counting from five to 40 by five. Then, as shown in the following code, the custom function trains the model as before, each time using the cost value and making predictions on the test dataset. Each model's accuracy is computed as the number of predictions that match the actual values divided by the total number of predictions. The result is visualized using the plot() function:

```
> cost_values <- c(1, seq(from = 5, to = 40, by = 5))
>
> accuracy_values <- sapply(cost_values, function(x) {
    set.seed(12345)
    m <- ksvm(letter ~ ., data = letters_train,
              kernel = "rbfdot", C = x)
    pred <- predict(m, letters_test)
    agree <- ifelse(pred == letters_test$letter, 1, 0)
    accuracy <- sum(agree) / nrow(letters_test)
    return (accuracy)
  })

> plot(cost_values, accuracy_values, type = "b")
```

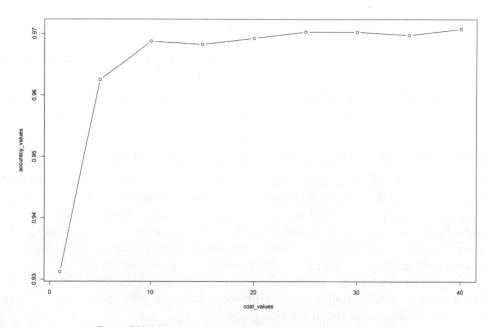

Figure 7.22: Mapping accuracy against SVM cost for the RBF kernel

As depicted in the visualization, with an accuracy of 93 percent, the default SVM cost parameter of C = 1 resulted in by far the least accurate model among the nine models evaluated. Instead, setting C to a value of 10 or higher results in an accuracy of around 97 percent, which is quite an improvement in performance! Perhaps this is close enough to perfect for the model to be deployed in a real-world environment, though it may still be worth experimenting further with various kernels to see if it is possible to get even closer to 100 percent accuracy. Each additional improvement in accuracy will result in fewer mistakes for the OCR software and a better overall experience for the end user.

Summary

In this chapter, we examined two machine learning methods that offer a great deal of potential but are often overlooked due to their complexity. Hopefully, you now see that this reputation is at least somewhat undeserved. The basic concepts that drive ANNs and SVMs are fairly easy to understand.

On the other hand, because ANNs and SVMs have been around for many decades, each of them has numerous variations. This chapter just scratches the surface of what is possible with these methods. By utilizing the terminology that you learned here, you should be capable of picking up the nuances that distinguish the many advancements that are being developed every day, including the ever-growing field of deep learning.

Now that we have spent some time learning about many different types of predictive models, from simple to sophisticated, in the next chapter, we will begin to consider methods for other types of learning tasks. These unsupervised learning techniques will bring to light fascinating patterns within the data.

8

Finding Patterns – Market Basket Analysis Using Association Rules

Think back to your last impulse purchase. Maybe in the grocery store checkout lane you bought a pack of chewing gum or a candy bar. Perhaps on a late-night trip for diapers and formula you picked up a caffeinated beverage or a six-pack of beer. You might have even bought this book on a bookseller's recommendation. These impulse buys are no coincidence, as retailers use sophisticated data analysis techniques to identify patterns that will drive retail behavior.

In years past, such recommendations were based on the subjective intuition of marketing professionals and inventory managers. Now, barcode scanners, inventory databases, and online shopping carts have generated transactional data that machine learning can use to learn purchasing patterns. The practice is commonly known as **market basket analysis** due to the fact that it has been so frequently applied to supermarket data.

Although the technique originated with shopping data, it is also useful in other contexts. By the time you finish this chapter, you will be able to apply market basket analysis techniques to your own tasks, whatever they may be. Generally, the work involves:

- Using simple performance measures to find associations in large databases
- Understanding the peculiarities of transactional data
- Knowing how to identify the useful and actionable patterns

The goal of a market basket analysis is to discover actionable patterns. Thus, as we apply the technique, you are likely to identify applications to your work even if you have no affiliation with a retail chain.

Understanding association rules

The building blocks of a market basket analysis are the items that may appear in any given transaction. Groups of one or more items are surrounded by brackets to indicate that they form a set, or more specifically, an **itemset** that appears in the data with some regularity. Transactions are specified in terms of itemsets, such as the following transaction that might be found in a typical grocery store:

$$\{\text{bread, peanut butter, jelly}\}$$

The result of a market basket analysis is a collection of **association rules** that specify patterns found in the relationships among items in the itemsets. Association rules are always composed from subsets of itemsets and are denoted by relating one itemset on the left-hand side (LHS) of the rule to another itemset on the right-hand side (RHS) of the rule. The LHS is the condition that needs to be met in order to trigger the rule, and the RHS is the expected result of meeting that condition. A rule identified from the preceding example transaction might be expressed in the form:

$$\{\text{peanut butter, jelly}\} \rightarrow \{\text{bread}\}$$

In plain language, this association rule states that if peanut butter and jelly are purchased together, then bread is also likely to be purchased. In other words, "peanut butter and jelly imply bread."

Developed in the context of retail transaction databases, association rules are not used for prediction, but rather for unsupervised knowledge discovery in large databases. This is unlike the classification and numeric prediction algorithms presented in previous chapters. Even so, you will find that association rule learners are closely related to and share many features of the classification rule learners presented in *Chapter 5, Divide and Conquer – Classification Using Decision Trees and Rules.*

Because association rule learners are unsupervised, there is no need for the algorithm to be trained; data does not need to be labeled ahead of time. The program is simply unleashed on a dataset in the hope that interesting associations are found. The downside, of course, is that there isn't an easy way to objectively measure the performance of a rule learner, aside from evaluating it for qualitative usefulness — typically, an eyeball test of some sort.

Although association rules are most often used for market basket analysis, they are helpful for finding patterns in many different types of data. Other potential applications include:

- Searching for interesting and frequently occurring patterns of DNA and protein sequences in cancer data
- Finding patterns of purchases or medical claims that occur in combination with fraudulent credit card or insurance use
- Identifying combinations of behavior that precede customers dropping their cellular phone service or upgrading their cable television package

Association rule analysis is used to search for interesting connections among a very large number of elements. Human beings are capable of such insight quite intuitively, but it often takes expert-level knowledge or a great deal of experience to do what a rule learning algorithm can do in minutes or even seconds. Additionally, some datasets are simply too large and complex for a human being to find the needle in the haystack.

The Apriori algorithm for association rule learning

Just as large transactional datasets create challenges for humans, these datasets also present challenges for machines. Transactional datasets can be large in both the number of transactions as well as the number of items or features that are recorded. The problem is that the number of potential itemsets grows exponentially with the number of features. Given k items that can appear or not appear in a set, there are 2^k possible itemsets that could be potential rules. A retailer that sells only 100 different items could have on the order of $2^{100} = 1.27e+30$ itemsets that an algorithm must evaluate—a seemingly impossible task.

Rather than evaluating each of these itemsets one by one, a smarter rule learning algorithm takes advantage of the fact that in reality, many of the potential combinations of items are rarely, if ever, found in practice. For instance, even if a store sells both automotive items and women's cosmetics, a set of *{motor oil, lipstick}* is likely to be extraordinarily uncommon. By ignoring these rare (and perhaps less important) combinations, it is possible to limit the scope of the search for rules to a more manageable size.

Much work has been done to identify heuristic algorithms for reducing the number of itemsets to search. Perhaps the most widely-used approach for efficiently searching large databases for rules is known as **Apriori**. Introduced in 1994 by Rakesh Agrawal and Ramakrishnan Srikant, the Apriori algorithm has since become somewhat synonymous with association rule learning. The name is derived from the fact that the algorithm utilizes a simple prior (that is, *a priori*) belief about the properties of frequent itemsets.

Before we discuss that in more depth, it's worth noting that this algorithm, like all learning algorithms, is not without its strengths and weaknesses. Some of these are listed as follows:

Strengths	Weaknesses
• Capable of working with large amounts of transactional data • Results in rules that are easy to understand • Useful for data mining and discovering unexpected knowledge in databases	• Not very helpful for small datasets • Takes effort to separate the true insight from the common sense • Easy to draw spurious conclusions from random patterns

As noted earlier, the Apriori algorithm employs a simple *a priori* belief as a guideline for reducing the association rule search space: all subsets of a frequent itemset must also be frequent. This heuristic is known as the **Apriori property**. Using this astute observation, it is possible to dramatically limit the number of rules to search. For example, the set *{motor oil, lipstick}* can only be frequent if both *{motor oil}* and *{lipstick}* occur frequently as well. Consequently, if either motor oil or lipstick is infrequent, then any set containing these items can be excluded from the search.

For additional details on the Apriori algorithm, refer to *Fast Algorithms for Mining Association Rules, Agrawal, R, Srikant, R, Proceedings of the 20th International Conference on Very Large Databases, 1994, pp. 487-499.*

To see how this principle can be applied in a more realistic setting, let's consider a simple transaction database. The following table shows five completed transactions at an imaginary hospital's gift shop:

transaction ID	items purchased
1	{flowers, get well card, soda}
2	{plush toy bear, flowers, balloons, candy bar}
3	{get well card, candy bar, flowers}
4	{plush toy bear, balloons, soda}
5	{flowers, get well card, soda}

Figure 8.1: Itemsets representing five transactions in a hypothetical hospital's gift shop

By looking at the sets of purchases, one can infer that there are a couple of typical buying patterns. A person visiting a sick friend or family member tends to buy a get well card and flowers, while visitors to new mothers tend to buy plush toy bears and balloons. Such patterns are notable because they appear frequently enough to catch our interest; we simply apply a bit of logic and subject matter experience to explain the rule.

In a similar fashion, the Apriori algorithm uses statistical measures of an itemset's "interestingness" to locate association rules in much larger transaction databases. In the sections that follow, we will discover how Apriori computes such measures of interest, and how they are combined with the Apriori property to reduce the number of rules to be learned.

Measuring rule interest – support and confidence

Whether or not an association rule is deemed interesting is determined by two statistical measures: support and confidence. By providing minimum thresholds for each of these metrics and applying the Apriori principle, it is easy to drastically limit the number of rules reported, perhaps even to the point where only the obvious or common sense rules are identified. For this reason, it is important to carefully understand the types of rules that are excluded under these criteria.

The **support** of an itemset or rule measures how frequently it occurs in the data. For instance, the itemset *{get well card, flowers}* has the support of 3 / 5 = 0.6 in the hospital gift shop data. Similarly, the support for *{get well card}* → *{flowers}* is also 0.6. Support can be calculated for any itemset or even a single item; for instance, the support for *{candy bar}* is 2/5 = 0.4, since candy bars appear in 40 percent of purchases. A function defining support for itemset X could be defined as:

$$\text{confidence}(X \to Y) = \frac{\text{support}(X,Y)}{\text{support}(X)}$$

Here, N is the number of transactions in the database and *count(X)* is the number of transactions containing itemset X.

A rule's **confidence** is a measurement of its predictive power or accuracy. It is defined as the support of the itemset containing both X and Y divided by the support of the itemset containing only X:

$$\text{confidence}(X \to Y) = \frac{\text{support}(X,Y)}{\text{support}(X)}$$

Essentially, the confidence tells us the proportion of transactions where the presence of item or itemset X results in the presence of item or itemset Y. Keep in mind that the confidence that X leads to Y is not the same as the confidence that Y leads to X. For example, the confidence of *{flowers}* → *{get well card}* is *0.6 / 0.8 = 0.75*. In comparison, the confidence of *{get well card}* → *{flowers}* is *0.6 / 0.6 = 1.0*. This means that a purchase of flowers also includes the purchase of a get well card 75 percent of the time, while a purchase of a get well card also includes flowers 100 percent of the time. This information could be quite useful to the gift shop management.

> You may have noticed similarities between support, confidence, and the Bayesian probability rules covered in *Chapter 4, Probabilistic Learning – Classification Using Naive Bayes*. In fact, *support(A, B)* is the same as $P(A \cap B)$ and *confidence(A → B)* is the same as $P(B \mid A)$. It is just the context that differs.

Rules like *{get well card}* → *{flowers}* are known as **strong rules** because they have both high support and confidence. One way to find more strong rules would be to examine every possible combination of items in the gift shop, measure the support and confidence, and report back only those rules that meet certain levels of interest. However, as noted before, this strategy is generally not feasible for anything but the smallest of datasets.

In the next section, you will see how the Apriori algorithm uses minimum levels of support and confidence with the Apriori principle to find strong rules quickly by reducing the number of rules to a more manageable level.

Building a set of rules with the Apriori principle

Recall that the Apriori principle states that all subsets of a frequent itemset must also be frequent. In other words, if *{A, B}* is frequent, then *{A}* and *{B}* must both be frequent. Recall also that, by definition, the support metric indicates how frequently an itemset appears in the data. Therefore, if we know that *{A}* does not meet a desired support threshold, there is no reason to consider *{A, B}* or any itemset containing *{A}*; it cannot possibly be frequent.

The Apriori algorithm uses this logic to exclude potential association rules prior to actually evaluating them. The process of creating rules then occurs in two phases:

1. Identifying all the itemsets that meet a minimum support threshold.
2. Creating rules from these itemsets using those meeting a minimum confidence threshold.

The first phase occurs in multiple iterations. Each successive iteration involves evaluating the support of a set of increasingly large itemsets. For instance, iteration one involves evaluating the set of 1-item itemsets (1-itemsets), iteration two evaluates the 2-itemsets, and so on. The result of each iteration *i* is a set of all the *i*-itemsets that meet the minimum support threshold.

All the itemsets from iteration *i* are combined in order to generate candidate itemsets for evaluation in iteration *i + 1*. But the Apriori principle can eliminate some of them even before the next round begins. If *{A}*, *{B}*, and *{C}* are frequent in iteration one, while *{D}* is not frequent, then iteration two will consider only *{A, B}*, *{A, C}*, and *{B, C}*. Thus, the algorithm needs to evaluate only three itemsets rather than the six that would have been evaluated if the sets containing *D* had not been eliminated *a priori*.

Continuing this thought, suppose during iteration two it is discovered that *{A, B}* and *{B, C}* are frequent, but *{A, C}* is not. Although iteration three would normally begin by evaluating the support for *{A, B, C}*, this step is not necessary. Why not? The Apriori principle states that *{A, B, C}* cannot possibly be frequent, since the subset *{A, C}* is not. Therefore, having generated no new itemsets in iteration three, the algorithm may stop.

iteration	must evaluate	frequent itemsets	infrequent itemsets
1	{A}, {B}, {C}, {D}	{A}, {B}, {C}	{D}
2	{A, B}, {A, C}, {B, C} ~~{A, D}, {B, D}, {C, D}~~	{A, B}, {B, C}	{A, C}
3	~~{A, B, C}~~		
4	~~{A, B, C, D}~~		

Figure 8.2: The Apriori algorithm only needs to evaluate seven of the 12 potential itemsets

At this point, the second phase of the Apriori algorithm may begin. Given the set of frequent itemsets, association rules are generated from all possible subsets. For instance, *{A, B}* would result in candidate rules for *{A}* → *{B}* and *{B}* → *{A}*. These are evaluated against a minimum confidence threshold, and any rule that does not meet the desired confidence level is eliminated.

Example – identifying frequently purchased groceries with association rules

As noted in this chapter's introduction, market basket analysis is used behind the scenes for the recommendation systems used in many brick-and-mortar and online retailers. The learned association rules indicate the combinations of items that are often purchased together. Knowledge of these patterns provides insight into new ways a grocery chain might optimize the inventory, advertise promotions, or organize the physical layout of the store. For instance, if shoppers frequently purchase coffee or orange juice with a breakfast pastry, it may be possible to increase profit by relocating pastries closer to coffee and juice.

In this tutorial, we will perform a market basket analysis of transactional data from a grocery store. However, the techniques could be applied to many different types of problems, from movie recommendations, to dating sites, to finding dangerous interactions among medications. In doing so, we will see how the Apriori algorithm is able to efficiently evaluate a potentially massive set of association rules.

Step 1 – collecting data

Our market basket analysis will utilize purchase data from one month of operation at a real-world grocery store. The data contains 9,835 transactions, or about 327 transactions per day (roughly 30 transactions per hour in a 12-hour business day), suggesting that the retailer is not particularly large, nor is it particularly small.

The dataset used here was adapted from the Groceries dataset in the arules R package. For more information, see *Implications of Probabilistic Data Modeling for Mining Association Rules, Hahsler, M, Hornik, K, Reutterer, T, 2005*. In *From Data and Information Analysis to Knowledge Engineering, Gaul W, Vichi M, Weihs C, Studies in Classification, Data Analysis, and Knowledge Organization, 2006, pp. 598–605.*

A typical grocery store offers a huge variety of items. There might be five brands of milk, a dozen types of laundry detergent, and three brands of coffee. Given the moderate size of the retailer in this example, we will assume that it is not terribly concerned with finding rules that apply only to a specific brand of milk or detergent. With this in mind, all brand names have been removed from the purchases. This reduced the number of groceries to a more manageable 169 types, using broad categories such as chicken, frozen meals, margarine, and soda.

If you hope to identify highly specific association rules — such as whether customers prefer grape or strawberry jelly with their peanut butter — you will need a tremendous amount of transactional data. Large chain retailers use databases of many millions of transactions in order to find associations among particular brands, colors, or flavors of items.

Do you have any guesses about which types of items might be purchased together? Will wine and cheese be a common pairing? Bread and butter? Tea and honey? Let's dig into this data and see if these guesses can be confirmed.

Step 2 – exploring and preparing the data

Transactional data is stored in a slightly different format than we have used previously. Most of our prior analyses utilized data in a matrix where rows indicated example instances and columns indicated features. In the matrix format, all examples must have exactly the same set of features.

In comparison, transactional data is more freeform. As usual, each row in the data specifies a single example — in this case, a transaction. However, rather than having a set number of features, each record comprises a comma-separated list of any number of items, from one to many. In essence, the features may differ from example to example.

To follow along with this analysis, download the `groceries.csv` file from the Packt Publishing website and save it in your R working directory.

The first five rows of the raw `groceries.csv` file are as follows:

```
citrus fruit,semi-finished bread,margarine,ready soups
tropical fruit,yogurt,coffee
whole milk
pip fruit,yogurt,cream cheese,meat spreads
other vegetables,whole milk,condensed milk,long life bakery product
```

These lines indicate five separate grocery store transactions. The first transaction included four items: citrus fruit, semi-finished bread, margarine, and ready soups. In comparison, the third transaction included only one item: whole milk.

Suppose we tried to load the data using the `read.csv()` function as we did in prior analyses. R would happily comply and read the data into matrix format as follows:

	V1	V2	V3	V4
1	citrus fruit	semi-finished bread	margarine	ready soups
2	tropical fruit	yogurt	coffee	
3	whole milk			
4	pip fruit	yogurt	cream cheese	meat spreads
5	other vegetables	whole milk	condensed milk	long life bakery product

Figure 8.3: Transactional data incorrectly loaded into matrix format

You will notice that R created four columns to store the items in the transactional data: V1, V2, V3, and V4. Although this may seem reasonable, if we use the data in this form, we will encounter problems later on. R chose to create four variables because the first line had exactly four comma-separated values. However, we know that grocery purchases can contain more than four items; in the four-column design, such transactions will be broken across multiple rows in the matrix. We could try to remedy this by putting the transaction with the largest number of items at the top of the file, but this ignores another more problematic issue.

By structuring the data this way, R has constructed a set of features that record not just the items in the transactions, but also the order they appear. If we imagine our learning algorithm as an attempt to find a relationship among V1, V2, V3, and V4, then the whole milk in V1 might be treated differently than the whole milk appearing in V2. Instead, we need a dataset that does not treat a transaction as a set of positions to be filled (or not filled) with specific items, but rather as a market basket that either contains or does not contain each particular item.

Data preparation – creating a sparse matrix for transaction data

The solution to this problem utilizes a data structure called a sparse matrix. You may recall that we used a sparse matrix for processing text data in *Chapter 4, Probabilistic Learning – Classification Using Naive Bayes*. Just as with the preceding dataset, each row in the sparse matrix indicates a transaction. However, the sparse matrix has a column (that is, feature) for every item that could possibly appear in someone's shopping bag. Since there are 169 different items in our grocery store data, our sparse matrix will contain 169 columns.

Why not just store this as a data frame as we did in most of our prior analyses? The reason is that as additional transactions and items are added, a conventional data structure quickly becomes too large to fit in the available memory. Even with the relatively small transactional dataset used here, the matrix contains nearly 1.7 million cells, most of which contain zeros (hence the name "sparse" matrix—there are very few non-zero values).

Since there is no benefit to storing all these zeros, a sparse matrix does not actually store the full matrix in memory; it only stores the cells that are occupied by an item. This allows the structure to be more memory efficient than an equivalently sized matrix or data frame.

In order to create the sparse matrix data structure from transactional data, we can use functionality provided by the `arules` (association rules) package. Install and load the package using the `install.packages("arules")` and `library(arules)` commands.

 For more information on the `arules` package, refer to: *arules - A Computational Environment for Mining Association Rules and Frequent Item Sets, Hahsler, M, Gruen, B, Hornik, K, Journal of Statistical Software, 2005, Vol. 14.*

Because we're loading transactional data, we cannot simply use the `read.csv()` function used previously. Instead, `arules` provides a `read.transactions()` function that is similar to `read.csv()` with the exception that it results in a sparse matrix suitable for transactional data. The parameter `sep = ","` specifies that items in the input file are separated by a comma. To read the `groceries.csv` data into a sparse matrix named `groceries`, type the following line:

```
> groceries <- read.transactions("groceries.csv", sep = ",")
```

To see some basic information about the `groceries` matrix we just created, use the `summary()` function on the object:

```
> summary(groceries)
transactions as itemMatrix in sparse format with
 9835 rows (elements/itemsets/transactions) and
 169 columns (items) and a density of 0.02609146
```

The first block of information in the output (as shown previously) provides a summary of the sparse matrix we created. The output `9835 rows` refers to the number of transactions, and `169 columns` indicates each of the 169 different items that might appear in someone's grocery basket. Each cell in the matrix is a `1` if the item was purchased for the corresponding transaction, or `0` otherwise.

The **density** value of `0.02609146` (2.6 percent) refers to the proportion of non-zero matrix cells. Since there are *9,835 * 169 = 1,662,115* positions in the matrix, we can calculate that a total of *1,662,115 * 0.02609146 = 43,367* items were purchased during the store's 30 days of operation (ignoring the fact that duplicates of the same items might have been purchased). With an additional step, we can determine that the average transaction contained *43,367 / 9,835 = 4.409* distinct grocery items. Of course, if we look a little further down the output, we'll see that the mean number of items per transaction has already been provided.

The next block of `summary()` output lists the items that were most commonly found in the transactional data. Since *2,513 / 9,835 = 0.2555*, we can determine that whole milk appeared in 25.6 percent of transactions. Other vegetables, rolls/buns, soda, and yogurt round out the list of other common items, as follows:

```
most frequent items:
     whole milk other vegetables      rolls/buns
           2513             1903            1809
           soda           yogurt         (Other)
           1715             1372           34055
```

Finally, we are presented with a set of statistics about the size of the transactions. A total of 2,159 transactions contained only a single item, while one transaction had 32 items. The first quartile and median purchase size are two and three items respectively, implying that 25 percent of the transactions contained two or fewer items and about half contained three items or fewer. The mean of 4.409 items per transaction matches the value we calculated by hand.

```
element (itemset/transaction) length distribution:
sizes
   1    2    3    4    5    6    7    8    9   10   11   12
2159 1643 1299 1005  855  645  545  438  350  246  182  117
  13   14   15   16   17   18   19   20   21   22   23   24
  78   77   55   46   29   14   14    9   11    4    6    1
  26   27   28   29   32
   1    1    1    3    1


  Min. 1st Qu.  Median    Mean 3rd Qu.    Max.
 1.000   2.000   3.000   4.409   6.000  32.000
```

The `arules` package includes some useful features for examining transaction data. To look at the contents of the sparse matrix, use the `inspect()` function in combination with R's vector operators. The first five transactions can be viewed as follows:

```
> inspect(groceries[1:5])
  items
1 {citrus fruit,
   margarine,
   ready soups,
   semi-finished bread}
```

```
2 {coffee,
   tropical fruit,
   yogurt}
3 {whole milk}
4 {cream cheese,
   meat spreads,
   pip fruit,
   yogurt}
5 {condensed milk,
   long life bakery product,
   other vegetables,
   whole milk}
```

These transactions match our look at the original CSV file. To examine a particular item (that is, a column of data), use the [row, column] matrix notion. Using this with the itemFrequency() function allows us to see the proportion of transactions that contain the specified item. For instance, to view the support level for the first three items in the grocery data, use the following command:

```
> itemFrequency(groceries[, 1:3])
abrasive cleaner artif. sweetener   baby cosmetics
     0.0035587189      0.0032536858      0.0006100661
```

Notice that the items in the sparse matrix are sorted in columns by alphabetical order. Abrasive cleaner and artificial sweeteners are found in about 0.3 percent of the transactions, while baby cosmetics are found in about 0.06 percent of the transactions.

Visualizing item support – item frequency plots

To present these statistics visually, use the itemFrequencyPlot() function. This creates a bar chart depicting the proportion of transactions containing specified items. Since transactional data contains a very large number of items, you will often need to limit those appearing in the plot in order to produce a legible chart.

If you would like to display items that appear in a minimum proportion of transactions, use itemFrequencyPlot() with the support parameter:

```
> itemFrequencyPlot(groceries, support = 0.1)
```

As shown in the following plot, this results in a histogram showing the eight items in the groceries data with at least 10 percent support:

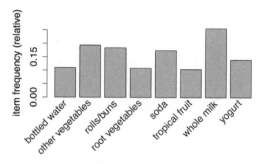

Figure 8.4: Support levels for all grocery items in at least 10 percent of transactions

If you would rather limit the plot to a specific number of items, use `itemFrequencyPlot()` with the `topN` parameter:

```
> itemFrequencyPlot(groceries, topN = 20)
```

The histogram is then sorted by decreasing support, as shown in the following diagram for the top 20 items in the groceries data:

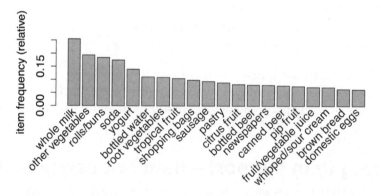

Figure 8.5: Support levels for the top 20 grocery items

Visualizing the transaction data – plotting the sparse matrix

In addition to looking at specific items, it's also possible to obtain a bird's-eye view of the entire sparse matrix using the `image()` function. Of course, because the matrix itself is very large, it is usually best to request a subset of the entire matrix. The command to display the sparse matrix for the first five transactions is as follows:

```
> image(groceries[1:5])
```

The resulting diagram depicts a matrix with five rows and 169 columns, indicating the five transactions and 169 possible items we requested. Cells in the matrix are filled with black for transactions (rows) where the item (column) was purchased.

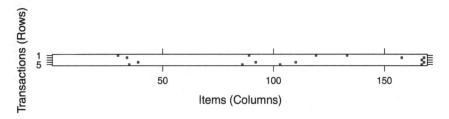

Figure 8.6: A visualization of the sparse matrix for the first five transactions

Although the preceding diagram is small and may be slightly hard to read, you can see that the first, fourth, and fifth transactions contained four items each, since their rows have four cells filled in. On the right side of the diagram, you can also see that rows three and five, and rows two and four, share an item in common.

This visualization can be a useful tool for exploring the transactional data. For one, it may help with the identification of potential data issues. Columns that are filled all the way down could indicate items that are purchased in every transaction—a problem that could arise, perhaps, if a retailer's name or identification number was inadvertently included in the transaction dataset.

Additionally, patterns in the diagram may help reveal interesting segments of transactions and items, particularly if the data is sorted in interesting ways. For example, if the transactions are sorted by date, patterns in the black dots could reveal seasonal effects in the number or types of items purchased. Perhaps around Christmas or Hanukkah, toys are more common; around Halloween, perhaps candies become popular. This type of visualization could be especially powerful if the items were also sorted into categories. In most cases, however, the plot will look fairly random, like static on a television screen.

Keep in mind that this visualization will not be as useful for extremely large transaction databases because the cells will be too small to discern. Still, by combining it with the `sample()` function, you can view the sparse matrix for a randomly sampled set of transactions. The command to create a random selection of 100 transactions is as follows:

```
> image(sample(groceries, 100))
```

This creates a matrix diagram with 100 rows and 169 columns:

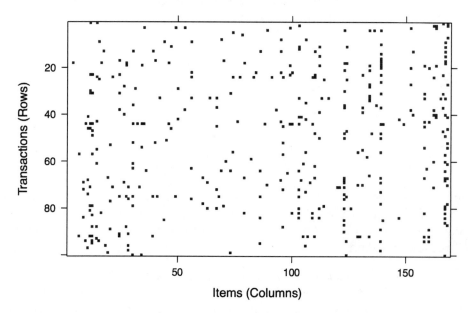

Figure 8.7: A visualization of the sparse matrix for 100 randomly-selected transactions

A few columns seem fairly heavily populated, indicating some very popular items at the store. However, the distribution of dots seems overall fairly random. Given nothing else of note, let's continue with our analysis.

Step 3 – training a model on the data

With data preparation complete, we can now work at finding associations among shopping cart items. We will use an implementation of the Apriori algorithm in the `arules` package we've been using for exploring and preparing the groceries data. You'll need to install and load this package if you have not done so already. The following table shows the syntax for creating sets of rules with the `apriori()` function:

Association rule syntax

using the `apriori()` function in the `arules` package

Finding association rules:

```
myrules <- apriori(data = mydata, parameter =
   list(support = 0.1, confidence = 0.8, minlen = 1))
```

- `data` is a sparse item matrix holding transactional data
- `support` specifies the minimum required rule support
- `confidence` specifies the minimum required rule confidence
- `minlen` specifies the minimum required rule items

The function will return a rules object storing all rules that meet the minimum criteria.

Examining association rules:

```
inspect(myrules)
```

- `myrules` is a set of association rules from the `apriori()` function

This will output the association rules to the screen. Vector operators can be used on myrules to choose a specific rule or rules to view.

Example:

```
groceryrules <- apriori(groceries, parameter =
   list(support = 0.01, confidence = 0.25, minlen = 2))
inspect(groceryrules[1:3])
```

Although running the `apriori()` function is straightforward, there can sometimes be a fair amount of trial and error needed to find the `support` and `confidence` parameters that produce a reasonable number of association rules. If you set these levels too high, then you might find no rules, or might find rules that are too generic to be very useful. On the other hand, a threshold too low might result in an unwieldy number of rules. Worse, the operation might take a very long time or run out of memory during the learning phase.

On the groceries data, using the default settings of `support = 0.1` and `confidence = 0.8` results in a set of zero rules:

```
> apriori(groceries)
set of 0 rules
```

Obviously, we need to widen the search a bit.

If you think about it, this outcome should not have been terribly surprising. Because `support = 0.1` by default, in order to generate a rule, an item must have appeared in at least *0.1 * 9,385 = 938.5* transactions. Since only eight items appeared this frequently in our data, it's no wonder we didn't find any rules.

One way to approach the problem of setting a minimum support is to think about the smallest number of transactions needed before you would consider a pattern interesting. For instance, you could argue that if an item is purchased twice a day (about 60 times in a month of data) then it may be important. From there, it is possible to calculate the support level needed to find only rules matching at least that many transactions. Since 60 out of 9,835 equals 0.006, we'll try setting the support there first.

Setting the minimum confidence involves a delicate balance. On the one hand, if the confidence is too low, then we might be overwhelmed with a large number of unreliable rules—such as dozens of rules indicating items commonly purchased with batteries. How would we know where to target our advertising budget then? On the other hand, if we set the confidence too high, then we will be limited to rules that are obvious or inevitable—like the fact that a smoke detector is always purchased in combination with batteries. In this case, moving the smoke detectors closer to the batteries is unlikely to generate additional revenue, since the two items were already almost always purchased together.

The appropriate minimum confidence level depends a great deal on the goals of your analysis. If you start with a conservative value, you can always reduce it to broaden the search if you aren't finding actionable intelligence.

We'll start with a confidence threshold of 0.25, which means that in order to be included in the results, the rule has to be correct at least 25 percent of the time. This will eliminate the most unreliable rules, while allowing some room for us to modify behavior with targeted promotions.

We are now ready to generate some rules. In addition to the minimum `support` and `confidence` parameters, it is helpful to set `minlen = 2` to eliminate rules that contain fewer than two items. This prevents uninteresting rules from being created simply because the item is purchased frequently, for instance, `{} => whole milk`. This rule meets the minimum support and confidence because whole milk is purchased in over 25 percent of transactions, but it isn't a very actionable insight.

The full command for finding a set of association rules using the Apriori algorithm is as follows:

```
> groceryrules <- apriori(groceries, parameter = list(support =
                    0.006, confidence = 0.25, minlen = 2))
```

This saves our rules in a rules object, which we can peek into by typing its name:

```
> groceryrules
set of 463 rules
```

Our `groceryrules` object contains a set of 463 association rules. To determine whether any of them are useful, we'll have to dig deeper.

Step 4 – evaluating model performance

To obtain a high-level overview of the association rules, we can use `summary()` as follows. The rule length distribution tells us how many rules have each count of items. In our rule set, 150 rules have only two items, while 297 have three, and 16 have four. The summary statistics associated with this distribution are also provided in the output:

```
> summary(groceryrules)
set of 463 rules


rule length distribution (lhs + rhs):sizes
  2   3   4
150 297  16


  Min. 1st Qu. Median   Mean 3rd Qu.   Max.
 2.000   2.000  3.000  2.711   3.000  4.000
```

 As noted in the previous output, the size of the rule is calculated as the total of both the left-hand side (`lhs`) and right-hand side (`rhs`) of the rule. This means that a rule like {bread} => {butter} is two items and {peanut butter, jelly} => {bread} is three.

Next, we see the summary statistics of the rule quality measures: `support`, `confidence`, and `lift`. The `support` and `confidence` measures should not be very surprising, since we used these as selection criteria for the rules. We might be alarmed if most or all of the rules had `support` and `confidence` very near the minimum thresholds, as this would mean that we may have set the bar too high.

This is not the case here, as there are many rules with much higher values of each:

```
summary of quality measures:
     support              confidence            lift
 Min.    :0.006101    Min.    :0.2500    Min.    :0.9932
 1st Qu.:0.007117     1st Qu.:0.2971     1st Qu.:1.6229
 Median :0.008744     Median :0.3554     Median :1.9332
 Mean    :0.011539    Mean    :0.3786    Mean    :2.0351
 3rd Qu.:0.012303     3rd Qu.:0.4495     3rd Qu.:2.3565
 Max.    :0.074835    Max.    :0.6600    Max.    :3.9565
```

The third column is a metric we have not considered yet. The **lift** of a rule measures how much more likely one item or itemset is to be purchased relative to its typical rate of purchase, given that you know another item or itemset has been purchased. This is defined by the following equation:

$$\text{lift}(X \to Y) = \frac{\text{confidence}(X \to Y)}{\text{support}(Y)}$$

Unlike confidence, where the item order matters, *lift(X → Y)* is the same as *lift(Y → X)*.

For example, suppose at a grocery store most people purchase milk and bread. By chance alone, we would expect to find many transactions with both milk and bread. However, if *lift(milk → bread)* is greater than one, this implies that the two items are found together more often than expected by chance alone. A large lift value is therefore a strong indicator that a rule is important and reflects a true connection between the items.

In the final section of the `summary()` output, we receive mining information, telling us about how the rules were chosen. Here, we see that the groceries data, which contained 9,835 transactions, was used to construct rules with a minimum support of 0.006 and minimum confidence of 0.25:

```
mining info:
      data    transactions support confidence
 groceries            9835    0.006        0.25
```

We can take a look at specific rules using the `inspect()` function. For instance, the first three rules in the `groceryrules` object can be viewed as follows:

```
> inspect(groceryrules[1:3])
```

	lhs		rhs	support	confidence	lift
1	{potted plants}	=>	{whole milk}	0.006914082	0.4000000	1.565460
2	{pasta}	=>	{whole milk}	0.006100661	0.4054054	1.586614
3	{herbs}	=>	{root vegetables}	0.007015760	0.4312500	3.956477

The first rule can be read in plain language as "if a customer buys potted plants, they will also buy whole milk." With a support of about 0.007 and confidence of 0.400, we can determine that this rule covers about 0.7 percent of transactions and is correct in 40 percent of purchases involving potted plants. The lift value tells us how much more likely a customer is to buy whole milk relative to the average customer, given that he or she bought a potted plant. Since we know that about 25.6 percent of customers bought whole milk (`support`), while 40 percent of customers buying a potted plant bought whole milk (`confidence`), we can compute the lift as *0.40 / 0.256 = 1.56*, which matches the value shown.

 Note that the column labeled `support` indicates the support for the rule, not the support for the `lhs` or `rhs` alone.

In spite of the fact that the confidence and lift are high, does *{potted plants}* → *{whole milk}* seem like a very useful rule? Probably not, as there doesn't seem to be a logical reason why someone would be more likely to buy milk with a potted plant. Yet our data suggests otherwise. How can we make sense of this fact?

A common approach is to take the association rules and divide them into the following three categories:

- Actionable
- Trivial
- Inexplicable

Obviously, the goal of a market basket analysis is to find **actionable** rules that provide a clear and useful insight. Some rules are clear and others are useful; it is less common to find a combination of both of these factors.

So-called **trivial** rules include any rules that are so obvious that they are not worth mentioning—they are clear, but not useful. Suppose you are a marketing consultant being paid large sums of money to identify new opportunities for cross-promoting items. If you report the finding that *{diapers}* → *{formula}*, you probably won't be invited back for another consulting job.

Trivial rules can also sneak in disguised as more interesting results. For instance, say you found an association between a particular brand of children's cereal and a certain DVD movie. This finding is not very insightful if the movie's main character is on the front of the cereal box.

Rules are **inexplicable** if the connection between the items is so unclear that figuring out how to use the information is impossible or nearly impossible. The rule may simply be a random pattern in the data, for instance, a rule stating that *{pickles}* → *{chocolate ice cream}* may be due to a single customer whose pregnant wife had regular cravings for strange combinations of foods.

The best rules are the hidden gems — the undiscovered insights that only seem obvious once discovered. Given enough time, one could evaluate each and every rule to find the gems. However, the data scientists working on the analysis may not be the best judge of whether a rule is actionable, trivial, or inexplicable. Consequently, better rules are likely to arise via collaboration with the domain experts responsible for managing the retail chain, who can help interpret the findings. In the next section, we'll facilitate such sharing by employing methods for sorting and exporting the learned rules so that the most interesting results float to the top.

Step 5 – improving model performance

Subject matter experts may be able to identify useful rules very quickly, but it would be a poor use of their time to ask them to evaluate hundreds or thousands of rules. Therefore, it's useful to be able to sort the rules according to different criteria, and get them out of R in a form that can be shared with marketing teams and examined in more depth. In this way, we can improve the performance of our rules by making the results more actionable.

Sorting the set of association rules

Depending upon the objectives of the market basket analysis, the most useful rules might be those with the highest support, confidence, or lift. The `arules` package includes a `sort()` function that can be used to reorder the list of rules so that those with the highest or lowest values of the quality measure come first.

To reorder the `groceryrules` object, we can `sort()` while specifying a value of `"support"`, `"confidence"`, or `"lift"` to the by parameter. By combining the sort with vector operators, we can obtain a specific number of interesting rules. For instance, the best five rules according to the `lift` statistic can be examined using the following command:

```
> inspect(sort(groceryrules, by = "lift")[1:5])
```

The output is as follows:

```
    lhs                    rhs                      support confidence      lift
1 {herbs}              => {root vegetables}     0.007015760 0.4312500 3.956477
2 {berries}            => {whipped/sour cream} 0.009049314 0.2721713 3.796886
3 {other vegetables,
   tropical fruit,
   whole milk}         => {root vegetables}     0.007015760 0.4107143 3.768074
4 {beef,
   other vegetables}   => {root vegetables}     0.007930859 0.4020619 3.688692
5 {other vegetables,
   tropical fruit}     => {pip fruit}           0.009456024 0.2634561 3.482649
```

These rules appear to be more interesting than the ones we looked at previously. The first rule, with a `lift` of about 3.96, implies that people who buy herbs are nearly four times more likely to buy root vegetables than the typical customer — perhaps for a stew of some sort? Rule two is also interesting. Whipped cream is over three times more likely to be found in a shopping cart with berries versus other carts, suggesting perhaps a dessert pairing?

> By default, the sort order is decreasing, meaning the largest values come first. To reverse this order, add an additional parameter `decreasing = FALSE`.

Taking subsets of association rules

Suppose that given the preceding rule, the marketing team is excited about the possibilities of creating an advertisement to promote berries, which are now in season. Before finalizing the campaign, however, they ask you to investigate whether berries are often purchased with other items. To answer this question, we'll need to find all the rules that include berries in some form.

The `subset()` function provides a method for searching for subsets of transactions, items, or rules. To use it to find any rules with berries appearing in the rule, use the following command. This will store the rules in a new object named `berryrules`:

```
> berryrules <- subset(groceryrules, items %in% "berries")
```

We can then inspect the rules as we had done with the larger set:

```
> inspect(berryrules)
```

The result is the following set of rules:

```
  lhs              rhs                          support confidence     lift
1 {berries} => {whipped/sour cream} 0.009049314  0.2721713 3.796886
2 {berries} => {yogurt}             0.010574479  0.3180428 2.279848
3 {berries} => {other vegetables}   0.010269446  0.3088685 1.596280
4 {berries} => {whole milk}         0.011794611  0.3547401 1.388328
```

There are four rules involving berries, two of which seem to be interesting enough to be called actionable. In addition to whipped cream, berries are also purchased frequently with yogurt—a pairing that could serve well for breakfast or lunch, as well as dessert.

The subset() function is very powerful. The criteria for choosing the subset can be defined with several keywords and operators:

- The keyword items, explained previously, matches an item appearing anywhere in the rule. To limit the subset to where the match occurs only on the left-hand side or right-hand side, use lhs or rhs instead.

- The operator %in% means that at least one of the items must be found in the list you defined. If you wanted any rules matching either berries or yogurt, you could write items %in% c("berries", "yogurt").

- Additional operators are available for partial matching (%pin%) and complete matching (%ain%). Partial matching allows you to find both citrus fruit and tropical fruit using one search: items %pin% "fruit". Complete matching requires that all listed items are present. For instance, items %ain% c("berries", "yogurt") finds only rules with both berries and yogurt.

- Subsets can also be limited by support, confidence, or lift. For instance, confidence > 0.50 would limit the rules to those with confidence greater than 50 percent.

- Matching criteria can be combined with standard R logical operators such as AND (&), OR (|), and NOT (!).

Using these options, you can limit the selection of rules to be as specific or general as you would like.

Saving association rules to a file or data frame

To share the results of your market basket analysis, you can save the rules to a CSV file with the write() function. This will produce a CSV file that can be used in most spreadsheet programs, including Microsoft Excel:

```
> write(groceryrules, file = "groceryrules.csv",
        sep = ",", quote = TRUE, row.names = FALSE)
```

Sometimes it is also convenient to convert the rules into an R data frame. This can be accomplished using the `as()` function, as follows:

```
> groceryrules_df <- as(groceryrules, "data.frame")
```

This creates a data frame with the rules in factor format, and numeric vectors for support, confidence, and lift:

```
> str(groceryrules_df)
'data.frame':     463 obs. of 4 variables:
 $ rules      : Factor w/ 463 levels "{baking powder} => {other
vegetables}",..: 340 302 207 206 208 341 402 21 139 140 ...
 $ support    : num   0.00691 0.0061 0.00702 0.00773 0.00773 ...
 $ confidence : num   0.4 0.405 0.431 0.475 0.475 ...
 $ lift       : num   1.57 1.59 3.96 2.45 1.86 ...
```

Saving the rules to a data frame may be useful if you want to perform additional processing on the rules or need to export them to another database.

Summary

Association rules are used to find useful insight in the massive transaction databases of large retailers. As an unsupervised learning process, association rule learners are capable of extracting knowledge from large databases without any prior knowledge of what patterns to seek. The catch is that it takes some effort to reduce the wealth of information into a smaller and more manageable set of results. The Apriori algorithm, which we studied in this chapter, does so by setting minimum thresholds of interestingness, and reporting only the associations meeting these criteria.

We put the Apriori algorithm to work while performing a market basket analysis for a month's worth of transactions at a modestly sized supermarket. Even in this small example, a wealth of associations was identified. Among these, we noted several patterns that may be useful for future marketing campaigns. The same methods we applied are used at much larger retailers on databases many times this size, and can also be applied to projects outside of a retail setting.

In the next chapter, we will examine another unsupervised learning algorithm. Just like association rules, it is intended to find patterns within data. But unlike association rules that seek groups of related items or features, the methods in the next chapter are concerned with finding connections and relationships among the examples.

9
Finding Groups of Data – Clustering with k-means

Have you ever spent time watching a crowd? If so, you are likely to have seen some recurring personalities. Perhaps a certain type of person, identified by a freshly pressed suit and a briefcase, comes to typify the "fat cat" business executive. A 20-something wearing skinny jeans, a flannel shirt, and sunglasses might be dubbed a "hipster," while a woman unloading children from a minivan may be labeled a "soccer mom."

Of course, these types of stereotypes are dangerous to apply to individuals, as no two people are exactly alike. Yet, understood as a way to describe a collective, the labels capture some underlying aspect of similarity shared among the individuals within the group.

As you will soon learn, the act of clustering, or spotting patterns in data, is not much different from spotting patterns in groups of people. This chapter describes:

- The ways clustering tasks differ from the classification tasks we examined previously
- How clustering defines a group and how such groups are identified by k-means, a classic and easy-to-understand clustering algorithm
- The steps needed to apply clustering to a real-world task of identifying marketing segments among teenage social media users

Before jumping into action, we'll begin by taking an in-depth look at exactly what clustering entails.

Understanding clustering

Clustering is an unsupervised machine learning task that automatically divides the data into **clusters**, or groups of similar items. It does this without having been told how the groups should look ahead of time. As we may not even know what we're looking for, clustering is used for knowledge discovery rather than prediction. It provides an insight into the natural groupings found within data.

Without advanced knowledge of what comprises a cluster, how can a computer possibly know where one group ends and another begins? The answer is simple: clustering is guided by the principle that items inside a cluster should be very similar to each other, but very different from those outside. The definition of similarity might vary across applications, but the basic idea is always the same: group the data such that related elements are placed together.

The resulting clusters can then be used for action. For instance, you might find clustering methods employed in applications such as:

- Segmenting customers into groups with similar demographics or buying patterns for targeted marketing campaigns
- Detecting anomalous behavior, such as unauthorized network intrusions, by identifying patterns of use falling outside known clusters
- Simplifying extremely large datasets by grouping similar feature values into a smaller number of homogeneous categories

Overall, clustering is useful whenever diverse and varied data can be exemplified by a much smaller number of groups. It results in meaningful and actionable data structures that reduce complexity and provide insight into patterns of relationships.

Clustering as a machine learning task

Clustering is somewhat different from the classification, numeric prediction, and pattern detection tasks we've examined so far. In each of these tasks, the goal is to build a model that relates features to an outcome, or to relate some features to other features. Each of these tasks describes existing patterns within data. In contrast, the goal of clustering is to create new data. In clustering, unlabeled examples are given a new cluster label that has been inferred entirely from the relationships within the data. For this reason, you will sometimes see a clustering task referred to as **unsupervised classification** because, in a sense, it classifies unlabeled examples.

The catch is that the class labels obtained from an unsupervised classifier are without intrinsic meaning. Clustering will tell you which groups of examples are closely related—for instance, it might return groups A, B, and C—but it's up to you to apply an actionable and meaningful label. To see how this impacts the clustering task, let's consider a hypothetical example.

Suppose you were organizing a conference on the topic of data science. To facilitate professional networking and collaboration, you planned to seat people in groups according to one of three research specialties: computer science, math and statistics, and machine learning. Unfortunately, after sending out the conference invitations, you realize that you had forgotten to include a survey asking which discipline the attendee would prefer to be seated with.

In a stroke of brilliance, you realize that you might be able to infer each scholar's research specialty by examining his or her publication history. To this end, you begin collecting data on the number of articles each attendee published in computer science-related journals and the number of articles published in math or statistics-related journals. Using the data collected for the scholars, you create a scatterplot:

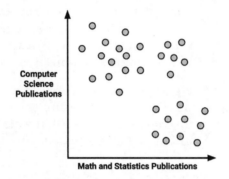

Figure 9.1: Visualizing scholars by their math and computer science publication data

As expected, there seems to be a pattern. We might guess that the upper-left corner, which represents people with many computer science publications but few articles on math, could be a cluster of computer scientists. Following this logic, the lower-right corner might be a group of mathematicians. Similarly, the upper-right corner, those with both math and computer science experience, may be machine learning experts. Applying these labels results in the following visualization:

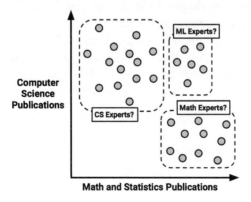

Figure 9.2: Clusters can be identified based on presumptions about the scholars in each group

Our groupings were formed visually; we simply identified clusters as closely grouped data points. Yet, in spite of the seemingly obvious groupings, without personally asking each scholar about his/her academic specialty, we have no way to know whether the groups are truly homogeneous. The labels are qualitative, presumptive judgments about the types of people in each group, based on a limited set of quantitative data.

Rather than defining the group boundaries subjectively, it would be nice to use machine learning to define them objectively. Given the axis-parallel splits in the previous figure, our problem seems like an obvious application for decision trees, as described in *Chapter 5, Divide and Conquer – Classification Using Decision Trees and Rules*. This would provide us with a clean rule like "if a scholar has few math publications, then he/she is a computer science expert." Unfortunately, there's a problem with this plan. Without data on the true class value for each point, a supervised learning algorithm would have no ability to learn such a pattern, as it would have no way of knowing what splits would result in homogenous groups.

On the other hand, clustering algorithms use a process very similar to what we did by visually inspecting the scatterplot. Using a measure of how closely the examples are related, homogeneous groups can be identified. In the next section, we'll start looking at how clustering algorithms are implemented.

This example highlights an interesting application of clustering. If you begin with unlabeled data, you can use clustering to create class labels. From there, you could apply a supervised learner such as decision trees to find the most important predictors of these classes. This is called **semi-supervised learning**.

The k-means clustering algorithm

The **k-means algorithm** is perhaps the most often used clustering method. Having been studied for several decades, it serves as the foundation for many more sophisticated clustering techniques. If you understand the simple principles it uses, you will have the knowledge needed to understand nearly any clustering algorithm in use today. Many such methods are listed on the following site, the CRAN task view for clustering:

```
http://cran.r-project.org/web/views/Cluster.html
```

As k-means has evolved over time, there are many implementations of the algorithm. One popular approach is described in *A k-means clustering algorithm, Hartigan, JA, Wong, MA, Applied Statistics, 1979, Vol. 28, pp. 100-108.*

Even though clustering methods have advanced since the inception of k-means, this is not to imply that k-means is obsolete. In fact, the method may be more popular now than ever. The following table lists some reasons why k-means is still used widely:

Strengths	Weaknesses
• Uses simple principles that can be explained in non-statistical terms	• Not as sophisticated as more modern clustering algorithms
• Highly flexible and can be adapted with simple adjustments to address nearly all of its shortcomings	• Because it uses an element of random chance, it is not guaranteed to find the optimal set of clusters
• Performs well enough under many real-world use cases	• Requires a reasonable guess as to how many clusters naturally exist in the data
	• Not ideal for non-spherical clusters or clusters of widely varying density

If the name k-means sounds familiar to you, you may be recalling the k-nearest neighbors (k-NN) algorithm presented in *Chapter 3, Lazy Learning – Classification Using Nearest Neighbors.* As you will soon see, k-means shares more in common with k-NN than just the letter k.

The k-means algorithm assigns each of the n examples to one of the k clusters, where k is a number that has been determined ahead of time. The goal is to minimize the differences within each cluster and maximize the differences between clusters.

Unless k and n are extremely small, it is not feasible to compute the optimal clusters across all possible combinations of examples. Instead, the algorithm uses a heuristic process that finds **locally optimal** solutions. Put simply, this means that it starts with an initial guess for the cluster assignments then modifies the assignments slightly to see if the changes improve the homogeneity within the clusters.

We will cover the process in depth shortly, but the algorithm essentially involves two phases. First, it assigns examples to an initial set of k clusters. Then, it updates the assignments by adjusting the cluster boundaries according to the examples that currently fall into the cluster. The process of updating and assigning occurs several times until changes no longer improve the cluster fit. At this point, the process stops and the clusters are finalized.

Due to the heuristic nature of k-means, you may end up with somewhat different final results by making only slight changes to the starting conditions. If the results vary dramatically, this could indicate a problem. For instance, the data may not have natural groupings, or the value of *k* has been poorly chosen. With this in mind, it's a good idea to try a cluster analysis more than once to test the robustness of your findings.

To see how the process of assigning and updating works in practice, let's revisit the case of the hypothetical data science conference. Though this is a simple example, it will illustrate the basics of how k-means operates under the hood.

Using distance to assign and update clusters

As with k-NN, k-means treats feature values as coordinates in a multidimensional feature space. For the conference data, there are only two features, so we can represent the feature space as a two-dimensional scatterplot as depicted previously.

The k-means algorithm begins by choosing *k* points in the feature space to serve as the cluster centers. These centers are the catalyst that spurs the remaining examples to fall into place. Often, the points are chosen by selecting *k* random examples from the training dataset. Because we hope to identify three clusters, using this method, *k* = 3 points will be selected at random. These points are indicated by the star, triangle, and diamond in the following diagram:

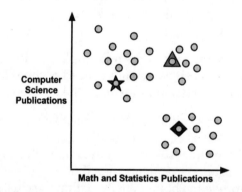

Figure 9.3: k-means clustering begins by selecting k random cluster centers

It's worth noting that although the three cluster centers in the preceding diagram happen to be widely spaced apart, this is not always necessarily the case. Since they are selected at random, the three centers could have just as easily been three adjacent points. Since the k-means algorithm is highly sensitive to the starting position of the cluster centers, this means that random chance may have a substantial impact on the final set of clusters.

To address this problem, k-means can be modified to use different methods for choosing the initial centers. For example, one variant chooses random values occurring anywhere in the feature space (rather than only selecting among values observed in the data). Another option is to skip this step altogether; by randomly assigning each example to a cluster, the algorithm can jump ahead immediately to the update phase. Each of these approaches adds a particular bias to the final set of clusters, which you may be able to use to improve your results.

> In 2007, an algorithm called **k-means++** was introduced, which proposes an alternative method for selecting the initial cluster centers. It purports to be an efficient way to get much closer to the optimal clustering solution while reducing the impact of random chance. For more information, see *k-means++: The advantages of careful seeding*, Arthur, D, Vassilvitskii, S, Proceedings of the eighteenth annual ACM-SIAM symposium on discrete algorithms, 2007, pp. 1027–1035.

After choosing the initial cluster centers, the other examples are assigned to the cluster center that is nearest according to the distance function. You will remember that we studied distance functions while learning about k-NN. Traditionally, k-means uses Euclidean distance, but Manhattan distance or Minkowski distance are also sometimes used.

Recall that if n indicates the number of features, the formula for Euclidean distance between example x and example y is as follows:

$$dist(x, y) = \sqrt{\sum_{i=1}^{n}(x_i - y_i)^2}$$

For instance, if we are comparing a guest with five computer science publications and one math publication to a guest with zero computer science papers and two math papers, we could compute this in R as:

```
> sqrt((5 - 0)^2 + (1 - 2)^2)
[1] 5.09902
```

Using this distance function, we find the distance between each example and each cluster center. The example is then assigned to the nearest cluster center.

> Keep in mind that because we are using distance calculations, all the features need to be numeric, and the values should be normalized to a standard range ahead of time. The methods presented in *Chapter 3, Lazy Learning – Classification Using Nearest Neighbors*, will prove helpful for this task.

As shown in the following figure, the three cluster centers partition the examples into three segments labeled **Cluster A**, **Cluster B**, and **Cluster C**. The dashed lines indicate the boundaries for the **Voronoi diagram** created by the cluster centers. The Voronoi diagram indicates the areas that are closer to one cluster center than any other; the vertex where all three boundaries meet is the maximal distance from all three cluster centers. Using these boundaries, we can easily see the regions claimed by each of the initial k-means seeds:

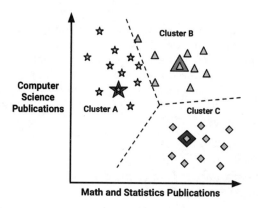

Figure 9.4: The initial cluster centers create three groups of "nearest" points

Now that the initial assignment phase has been completed, the k-means algorithm proceeds to the update phase. The first step of updating the clusters involves shifting the initial centers to a new location, known as the **centroid**, which is calculated as the average position of the points currently assigned to that cluster. The following figure illustrates how as the cluster centers shift to the new centroids, the boundaries in the Voronoi diagram also shift, and a point that was once in **Cluster B** (indicated by an arrow) is added to **Cluster A**:

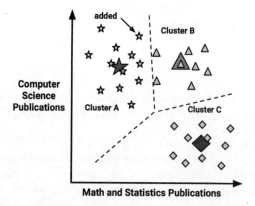

Figure 9.5: The update phase shifts the cluster centers, which causes the reassignment of one point

As a result of this reassignment, the k-means algorithm will continue through another update phase. After shifting the cluster centroids, updating the cluster boundaries, and reassigning points into new clusters (as indicated by arrows), the figure looks like this:

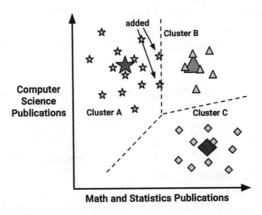

Figure 9.6: After another update, two more points are reassigned to the nearest cluster center

Because two more points were reassigned, another update must occur, which moves the centroids and updates the cluster boundaries. However, because these changes result in no reassignments, the k-means algorithm stops. The cluster assignments are now final:

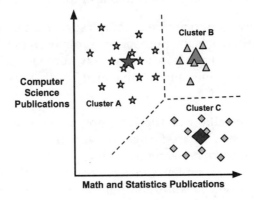

Figure 9.7: Clustering stops after the update phase results in no new cluster assignments

The final clusters can be reported in one of two ways. First, you might simply report the cluster assignments of A, B, or C for each example. Alternatively, you could report the coordinates of the cluster centroids after the final update. Given either reporting method, you are able to compute the other by calculating the centroids using the coordinates of each cluster's examples, or assigning each example to its nearest cluster centroid.

Choosing the appropriate number of clusters

In the introduction to k-means, we learned that the algorithm is sensitive to the randomly-chosen cluster centers. Indeed, if we had selected a different combination of three starting points in the previous example, we may have found clusters that split the data differently from what we had expected. Similarly, k-means is sensitive to the number of clusters; the choice requires a delicate balance. Setting the k to be very large will improve the homogeneity of the clusters and, at the same time, it risks overfitting the data.

Ideally, you will have *a priori* knowledge (a prior belief) about the true groupings and you can apply this information to choosing the number of clusters. For instance, if you were clustering movies, you might begin by setting k equal to the number of genres considered for the Academy Awards. In the data science conference seating problem that we worked through previously, k might reflect the number of academic fields of study that were invited.

Sometimes the number of clusters is dictated by business requirements or the motivation for the analysis. For example, the number of tables in the meeting hall might dictate how many groups of people should be created from the data science attendee list.

Extending this idea to another business case, if the marketing department only has resources to create three distinct advertising campaigns, it might make sense to set $k = 3$ to assign all the potential customers to one of the three appeals.

Without any prior knowledge, one rule of thumb suggests setting k equal to the square root of $(n / 2)$, where n is the number of examples in the dataset. However, this rule of thumb is likely to result in an unwieldy number of clusters for large datasets. Luckily, there are other quantitative methods that can assist in finding a suitable k-means cluster set.

A technique known as the **elbow method** attempts to gauge how the homogeneity or heterogeneity within the clusters changes for various values of k. As illustrated in the following diagrams, the homogeneity within clusters is expected to increase as additional clusters are added; similarly, heterogeneity will also continue to decrease with more clusters. Because you could continue to see improvements until each example is in its own cluster, the goal is not to maximize homogeneity or minimize heterogeneity endlessly, but rather to find k such that there are diminishing returns beyond that value. This value of k is known as the **elbow point**, because it looks like an elbow.

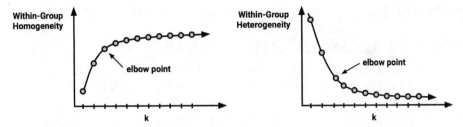

Figure 9.8: The elbow is the point at which increasing k results in relatively small improvements

There are numerous statistics to measure homogeneity and heterogeneity within clusters that can be used with the elbow method (the following information box provides a citation for more detail). Still, in practice, it is not always feasible to iteratively test a large number of *k* values. This is in part because clustering large datasets can be fairly time consuming; clustering the data repeatedly is even worse. Furthermore, applications requiring the exact optimal set of clusters are fairly rare. In most clustering applications, it suffices to choose a *k* value based on convenience rather than strict performance requirements.

 For a very thorough review of the vast assortment of cluster performance measures, refer to *On Clustering Validation Techniques, Halkidi, M, Batistakis, Y, Vazirgiannis, M, Journal of Intelligent Information Systems, 2001, Vol. 17, pp. 107-145.*

The process of setting *k* itself can sometimes lead to interesting insights. By observing how the characteristics of the clusters change as *k* is varied, one might infer where the data have naturally defined boundaries. Groups that are more tightly clustered will change very little, while less homogeneous groups will form and disband over time.

In general, it may be wise to spend little time worrying about getting *k* exactly right. The next example will demonstrate how even a tiny bit of subject-matter knowledge borrowed from a Hollywood film can be used to set *k* such that actionable and interesting clusters are found. As clustering is unsupervised, the task is really about what you make of it; the value is in the insights you take away from the algorithm's findings.

Finding teen market segments using k-means clustering

Interacting with friends on a **social networking service** (**SNS**) such as Facebook, Tumblr, and Instagram has become a rite of passage for teenagers around the world. Having a relatively large amount of disposable income, these adolescents are a coveted demographic for businesses hoping to sell snacks, beverages, electronics, and hygiene products.

The many millions of teenage consumers using such sites have attracted the attention of marketers struggling to find an edge in an increasingly competitive market. One way to gain this edge is to identify segments of teenagers who share similar tastes, so that clients can avoid targeting advertisements to teens with no interest in the product being sold. For instance, sporting apparel is likely to be a difficult sell to teens with no interest in sports.

Given the text of teenagers' SNS pages, we can identify groups that share common interests such as sports, religion, or music. Clustering can automate the process of discovering the natural segments in this population. However, it will be up to us to decide whether or not the clusters are interesting and how we can use them for advertising. Let's try this process from start to finish.

Step 1 – collecting data

For this analysis, we will be using a dataset representing a random sample of 30,000 US high school students who had profiles on a well-known SNS in 2006. To protect the users' anonymity, the SNS will remain unnamed. However, at the time the data was collected, the SNS was a popular web destination for US teenagers. Therefore, it is reasonable to assume that the profiles represent a fairly wide cross section of American adolescents in 2006.

I compiled this dataset while conducting my own sociological research on teenage identities at the University of Notre Dame. If you use the data for research purposes, please cite this book chapter. The full dataset is available at the Packt Publishing website with the filename snsdata. csv. To follow along interactively, this chapter assumes you have saved this file to your R working directory.

The data was sampled evenly across four high school graduation years (2006 through to 2009) representing the senior, junior, sophomore, and freshman classes at the time of data collection. Using an automated web crawler, the full text of the SNS profiles were downloaded, and each teen's gender, age, and number of SNS friends was recorded.

A text mining tool was used to divide the remaining SNS page content into words. From the top 500 words appearing across all pages, 36 words were chosen to represent five categories of interests: extracurricular activities, fashion, religion, romance, and antisocial behavior. The 36 words include terms such as *football*, *sexy*, *kissed*, *bible*, *shopping*, *death*, and *drugs*. The final dataset indicates, for each person, how many times each word appeared in the person's SNS profile.

Step 2 – exploring and preparing the data

We can use the default settings of read.csv() to load the data into a data frame:

```
> teens <- read.csv("snsdata.csv")
```

Let's also take a quick look at the specifics of the data. The first several lines of the str() output are as follows:

```
> str(teens)
'data.frame':   30000 obs. of  40 variables:
 $ gradyear    : int  2006 2006 2006 2006 2006 2006 2006 2006 ...
 $ gender      : Factor w/ 2 levels "F","M": 2 1 2 1 NA 1 1 2 ...
 $ age         : num  19 18.8 18.3 18.9 19 ...
 $ friends     : int  7 0 69 0 10 142 72 17 52 39 ...
 $ basketball  : int  0 0 0 0 0 0 0 0 0 0 ...
```

As we had expected, the data include 30,000 teenagers with four variables indicating personal characteristics and 36 words indicating interests.

Do you notice anything strange around the gender row? If you were looking carefully, you may have noticed the NA value, which is out of place compared to the 1 and 2 values. The NA is R's way of telling us that the record has a **missing value**—we do not know the person's gender. Until now, we haven't dealt with missing data, but it can be a significant problem for many types of analyses.

Let's see how substantial this problem is. One option is to use the table() command, as follows:

```
> table(teens$gender)

    F     M
22054  5222
```

Although this tells us how many F and M values are present, the table() function excluded the NA values rather than treating it as a separate category. To include the NA values (if there are any), we simply need to add an additional parameter:

```
> table(teens$gender, useNA = "ifany")

    F     M  <NA>
22054  5222  2724
```

Here, we see that 2,724 records (nine percent) have missing gender data. Interestingly, there are over four times as many females as males in the SNS data, suggesting that males are not as inclined to use this SNS as females.

If you examine the other variables in the data frame, you will find that besides gender, only age has missing values. For numeric data, the summary() command tells us the number of missing values:

```
> summary(teens$age)
   Min. 1st Qu.  Median    Mean 3rd Qu.    Max.    NA's
  3.086  16.310  17.290  17.990  18.260 106.900    5086
```

A total of 5,086 records (17 percent) have missing ages. Also concerning is the fact that the minimum and maximum values seem to be unreasonable: it is unlikely that a three-year-old or a 106-year-old is attending high school. To ensure that these extreme values don't cause problems for the analysis, we'll need to clean them up before moving on.

A more reasonable range of ages for high school students includes those who are at least 13 years old and not yet 20 years old. Any age value falling outside this range should be treated the same as missing data—we cannot trust the age provided. To recode the age variable, we can use the ifelse() function, assigning teen$age the value of teen$age if the age is at least 13 and less than 20 years; otherwise, it will receive the value NA:

```
> teens$age <- ifelse(teens$age >= 13 & teens$age < 20,
                      teens$age, NA)
```

By rechecking the summary() output, we see that the age range now follows a distribution that looks much more like an actual high school:

```
> summary(teens$age)
   Min. 1st Qu.  Median    Mean 3rd Qu.    Max.    NA's
  13.03   16.30   17.26   17.25   18.22   20.00    5523
```

Unfortunately, now we've created an even larger missing data problem. We'll need to find a way to deal with these values before continuing with our analysis.

Data preparation – dummy coding missing values

An easy solution for handling missing values is to exclude any record with a missing value. However, if you think through the implications of this practice, you might think twice before doing so—just because it is easy does not mean it is a good idea! The problem with this approach is that even if the missingness is not extensive, you can easily exclude large portions of the data.

For example, suppose that in our data the people with NA values for gender are completely different from those with missing age data. This would imply that by excluding those missing either gender or age, you would exclude *9% + 17% = 26%* of the data, or over 7,500 records. And this is for missing data on only two variables! The larger the number of missing values present in a dataset, the more likely it is that any given record will be excluded. Fairly soon, you will be left with a tiny subset of data, or worse, the remaining records will be systematically different or non-representative of the full population.

An alternative solution for categorical data like gender is to treat a missing value as a separate category. For instance, rather than limiting to female and male, we can add an additional category for unknown gender. This allows us to utilize dummy coding, which was covered in *Chapter 3, Lazy Learning – Classification Using Nearest Neighbors.*

If you recall, dummy coding involves creating a separate binary (1 or 0) valued dummy variable for each level of a nominal feature except one, which is held out to serve as the reference group. The reason one category can be excluded is because its status can be inferred from the other categories. For instance, if someone is not female and not unknown gender, they must be male. Therefore, in this case, we need to only create dummy variables for female and unknown gender:

```
> teens$female <- ifelse(teens$gender == "F" &
                          !is.na(teens$gender), 1, 0)
> teens$no_gender <- ifelse(is.na(teens$gender), 1, 0)
```

As you might expect, the is.na() function tests whether the gender is equal to NA. Therefore, the first statement assigns teens$female the value 1 if the gender is equal to F and the gender is not equal to NA, otherwise it assigns the value 0. In the second statement, if is.na() returns TRUE, meaning the gender is missing, then the teens$no_gender variable is assigned 1, otherwise it is assigned the value 0. To confirm that we did the work correctly, let's compare our constructed dummy variables to the original gender variable:

```
> table(teens$gender, useNA = "ifany")
    F     M   <NA>
22054  5222   2724
> table(teens$female, useNA = "ifany")
    0     1
 7946 22054
> table(teens$no_gender, useNA = "ifany")
    0     1
27276  2724
```

The number of 1 values for teens$female and teens$no_gender matches the number of F and NA values respectively, so we should be able to trust our work.

Data preparation – imputing the missing values

Next, let's eliminate the 5,523 missing age values. As age is numeric, it doesn't make sense to create an additional category for unknown values—where would you rank "unknown" relative to the other ages? Instead, we'll use a different strategy known as **imputation**, which involves filling in the missing data with a guess as to the true value.

Can you think of a way we might be able to use the SNS data to make an informed guess about a teenager's age? If you are thinking of using the graduation year, you've got the right idea. Most people in a graduation cohort were born within a single calendar year. If we can identify the typical age for each cohort, then we will have a fairly reasonable estimate of the age of a student in that graduation year.

One way to find a typical value is by calculating the average, or mean, value. If we try to apply the mean() function as we have done for previous analyses, there's a problem:

```
> mean(teens$age)
[1] NA
```

The issue is that the mean value is undefined for a vector containing missing data. As our age data contains missing values, mean(teens$age) returns a missing value. We can correct this by adding an additional parameter to remove the missing values before calculating the mean:

```
> mean(teens$age, na.rm = TRUE)
[1] 17.25243
```

This reveals that the average student in our data is about 17 years old. This only gets us part of the way there; we actually need the average age for each graduation year. You might first attempt to calculate the mean four times, but one of the benefits of R is that there's usually a way to avoid repeating oneself. In this case, the aggregate() function is the tool for the job. It computes statistics for subgroups of data. Here, it calculates the mean age by graduation year after removing the NA values:

```
> aggregate(data = teens, age ~ gradyear, mean, na.rm = TRUE)
  gradyear        age
1    2006 18.65586
2    2007 17.70617
3    2008 16.76770
4    2009 15.81957
```

The mean age differs by roughly one year per change in graduation year. This is not at all surprising, but is a nice confirmation that our data is reasonable.

The `aggregate()` output is in a data frame. This would require extra work to merge back onto our original data. As an alternative, we can use the `ave()` function, which returns a vector with the group means repeated such that the result is equal in length to the original vector:

```
> ave_age <- ave(teens$age, teens$gradyear, FUN =
                 function(x) mean(x, na.rm = TRUE))
```

To impute these means onto the missing values, we need one more `ifelse()` call to use the `ave_age` value only if the original age value was `NA`:

```
> teens$age <- ifelse(is.na(teens$age), ave_age, teens$age)
```

The `summary()` results show that the missing values have now been eliminated:

```
> summary(teens$age)
   Min. 1st Qu.  Median    Mean 3rd Qu.    Max.
  13.03   16.28   17.24   17.24   18.21   20.00
```

With the data ready for analysis, we are ready to dive into the interesting part of this project. Let's see if our efforts have paid off.

Step 3 – training a model on the data

To cluster the teenagers into marketing segments, we will use an implementation of k-means in the `stats` package, which should be included in your R installation by default. If by chance you do not have this package, you can install it as you would any other package and load it using the `library(stats)` command.

Although there is no shortage of k-means functions available in various R packages, the kmeans() function in the stats package is widely used and provides a vanilla implementation of the algorithm.

Clustering syntax

using the **kmeans()** function in the **stats** package

Finding clusters:

```
myclusters <- kmeans(mydata, k)
```

- **mydata** is a matrix or data frame with the examples to be clustered
- **k** specifies the desired number of clusters

The function will return a cluster object that stores information about the clusters.

Examining clusters:

- **myclusters$cluster** is a vector of cluster assignments from the **kmeans()** function
- **myclusters$centers** is a matrix indicating the mean values for each feature and cluster combination
- **myclusters$size** lists the number of examples assigned to each cluster

Example:

```
teen_clusters <- kmeans(teens, 5)
teens$cluster_id <- teen_clusters$cluster
```

The kmeans() function requires a data frame containing only numeric data and a parameter specifying the desired number of clusters. If you have these two things ready, the actual process of building the model is simple. The trouble is that choosing the right combination of data and clusters can be a bit of an art; sometimes a great deal of trial and error is involved.

We'll start our cluster analysis by considering only the 36 features that represent the number of times various interests appeared on the teenager SNS profiles. For convenience, let's make a data frame containing only these features:

```
> interests <- teens[5:40]
```

If you recall from *Chapter 3, Lazy Learning – Classification Using Nearest Neighbors*, a common practice employed prior to any analysis using distance calculations is to normalize or z-score standardize the features such that each utilizes the same range. By doing so, you can avoid a problem in which some features come to dominate solely because they have a larger range of values than the others.

The process of z-score standardization rescales features such that they have a mean of zero and a standard deviation of one. This transformation changes the interpretation of the data in a way that may be useful here.

Specifically, if someone mentions basketball three times on their profile, without additional information, we have no idea whether this implies they like basketball more or less than their peers. On the other hand, if the z-score is three, we know that they mentioned basketball many more times than the average teenager.

To apply z-score standardization to the `interests` data frame, we can use the `scale()` function with `lapply()`. Since `lapply()` returns a matrix, it must be coerced back to data frame form using the `as.data.frame()` function, as follows:

```
> interests_z <- as.data.frame(lapply(interests, scale))
```

To confirm that the transformation worked correctly, we can compare the summary statistics of the `basketball` column in the old and new interests data:

```
> summary(interests$basketball)
   Min. 1st Qu.  Median    Mean 3rd Qu.    Max.
 0.0000  0.0000  0.0000  0.2673  0.0000 24.0000
> summary(interests_z$basketball)
   Min. 1st Qu.  Median    Mean 3rd Qu.    Max.
-0.3322 -0.3322 -0.3322  0.0000 -0.3322 29.4923
```

As expected, the `interests_z` dataset transformed the `basketball` feature to have a mean of zero and a range that spans above and below zero. Now, a value less than zero can be interpreted as a person having fewer-than-average mentions of basketball in their profile. A value greater than zero implies that the person mentioned basketball more frequently than the average.

Our last decision involves deciding how many clusters to use for segmenting the data. If we use too many clusters, we may find them too specific to be useful; conversely, choosing too few may result in heterogeneous groupings. You should feel comfortable experimenting with the value of *k*. If you don't like the result, you can easily try another value and start over.

 Choosing the number of clusters is easier if you are familiar with the analysis population. Having a hunch about the true number of natural groupings can save you some trial and error.

To help us choose the number of clusters in the data, I'll defer to one of my favorite films, *The Breakfast Club*, a coming-of-age comedy released in 1985 and directed by John Hughes. The teenage characters in this movie are identified in terms of five stereotypes: a brain, an athlete, a basket case, a princess, and a criminal. Given that these identities prevail throughout popular teenage fiction, five seems like a reasonable starting point for *k*.

To use the k-means algorithm to divide the teenagers' interest data into five clusters, we use the kmeans() function on the interests data frame. Because the k-means algorithm utilizes random starting points, the set.seed() function is used to ensure that the results match the output in the examples that follow. If you recall from previous chapters, this command initializes R's random number generator to a specific sequence. In the absence of this statement, the results may vary each time the k-means algorithm is run.

```
> RNGversion("3.5.2")
> set.seed(2345)
> teen_clusters <- kmeans(interests_z, 5)
```

The result of the k-means clustering process is a list named teen_clusters that stores the properties of each of the five clusters. Let's dig in and see how well the algorithm has divided the teenagers' interest data.

> If you find that your results differ from those shown here, ensure that the RNGversion("3.5.2") and set.seed(2345) commands are run immediately prior to the kmeans() function.

Step 4 – evaluating model performance

Evaluating clustering results can be somewhat subjective. Ultimately, the success or failure of the model hinges on whether the clusters are useful for their intended purpose. As the goal of this analysis was to identify clusters of teenagers with similar interests for marketing purposes, we will largely measure our success in qualitative terms. For other clustering applications, more quantitative measures of success may be needed.

One of the most basic ways to evaluate the utility of a set of clusters is to examine the number of examples falling in each of the groups. If the groups are too large or too small, then they are not likely to be very useful. To obtain the size of the kmeans() clusters, use the teen_clusters$size component as follows:

```
> teen_clusters$size
[1]    871    600   5981   1034 21514
```

Here we see the five clusters we requested. The smallest cluster has 600 teenagers (two percent) while the largest has 21,514 (72 percent). Although the large gap between the number of people in the largest and smallest clusters is slightly concerning, without examining these groups more carefully, we will not know whether or not this indicates a problem.

It may be the case that the clusters' size disparity indicates something real, such as a big group of teenagers who share similar interests, or it may be a random fluke caused by the initial k-means cluster centers. We'll know more as we start to look at each cluster's homogeneity.

Sometimes k-means may find extremely small clusters—occasionally as small as a single point. This can happen if one of the initial cluster centers happens to fall on an outlier far from the rest of the data. It is not always clear whether to treat such small clusters as a true finding that represents a cluster of extreme cases, or a problem caused by random chance. If you encounter this issue, it may be worth re-running the k-means algorithm with a different random seed to see whether the small cluster is robust to different starting points.

For a more in-depth look at the clusters, we can examine the coordinates of the cluster centroids using the `teen_clusters$centers` component, which is as follows for the first four interests:

```
> teen_clusters$centers
    basketball    football      soccer      softball
1   0.16001227   0.2364174   0.10385512   0.07232021
2  -0.09195886   0.0652625  -0.09932124  -0.01739428
3   0.52755083   0.4873480   0.29778605   0.37178877
4   0.34081039   0.3593965   0.12722250   0.16384661
5  -0.16695523  -0.1641499  -0.09033520  -0.11367669
```

The rows of the output (labeled 1 to 5) refer to the five clusters, while the numbers across each row indicate the cluster's average value for the interest listed at the top of the column. Because the values are z-score standardized, positive values are above the overall mean level for all teenagers and negative values are below the overall mean. For example, the third row has the highest value in the `basketball` column, which means that cluster 3 has the highest average interest in basketball among all the clusters.

By examining whether clusters fall above or below the mean level for each interest category, we can begin to notice patterns that distinguish the clusters from one another. In practice, this involves printing the cluster centers and searching through them for any patterns or extreme values, much like a word search puzzle but with numbers. The following annotated screenshot shows a highlighted pattern for each of the five clusters, for 19 of the 36 teenager interests:

```
> teen_clusters$centers
    basketball   football    soccer    softball  volleyball    swimming
1   0.16001227  0.2364174  0.10385512  0.07232021  0.18897158  0.23970234
2  -0.09195886  0.0652625 -0.09932124 -0.01739428 -0.06219308  0.03339844
3   0.52755083  0.4873480  0.29778605  0.37178877  0.37986175  0.29628671
4   0.34081039  0.3593965  0.12722250  0.16384661  0.11032200  0.26943332
5  -0.16695523 -0.1641499 -0.09033520 -0.11367669 -0.11682181 -0.10595448
    cheerleading    baseball      tennis      sports        cute         sex
1      0.3931445  0.02993479  0.13532387  0.10257837  0.37884271  0.020042068
2     -0.1101103 -0.11487510  0.04062204 -0.09899231 -0.03265037 -0.042486141
3      0.3303485  0.35231971  0.14057808  0.32967130  0.54442929  0.002913623
4      0.1856664  0.27527088  0.10980958  0.79711920  0.47866008  2.028471066
5     -0.1136077 -0.10918483 -0.05097057 -0.13135334 -0.18878627 -0.097928345
         sexy         hot      kissed        dance        band    marching       music
1  0.11740551  0.41389104  0.06787768  0.22780899 -0.10257102 -0.10942590  0.1378306
2 -0.04329091 -0.03812345 -0.04554933  0.04573186  4.06726666  5.25757242  0.4981238
3  0.24040196  0.38551819 -0.03356121  0.45662534 -0.02120728 -0.10880541  0.2844999
4  0.51266080  0.31708549  2.97973077  0.45535061  0.38053621 -0.02014608  1.1367885
5 -0.09501817 -0.13810894 -0.13535855 -0.15932739 -0.12167214 -0.11098063 -0.1532006
```

Figure 9.9: To distinguish clusters, it can be helpful to highlight patterns in their coordinates

Given this snapshot of the interest data, we can already infer some characteristics of the clusters. Cluster three is substantially above the mean interest level on all the sports, which suggests that this may be a group of *athletes* per the *Breakfast Club* stereotype. Cluster one includes the most mentions of "cheerleading" and the word "hot." Are these the so-called *princesses*?

By continuing to examine the clusters in this way, it is possible to construct a table listing the dominant interests of each of the groups. In the following table, each cluster is shown with the features that most distinguish it from the other clusters, and the *Breakfast Club* identity that most accurately captures the group's characteristics.

Interestingly, cluster five is distinguished by the fact that it is unexceptional: its members had lower-than-average levels of interest in every measured activity. It is also the single largest group in terms of the number of members. One potential explanation is that these users created a profile on the website but never posted any interests.

Cluster 1 (N = 3,376)	Cluster 2 (N = 601)	Cluster 3 (N = 1,036)	Cluster 4 (N = 3,279)	Cluster 5 (N = 21,708)
swimming cheerleading cute sexy hot dance dress hair mall hollister abercrombie shopping clothes	band marching music rock	sports sex sexy hot kissed dance music band die death drunk drugs	basketball football soccer softball volleyball baseball sports god church Jesus bible	???
Princesses	Brains	Criminals	Athletes	Basket Cases

Figure 9.10: A table can be used to list important dimensions of each cluster

When sharing the results of a segmentation analysis, it is often helpful to apply informative labels that simplify and capture the essence of the groups, such as the *Breakfast Club* typology applied here. The risk in adding such labels is that they can obscure the groups' nuances by stereotyping the group members. Because such labels can bias our thinking, important patterns can be missed if labels are taken as the whole truth.

Given the table, a marketing executive would have a clear depiction of five types of teenage visitors to the social networking website. Based on these profiles, the executive could sell targeted advertising impressions to businesses with products relevant to one or more of the clusters. In the next section, we will see how the cluster labels can be applied back to the original population for such uses.

Step 5 – improving model performance

Because clustering creates new information, the performance of a clustering algorithm depends at least somewhat on both the quality of the clusters themselves and what is done with that information. In the preceding section, we demonstrated that the five clusters provided useful and novel insights into the interests of teenagers. By that measure, the algorithm appears to be performing quite well. Therefore, we can now focus our effort on turning these insights into action.

We'll begin by applying the clusters back onto the full dataset. The `teen_clusters` object created by the `kmeans()` function includes a component named `cluster` that contains the cluster assignments for all 30,000 individuals in the sample. We can add this as a column on the `teens` data frame with the following command:

```
> teens$cluster <- teen_clusters$cluster
```

Given this new data, we can start to examine how the cluster assignment relates to individual characteristics. For example, here's the personal information for the first five teenagers in the SNS data:

```
> teens[1:5, c("cluster", "gender", "age", "friends")]
  cluster gender     age friends
1       5        M 18.982       7
2       3        F 18.801       0
3       5        M 18.335      69
4       5        F 18.875       0
5       4     <NA> 18.995      10
```

Using the `aggregate()` function, we can also look at the demographic characteristics of the clusters. The mean age does not vary much by cluster, which is not too surprising, as these teenager identities are often determined before high school. This is depicted as follows:

```
> aggregate(data = teens, age ~ cluster, mean)
  cluster       age
1       1 16.86497
2       2 17.39037
3       3 17.07656
4       4 17.11957
5       5 17.29849
```

On the other hand, there are some substantial differences in the proportion of females by cluster. This is a very interesting finding, as we didn't use gender data to create the clusters, yet the clusters are still predictive of gender:

```
> aggregate(data = teens, female ~ cluster, mean)
  cluster    female
1       1 0.8381171
2       2 0.7250000
3       3 0.8378198
4       4 0.8027079
5       5 0.6994515
```

Recall that overall about 74 percent of the SNS users are female. Cluster one, the so-called *princesses*, is nearly 84 percent female, while clusters two and five are only about 70 percent female. These disparities imply that there are differences in the interests that teenage boys and girls discuss on their social networking pages.

Given our success in predicting gender, you might suspect that the clusters are also predictive of the number of friends the users have. This hypothesis seems to be supported by the data, which is as follows:

```
> aggregate(data = teens, friends ~ cluster, mean)
   cluster   friends
1        1 41.43054
2        2 32.57333
3        3 37.16185
4        4 30.50290
5        5 27.70052
```

On average, *princesses* have the most friends (41.4), followed by *athletes* (37.2) and *brains* (32.6). On the low end are *criminals* (30.5) and *basket cases* (27.7). As with gender, the connection between a teenager's number of friends and their predicted cluster is remarkable given that we did not use the friendship data as an input to the clustering algorithm. Also interesting is the fact that the number of friends seems to be related to the stereotype of each cluster's high school popularity: the stereotypically popular groups tend to have more friends.

The association among group membership, gender, and number of friends suggests that the clusters can be useful predictors of behavior. Validating their predictive ability in this way may make the clusters an easier sell when they are pitched to the marketing team, ultimately improving the performance of the algorithm.

Summary

Our findings support the popular adage that "birds of a feather flock together." By using machine learning methods to cluster teenagers with others who have similar interests, we were able to develop a typology of teenage identities that was predictive of personal characteristics such as gender and number of friends. These same methods can be applied to other contexts with similar results.

This chapter covered only the fundamentals of clustering. There are many variants of the k-means algorithm, as well as many other clustering algorithms that bring unique biases and heuristics to the task. Based on the foundation in this chapter, you will be able to understand these clustering methods and apply them to new problems.

In the next chapter, we will begin to look at methods for measuring the success of a learning algorithm that are applicable across many machine learning tasks. While our process has always devoted some effort to evaluating the success of learning, in order to obtain the highest degree of performance, it is crucial to be able to define and measure it in the strictest terms.

10

Evaluating Model Performance

When only the wealthy could afford education, tests and exams did not evaluate the students. Instead, tests judged teachers for parents who wanted to know whether their children learned enough to justify the instructors' wages. Obviously, this is different today. Now, such evaluations are used to distinguish between high-achieving and low-achieving students, filtering them into careers and other opportunities.

Given the significance of this process, a great deal of effort is invested in developing accurate student assessments. Fair assessments have a large number of questions that cover a wide breadth of topics and reward true knowledge over lucky guesses. A good assessment also requires students to think about problems they have never faced before. Correct responses therefore reflect an ability to generalize knowledge more broadly.

The process of evaluating machine learning algorithms is very similar to the process of evaluating students. Since algorithms have varying strengths and weaknesses, tests should distinguish among the learners. It is also important to understand how a learner will perform on future data.

This chapter provides the information needed to assess machine learners, such as:

- The reasons why predictive accuracy is not sufficient to measure performance, and the performance measures you might use instead
- Methods to ensure that the performance measures reasonably reflect a model's ability to predict or forecast unseen cases
- How to use R to apply these more useful measures and methods to the predictive models covered in previous chapters

Just as the best way to learn a topic is to attempt to teach it to someone else, the process of teaching and evaluating machine learners will provide you with greater insight into the methods you've learned so far.

Measuring performance for classification

In the previous chapters, we measured classifier accuracy by dividing the number of correct predictions by the total number of predictions. This finds the proportion of cases in which the learner is right or wrong. For example, suppose that a classifier correctly predicted for 99,990 out of 100,000 newborn babies whether they were a carrier of a treatable but potentially fatal genetic defect. This would imply an accuracy of 99.99 percent and an error rate of only 0.01 percent.

At first glance, this appears to be an extremely valuable classifier. However, it would be wise to collect additional information before trusting a child's life to the test. What if the genetic defect is found in only 10 out of every 100,000 babies? A test that invariably predicts *no defect* will be correct for 99.99 percent of all cases, but incorrect for 100 percent of the cases that matter most. In other words, even though the classifier is extremely accurate, it is not very useful for preventing treatable birth defects.

 This is one consequence of the **class imbalance problem**, which refers to the trouble associated with data having a large majority of records belonging to a single class.

Though there are many ways to measure a classifier's performance, the best measure is always that which captures whether the classifier is successful at its intended purpose. It is crucial to define performance measures for utility rather than raw accuracy. To this end, we will explore a variety of alternative performance measures derived from the confusion matrix. Before we get started, however, we need to consider how to prepare a classifier for evaluation.

Understanding a classifier's predictions

The goal of evaluating a classification model is to better understand how its performance will extrapolate to future cases. Since it is usually infeasible to test an unproven model in a live environment, we typically simulate future conditions by asking the model to classify cases in a dataset made of cases that resemble what it will be asked to do in the future. By observing the learner's responses to this examination, we can learn about its strengths and weaknesses.

Though we've evaluated classifiers in prior chapters, it's worth reflecting on the types of data at our disposal:

- Actual class values
- Predicted class values
- Estimated probability of the prediction

The actual and predicted class values may be self-evident, but they are the key to the evaluation. Just like a teacher uses an answer key—a list of correct answers—to assess the student's answers, we need to know the correct answer for a machine learner's predictions. The goal is to maintain two vectors of data: one holding the correct or actual class values, and the other holding the predicted class values. Both vectors must have the same number of values stored in the same order. The predicted and actual values may be stored as separate R vectors or as columns in a single R data frame.

Obtaining this data is easy. The actual class values come directly from the target in the test dataset. Predicted class values are obtained from the classifier built upon the training data, which is then applied to the test data. For most machine learning packages, this involves applying the `predict()` function to a model object and a data frame of test data, such as: `predictions <- predict(model, test_data)`.

Until now, we have only examined classification predictions using these two vectors of data, but most models can supply another piece of useful information. Even though the classifier makes a single prediction about each example, it may be more confident about some decisions than others. For instance, a classifier may be 99 percent certain that a SMS with the words "free" and "ringtones" is spam, but only 51 percent certain that a SMS with the word "tonight" is spam. In both cases, the classifier classifies the message as spam, but it is far more certain about one decision than the other.

Figure 10.1: Learners may differ in their prediction confidence even when trained on the same data

Studying these internal prediction probabilities provides useful data to evaluate a model's performance. If two models make the same number of mistakes, but one is more able to accurately assess its uncertainty, then it is a smarter model. It's ideal to find a learner that is extremely confident when making a correct prediction, but timid in the face of doubt. The balance between confidence and caution is a key part of model evaluation.

The function call to obtain the internal prediction probabilities varies across R packages. In general, for most classifiers, the predict() function allows an additional parameter to specify the desired type of prediction. To obtain a single predicted class, such as spam or ham, you typically set the type = "class" parameter. To obtain the prediction probability, the type parameter should be set to one of "prob", "posterior", "raw", or "probability" depending on the classifier used.

All of the classifiers presented in this book will provide prediction probabilities. The type parameter is included in the syntax box introducing each model.

For example, to output the predicted probabilities for the C5.0 classifier built in *Chapter 5, Divide and Conquer – Classification Using Decision Trees and Rules*, use the predict() function with type = "prob" as follows:

```
> predicted_prob <- predict(credit_model, credit_test,
    type = "prob")
```

To output the Naive Bayes predicted probabilities for the SMS spam classification model developed in *Chapter 4, Probabilistic Learning – Classification Using Naive Bayes*, use predict() with type = "raw" as follows:

```
> sms_test_prob <- predict(sms_classifier, sms_test, type = "raw")
```

In most cases, the predict() function returns a probability for each category of the outcome. For example, in the case of a two-outcome model like the SMS classifier, the predicted probabilities might be stored in a matrix or data frame, as shown here:

```
> head(sms_test_prob)
            ham           spam
[1,] 9.999995e-01  4.565938e-07
[2,] 9.999995e-01  4.540489e-07
[3,] 9.998418e-01  1.582360e-04
[4,] 9.999578e-01  4.223125e-05
[5,] 4.816137e-10  1.000000e+00
[6,] 9.997970e-01  2.030033e-04
```

Each line in this output shows the classifier's predicted probability of spam and ham. According to probability rules, the sum of each line is one because these are mutually exclusive and exhaustive outcomes. Given this data, when constructing the model evaluation dataset, it is important to ensure that you select only the probability for the class level of interest. For convenience during the evaluation process, it can be helpful to construct a data frame collecting the predicted class, the actual class, and the predicted probability of the class level of interest.

> The steps required to construct the evaluation dataset have been omitted for brevity but are included in this chapter's code on the Packt Publishing website. To follow along with the examples here, download the sms_results.csv file and load it to a data frame using the sms_results <- read.csv("sms_results.csv") command.

The sms_results data frame is simple. It contains four vectors of 1,390 values. One vector contains values indicating the actual type of SMS message (spam or ham), one vector indicates the Naive Bayes model's predicted message type, and the third and fourth vectors indicate the probability that the message was spam or ham, respectively:

```
> head(sms_results)
  actual_type predict_type prob_spam prob_ham
1         ham          ham   0.00000  1.00000
2         ham          ham   0.00000  1.00000
3         ham          ham   0.00016  0.99984
4         ham          ham   0.00004  0.99996
5        spam         spam   1.00000  0.00000
6         ham          ham   0.00020  0.99980
```

For these six test cases, the predicted and actual SMS message types agree; the model predicted their status correctly. Furthermore, the prediction probabilities suggest that the model was extremely confident about these predictions because they all fall close to zero or one.

What happens when the predicted and actual values are further from zero and one? Using the subset() function, we can identify a few of these records. The following output shows test cases where the model estimated the probability of spam between 40 and 60 percent:

```
> head(subset(sms_results, prob_spam > 0.40 & prob_spam < 0.60))
      actual_type predict_type prob_spam prob_ham
377          spam          ham   0.47536  0.52464
717           ham         spam   0.56188  0.43812
1311          ham         spam   0.57917  0.42083
```

By the model's own estimation, these were cases in which a correct prediction was virtually a coin flip. Yet all three predictions were wrong—an unlucky result. Let's look at a few more cases where the model was wrong:

```
> head(subset(sms_results, actual_type != predict_type))
    actual_type predict_type prob_spam prob_ham
53          spam          ham   0.00071  0.99929
59          spam          ham   0.00156  0.99844
73          spam          ham   0.01708  0.98292
76          spam          ham   0.00851  0.99149
184         spam          ham   0.01243  0.98757
332         spam          ham   0.00003  0.99997
```

These cases illustrate the important fact that a model can be extremely confident and yet it can still be extremely wrong. All six of these test cases were spam messages that the classifier believed to have no less than a 98 percent chance of being ham.

In spite of such mistakes, is the model still useful? We can answer this question by applying various error metrics to this evaluation data. In fact, many such metrics are based on a tool we've already used extensively in previous chapters.

A closer look at confusion matrices

A **confusion matrix** is a table that categorizes predictions according to whether they match the actual value. One of the table's dimensions indicates the possible categories of predicted values, while the other dimension indicates the same for actual values. Although we have only seen 2x2 confusion matrices so far, a matrix can be created for models that predict any number of class values. The following figure depicts the familiar confusion matrix for a two-class binary model, as well as the 3x3 confusion matrix for a three-class model.

When the predicted value is the same as the actual value, this is a correct classification. Correct predictions fall on the diagonal in the confusion matrix (denoted by **O**). The off-diagonal matrix cells (denoted by **X**) indicate the cases where the predicted value differs from the actual value. These are incorrect predictions. Performance measures for classification models are based on the counts of predictions falling on and off the diagonal in these tables:

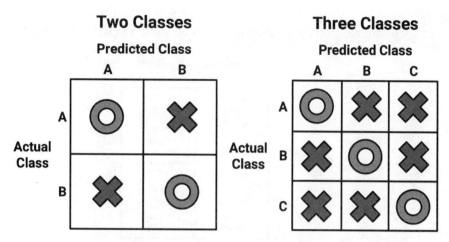

Figure 10.2: Confusion matrices count cases where the predicted class agrees or disagrees with the actual value

The most common performance measures consider the model's ability to discern one class versus all others. The class of interest is known as the **positive** class, while all others are known as **negative**.

 The use of the terms positive and negative is not intended to imply any value judgment (that is, good versus bad), nor does it necessarily suggest that the outcome is present or absent (such as birth defect versus none). The choice of the positive outcome can even be arbitrary, as in cases where a model is predicting categories such as sunny versus rainy, or dog versus cat.

The relationship between positive class and negative class predictions can be depicted as a 2x2 confusion matrix that tabulates whether predictions fall into one of four categories:

- **True positive (TP)**: Correctly classified as the class of interest
- **True negative (TN)**: Correctly classified as not the class of interest
- **False positive (FP)**: Incorrectly classified as the class of interest
- **False negative (FN)**: Incorrectly classified as not the class of interest

For the spam classifier, the positive class is spam, as this is the outcome we hope to detect. We then can imagine the confusion matrix as shown in the following diagram:

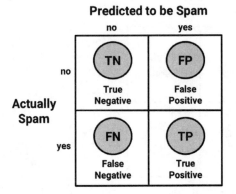

Figure 10.3: Distinguishing between positive and negative classes adds detail to the confusion matrix

The confusion matrix presented in this way is the basis for many of the most important measures of model performance. In the next section, we'll use this matrix to better understand exactly what is meant by accuracy.

Using confusion matrices to measure performance

With the 2x2 confusion matrix, we can formalize our definition of prediction accuracy (sometimes called the **success rate**) as:

$$accuracy = \frac{TP + TN}{TP + TN + FP + FN}$$

In this formula, the terms *TP, TN, FP,* and *FN* refer to the number of times the model's predictions fell into each of these categories. The accuracy is therefore a proportion that represents the number of true positives and true negatives divided by the total number of predictions.

The **error rate**, or the proportion of incorrectly classified examples, is specified as:

$$error\ rate = \frac{FP + FN}{TP + TN + FP + FN} = 1 - accuracy$$

Notice that the error rate can be calculated as one minus the accuracy. Intuitively, this makes sense; a model that is correct 95 percent of the time is incorrect five percent of the time.

An easy way to tabulate a classifier's predictions into a confusion matrix is to use R's `table()` function. The command for creating a confusion matrix for the SMS data is shown as follows. The counts in this table could then be used to calculate accuracy and other statistics:

```
> table(sms_results$actual_type, sms_results$predict_type)
```

```
        ham spam
  ham  1203    4
 spam    31  152
```

If you would like to create a confusion matrix with more informative output, the `CrossTable()` function in the `gmodels` package offers a customizable solution. If you recall, we first used this function in *Chapter 2, Managing and Understanding Data*. If you didn't install the package at that time, you will need to do so using the `install.packages("gmodels")` command.

By default, the `CrossTable()` output includes proportions in each cell that indicate the cell count as a percentage of the table's row, column, and overall total counts. The output also includes row and column totals. As shown in the following code, the syntax is similar to the `table()` function:

```
> library(gmodels)
> CrossTable(sms_results$actual_type, sms_results$predict_type)
```

The result is a confusion matrix with a wealth of additional detail:

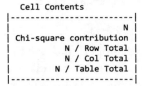

```
   Cell Contents
|-------------------------|
|                       N |
| Chi-square contribution |
|         N / Row Total   |
|         N / Col Total   |
|         N / Table Total |
|-------------------------|

Total Observations in Table:  1390

                        | sms_results$predict_type
  sms_results$actual_type |       ham |      spam | Row Total |
 -----------------------|-----------|-----------|-----------|
                    ham |      1203 |         4 |      1207 |
                        |    16.128 |   127.580 |           |
                        |     0.997 |     0.003 |     0.868 |
                        |     0.975 |     0.026 |           |
                        |     0.865 |     0.003 |           |
 -----------------------|-----------|-----------|-----------|
                   spam |        31 |       152 |       183 |
                        |   106.377 |   841.470 |           |
                        |     0.169 |     0.831 |     0.132 |
                        |     0.025 |     0.974 |           |
                        |     0.022 |     0.109 |           |
 -----------------------|-----------|-----------|-----------|
           Column Total |      1234 |       156 |      1390 |
                        |     0.888 |     0.112 |           |
 -----------------------|-----------|-----------|-----------|
```

We've used `CrossTable()` in several previous chapters, so by now you should be familiar with the output. If you ever forget how to interpret the output, simply refer to the key (labeled `Cell Contents`), which provides the definition of each number in the table cells.

We can use the confusion matrix to obtain the accuracy and error rate. Since accuracy is *(TP + TN) / (TP + TN + FP + FN)*, we can calculate it as follows:

```
> (152 + 1203) / (152 + 1203 + 4 + 31)
[1] 0.9748201
```

We can also calculate the error rate *(FP + FN) / (TP + TN + FP + FN)* as:

```
> (4 + 31) / (152 + 1203 + 4 + 31)
[1] 0.02517986
```

This is the same as one minus accuracy:

```
> 1 - 0.9748201
[1] 0.0251799
```

Although these calculations may seem simple, it is important to practice thinking about how the components of the confusion matrix relate to one another. In the next section, you will see how these same pieces can be combined in different ways to create a variety of additional performance measures.

Beyond accuracy – other measures of performance

Countless performance measures have been developed and used for specific purposes in disciplines as diverse as medicine, information retrieval, marketing, and signal detection theory, among others. To cover all of them could fill hundreds of pages, which makes a comprehensive description infeasible here. Instead, we'll consider only some of the most useful and most commonly cited measures in machine learning literature.

The **Classification and Regression Training** package `caret` by Max Kuhn includes functions for computing many such performance measures. This package provides a large number of tools for preparing, training, evaluating, and visualizing machine learning models and data. In addition to its use here, we will also employ `caret` extensively in *Chapter 11, Improving Model Performance*. Before proceeding, you will need to install the package using the `install.packages("caret")` command.

For more information on `caret`, please refer to *Building Predictive Models in R Using the caret Package, Kuhn, M, Journal of Statistical Software, 2008, Vol. 28.*

The `caret` package adds yet another function to create a confusion matrix. As shown in the following command, the syntax is similar to `table()`, but with a minor difference. Because `caret` computes measures of model performance that reflect the ability to classify the positive class, a `positive` parameter should be specified. In this case, since the SMS classifier is intended to detect spam, we will set `positive = "spam"` as follows:

```
> library(caret)
> confusionMatrix(sms_results$predict_type,
    sms_results$actual_type, positive = "spam")
```

This results in the following output:

```
              Confusion Matrix and Statistics

                Reference
    Prediction  ham  spam
          ham  1203   31
         spam     4  152

               Accuracy : 0.9748
                 95% CI : (0.9652, 0.9824)
    No Information Rate : 0.8683
    P-Value [Acc > NIR] : < 2.2e-16

                  Kappa : 0.8825
 Mcnemar's Test P-Value : 1.109e-05

            Sensitivity : 0.8306
            Specificity : 0.9967
         Pos Pred Value : 0.9744
         Neg Pred Value : 0.9749
             Prevalence : 0.1317
         Detection Rate : 0.1094
   Detection Prevalence : 0.1122
      Balanced Accuracy : 0.9136

       'Positive' Class : spam
```

At the top of the output is a confusion matrix much like the one produced by the `table()` function, but transposed. The output also includes a set of performance measures. Some of these, like accuracy, are familiar, while many others are new. Let's take a look at some of the most important metrics.

The kappa statistic

The **kappa statistic** (labeled Kappa in the previous output) adjusts accuracy by accounting for the possibility of a correct prediction by chance alone. This is especially important for datasets with severe class imbalance because a classifier can obtain high accuracy simply by always guessing the most frequent class. The kappa statistic will only reward the classifier if it is correct more often than this simplistic strategy.

Kappa values range from zero to a maximum of one, which indicates perfect agreement between the model's predictions and the true values. Values less than one indicate imperfect agreement. Depending on how a model is to be used, the interpretation of the kappa statistic might vary. One common interpretation is shown as follows:

- Poor agreement = less than 0.20
- Fair agreement = 0.20 to 0.40
- Moderate agreement = 0.40 to 0.60
- Good agreement = 0.60 to 0.80
- Very good agreement = 0.80 to 1.00

It's important to note that these categories are subjective. While "good agreement" may be more than adequate for predicting someone's favorite ice cream flavor, "very good agreement" may not suffice if your goal is to identify birth defects.

For more information on the previous scale, refer to *The measurement of observer agreement for categorical data, Landis, JR, Koch, GG. Biometrics, 1997, Vol. 33, pp. 159-174.*

The following is the formula for calculating the kappa statistic. In this formula, *Pr(a)* refers to the proportion of actual agreement and *Pr(e)* refers to the expected agreement between the classifier and the true values, under the assumption that they were chosen at random:

$$\kappa = \frac{\Pr(a) - \Pr(e)}{1 - \Pr(e)}$$

There is more than one way to define the kappa statistic. The most common method, described here, uses **Cohen's kappa coefficient** as described in the paper *A coefficient of agreement for nominal scales, Cohen, J, Education and Psychological Measurement, 1960, Vol. 20, pp. 37-46.*

These proportions are easy to obtain from a confusion matrix once you know where to look. Let's consider the confusion matrix for the SMS classification model created with the `CrossTable()` function, which is repeated here for convenience:

```
                              |  sms_results$predict_type
     sms_results$actual_type  |      ham  |     spam | Row Total |
    --------------------------|----------|----------|-----------|
                         ham  |    1203  |        4 |      1207 |
                              |  16.128  |  127.580 |           |
                              |   0.997  |    0.003 |     0.868 |
                              |   0.975  |    0.026 |           |
                              |   0.865  |    0.003 |           |
    --------------------------|----------|----------|-----------|
                        spam  |      31  |      152 |       183 |
                              | 106.377  |  841.470 |           |
                              |   0.169  |    0.831 |     0.132 |
                              |   0.025  |    0.974 |           |
                              |   0.022  |    0.109 |           |
    --------------------------|----------|----------|-----------|
                Column Total  |    1234  |      156 |      1390 |
                              |   0.888  |    0.112 |           |
    --------------------------|----------|----------|-----------|
```

Remember that the bottom value in each cell indicates the proportion of all instances falling into that cell. Therefore, to calculate the observed agreement Pr(a), we simply add the proportion of all instances where the predicted type and actual SMS type agree. Thus, we can calculate Pr(a) as:

```
> pr_a <- 0.865 + 0.109
> pr_a
[1] 0.974
```

For this classifier, the observed and actual values agree 97.4 percent of the time—you will note that this is the same as the accuracy. The kappa statistic adjusts the accuracy relative to the expected agreement, Pr(e), which is the probability that chance alone would lead the predicted and actual values to match, under the assumption that both are selected randomly according to the observed proportions.

To find these observed proportions, we can use the probability rules we learned in *Chapter 4, Probabilistic Learning – Classification Using Naive Bayes*. Assuming two events are independent (meaning one does not affect the other), probability rules note that the probability of both occurring is equal to the product of the probabilities of each one occurring. For instance, we know that the probability of both choosing ham is:

*Pr(actual_type is ham) * Pr(predicted_type is ham)*

And the probability of both choosing spam is:

*Pr(actual_type is spam) * Pr(predicted_type is spam)*

The probability that the predicted or actual type is spam or ham can be obtained from the row or column totals. For instance, *Pr(actual_type is ham) = 0.868* and *Pr(predicted type is ham) = 0.888*.

Pr(e) can be calculated as the sum of the probabilities that either the predicted and actual values agree that the message is spam or ham. Recall that for mutually exclusive events (events that cannot happen simultaneously), the probability of either occurring is equal to the sum of their probabilities. Therefore, to obtain the final *Pr(e)*, we simply add both products, as follows:

```
> pr_e <- 0.868 * 0.888 + 0.132 * 0.112
> pr_e
[1] 0.785568
```

Since *Pr(e)* is 0.786, by chance alone we would expect the observed and actual values to agree about 78.6 percent of the time.

This means that we now have all the information needed to complete the kappa formula. Plugging the *Pr(a)* and *Pr(e)* values into the kappa formula, we find:

```
> k <- (pr_a - pr_e) / (1 - pr_e)
> k
[1] 0.8787494
```

The kappa is about 0.88, which agrees with the previous `confusionMatrix()` output from `caret` (the small difference is due to rounding). Using the suggested interpretation, we note that there is very good agreement between the classifier's predictions and the actual values.

There are a couple of R functions to calculate kappa automatically. The `Kappa()` function (be sure to note the capital "K") in the **Visualizing Categorical Data** (vcd) package uses a confusion matrix of predicted and actual values. After installing the package by typing `install.packages("vcd")`, the following commands can be used to obtain kappa:

```
> Kappa(table(sms_results$actual_type, sms_results$predict_type))
              value       ASE
Unweighted 0.8825203 0.01949315
Weighted   0.8825203 0.01949315
```

We're interested in the unweighted kappa. The value of 0.88 matches what we computed by hand.

The weighted kappa is used when there are varying degrees of agreement. For example, using a scale of cold, cool, warm, and hot, a value of warm agrees more with hot than it does with the value of cold. In the case of a two-outcome event, such as spam and ham, the weighted and unweighted kappa statistics will be identical.

The `kappa2()` function in the **Interrater Reliability** (`irr`) package can be used to calculate kappa from vectors of predicted and actual values stored in a data frame. After installing the package using `install.packages("irr")`, the following commands can be used to obtain kappa:

```
> kappa2(sms_results[1:2])
 Cohen's Kappa for 2 Raters (Weights: unweighted)

 Subjects = 1390
   Raters = 2
    Kappa = 0.883

        z = 33
  p-value = 0
```

The `Kappa()` and `kappa2()` functions report the same kappa statistic, so use whichever option you are more comfortable with.

Be careful not to use the built-in `kappa()` function. It is completely unrelated to the kappa statistic reported previously!

Sensitivity and specificity

Finding a useful classifier often involves a balance between predictions that are overly conservative and overly aggressive. For example, an email filter could guarantee to eliminate every spam message by aggressively eliminating nearly every ham message. On the other hand, to guarantee that no ham messages will be inadvertently filtered might require us to allow an unacceptable amount of spam to pass through the filter. A pair of performance measures captures this tradeoff: sensitivity and specificity.

The **sensitivity** of a model (also called the **true positive rate**), measures the proportion of positive examples that were correctly classified. Therefore, as shown in the following formula, it is calculated as the number of true positives divided by the total number of positives, both those correctly classified (the true positives), as well as those incorrectly classified (the false negatives):

$$\text{sensitivity} = \frac{TP}{TP + FN}$$

The **specificity** of a model (also called the **true negative rate**), measures the proportion of negative examples that were correctly classified. As with sensitivity, this is computed as the number of true negatives divided by the total number of negatives — the true negatives plus the false positives.

$$\text{specificity} = \frac{TN}{TN + FP}$$

Given the confusion matrix for the SMS classifier, we can easily calculate these measures by hand. Assuming that spam is the positive class, we can confirm that the numbers in the `confusionMatrix()` output are correct. For example, the calculation for sensitivity is:

```
> sens <- 152 / (152 + 31)
> sens
[1] 0.8306011
```

Similarly, for specificity we can calculate:

```
> spec <- 1203 / (1203 + 4)
> spec
[1] 0.996686
```

The `caret` package provides functions for calculating sensitivity and specificity directly from vectors of predicted and actual values. Be careful to specify the `positive` or `negative` parameter appropriately, as shown in the following lines:

```
> library(caret)
> sensitivity(sms_results$predict_type, sms_results$actual_type,
            positive = "spam")
[1] 0.8306011

> specificity(sms_results$predict_type, sms_results$actual_type,
            negative = "ham")
[1] 0.996686
```

Sensitivity and specificity range from zero to one, with values close to one being more desirable. Of course, it is important to find an appropriate balance between the two—a task that is often quite context-specific.

For example, in this case, the sensitivity of 0.831 implies that 83.1 percent of the spam messages were correctly classified. Similarly, the specificity of 0.997 implies that 99.7 percent of non-spam messages were correctly classified, or alternatively, 0.3 percent of valid messages were rejected as spam. The idea of rejecting 0.3 percent of valid SMS messages may be unacceptable, or it may be a reasonable tradeoff given the reduction in spam.

Sensitivity and specificity provide tools for thinking about such tradeoffs. Typically, changes are made to the model and different models are tested until you find one that meets a desired sensitivity and specificity threshold. Visualizations, such as those discussed later in this chapter, can also assist with understanding the balance between sensitivity and specificity.

Precision and recall

Closely related to sensitivity and specificity are two other performance measures related to compromises made in classification: precision and recall. Used primarily in the context of information retrieval, these statistics are intended to indicate how interesting and relevant a model's results are, or whether the predictions are diluted by meaningless noise.

The **precision** (also known as the **positive predictive value**) is defined as the proportion of positive examples that are truly positive; in other words, when a model predicts the positive class, how often is it correct? A precise model will only predict the positive class in cases very likely to be positive. It will be very trustworthy.

Consider what would happen if the model was very imprecise. Over time, the results would be less likely to be trusted. In the context of information retrieval, this would be similar to a search engine such as Google returning unrelated results. Eventually, users would switch to a competitor like Bing. In the case of the SMS spam filter, high precision means that the model is able to carefully target only the spam while ignoring the ham.

$$\text{precision} = \frac{TP}{TP + FP}$$

On the other hand, **recall** is a measure of how complete the results are. As shown in the following formula, this is defined as the number of true positives over the total number of positives. You may have already recognized this as the same as sensitivity; however, the interpretation differs slightly.

A model with high recall captures a large portion of the positive examples, meaning that it has wide breadth. For example, a search engine with high recall returns a large number of documents pertinent to the search query. Similarly, the SMS spam filter has high recall if the majority of spam messages are correctly identified.

$$recall = \frac{TP}{TP + FN}$$

We can calculate precision and recall from the confusion matrix. Again, assuming that spam is the positive class, the precision is:

```
> prec <- 152 / (152 + 4)
> prec
[1] 0.974359
```

And the recall is:

```
> rec <- 152 / (152 + 31)
> rec
[1] 0.8306011
```

The caret package can be used to compute either of these measures from vectors of predicted and actual classes. Precision uses the posPredValue() function:

```
> library(caret)
> posPredValue(sms_results$predict_type, sms_results$actual_type,
              positive = "spam")
[1] 0.974359
```

Recall uses the sensitivity() function that we used earlier:

```
> sensitivity(sms_results$predict_type, sms_results$actual_type,
              positive = "spam")
[1] 0.8306011
```

Similar to the inherent tradeoff between sensitivity and specificity, for most real-world problems, it is difficult to build a model with both high precision and high recall. It is easy to be precise if you target only the low-hanging fruit—the easy-to-classify examples. Similarly, it is easy for a model to have high recall by casting a very wide net, meaning that the model is overly aggressive at identifying the positive cases. In contrast, having both high precision and recall at the same time is very challenging. It is therefore important to test a variety of models in order to find the combination of precision and recall that meets the needs of your project.

The F-measure

A measure of model performance that combines precision and recall into a single number is known as the **F-measure** (also sometimes called the **F$_1$ score** or the **F-score**). The F-measure combines precision and recall using the **harmonic mean**, a type of average that is used for rates of change. The harmonic mean is used rather than the more common arithmetic mean since both precision and recall are expressed as proportions between zero and one, which can be interpreted as rates. The following is the formula for the F-measure:

$$\text{F-measure} = \frac{2 \times \text{precision} \times \text{recall}}{\text{recall} + \text{precision}} = \frac{2 \times \text{TP}}{2 \times \text{TP} + \text{FP} + \text{FN}}$$

To calculate the F-measure, use the precision and recall values computed previously:

```
> f <- (2 * prec * rec) / (prec + rec)
> f
[1] 0.8967552
```

This comes out exactly the same as using the counts from the confusion matrix:

```
> f <- (2 * 152) / (2 * 152 + 4 + 31)
> f
[1] 0.8967552
```

Since the F-measure describes model performance in a single number, it provides a convenient way to compare several models side-by-side. However, this assumes that equal weight should be assigned to precision and recall, an assumption that is not always valid. It is possible to calculate F-scores using different weights for precision and recall, but choosing the weights can be tricky at best and arbitrary at worst. A better practice is to use measures such as the F-score in combination with methods that consider a model's strengths and weaknesses more globally, such as those described in the next section.

Visualizing performance tradeoffs with ROC curves

Visualizations are helpful for understanding the performance of machine learning algorithms in greater detail. Where statistics such as sensitivity and specificity, or precision and recall, attempt to boil model performance down to a single number, visualizations depict how a learner performs across a wide range of conditions.

Because learning algorithms have different biases, it is possible that two models with similar accuracy could have drastic differences in how they achieve their accuracy. Some models may struggle with certain predictions that others make with ease, while breezing through cases that others cannot get right. Visualizations provide a method for understanding these tradeoffs by comparing learners side-by-side in a single chart.

The **receiver operating characteristic (ROC) curve** is commonly used to examine the tradeoff between the detection of true positives while avoiding the false positives. As you might suspect from the name, ROC curves were developed by engineers in the field of communications. Around the time of World War II, radar and radio operators used ROC curves to measure a receiver's ability to discriminate between true signals and false alarms. The same technique is useful today for visualizing the efficacy of machine learning models.

The characteristics of a typical ROC diagram are depicted in the following plot. The figure is drawn using the proportion of true positives on the vertical axis and the proportion of false positives on the horizontal axis. Because these values are equivalent to sensitivity and (1 – specificity) respectively, the diagram is also known as a sensitivity/specificity plot.

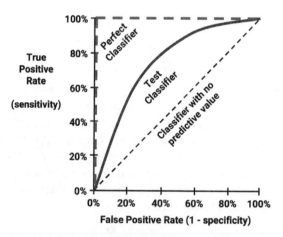

Figure 10.4: The ROC curve depicts classifier shapes relative to perfect and useless classifiers

The points comprising ROC curves indicate the true positive rate at varying false positive thresholds. To create the curves, a classifier's predictions are sorted by the model's estimated probability of the positive class, with the largest values first. Beginning at the origin, each prediction's impact on the true positive rate and false positive rate will result in a curve tracing vertically (for a correct prediction) or horizontally (for an incorrect prediction).

To illustrate this concept, three hypothetical classifiers are contrasted in the previous plot. First, the diagonal line from the bottom-left to the top-right corner of the diagram represents a *classifier with no predictive value*. This type of classifier detects true positives and false positives at exactly the same rate, implying that the classifier cannot discriminate between the two. This is the baseline by which other classifiers may be judged. ROC curves falling close to this line indicate models that are not very useful. Similarly, *the perfect classifier* has a curve that passes through the point at 100 percent true positive rate and zero percent false positive rate. It is able to correctly identify all of the true positives before it incorrectly classifies any negative result. Most real-world classifiers are similar to the *test classifier* in that they fall somewhere in the zone between perfect and useless.

The closer the curve is to the perfect classifier, the better it is at identifying positive values. This can be measured using a statistic known as the **area under the ROC curve** (**AUC**). The AUC treats the ROC diagram as a two-dimensional square and measures the total area under the ROC curve. AUC ranges from 0.5 (for a classifier with no predictive value), to 1.0 (for a perfect classifier). A convention for interpreting AUC scores uses a system similar to academic letter grades:

- **A**: Outstanding = 0.9 to 1.0
- **B**: Excellent/Good = 0.8 to 0.9
- **C**: Acceptable/Fair = 0.7 to 0.8
- **D**: Poor = 0.6 to 0.7
- **E**: No Discrimination = 0.5 to 0.6

As with most scales like this, the levels may work better for some tasks than others; the categorization is somewhat subjective.

As illustrated by the following figure, it's also worth noting that two ROC curves may be shaped very differently, yet have identical AUC. For this reason, AUC alone is insufficient to identify a "best" model. The safest practice is to use AUC in combination with qualitative examination of the ROC curve.

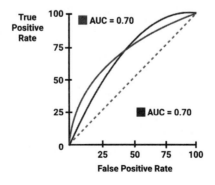

Figure 10.5: ROC curves may have different performance despite having the same AUC

The pROC package provides an easy-to-use set of functions for creating ROC curves and computing AUC. The pROC website at `https://web.expasy.org/pROC/` includes a list of the full set of features, as well as several examples of the visualization capabilities. Before continuing, be sure that you have installed the package using the `install.packages("pROC")` command.

> For more information on the pROC package, see *pROC: an open-source package for R and S+ to analyze and compare ROC curves, Robin, X, Turck, N, Hainard, A, Tiberti, N, Lisacek, F, Sanchez, JC,* and *Mueller M, BMC Bioinformatics, 2011, pp. 12-77.*

To create visualizations with pROC, two vectors of data are needed. The first must contain the estimated probability of the positive class and the second must contain the predicted class values. For the SMS classifier, we'll supply estimated spam probabilities and the actual class labels to the `roc()` function as follows:

```
> library(pROC)
> sms_roc <- roc(sms_results$prob_spam, sms_results$actual_type)
```

Using the `sms_roc` object, we can visualize the ROC curve with R's `plot()` function. As shown in the following code lines, many of the standard parameters for adjusting the visualization can be used, such as `main` (for adding a title), `col` (for changing the line color), and `lwd` (for adjusting the line width). The `legacy.axes` parameter instructs pROC to use an x axis of 1 – specificity, which is a popular convention:

```
> plot(sms_roc, main = "ROC curve for SMS spam filter",
        col = "blue", lwd = 2, legacy.axes = TRUE)
```

The end result is a ROC plot with a diagonal reference line representing a baseline classifier with no predictive value:

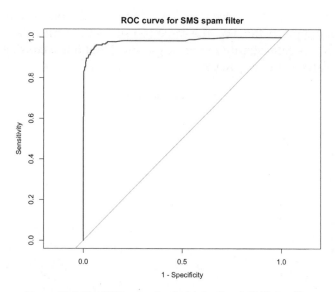

Figure 10.6: The ROC curve for the Naive Bayes SMS classifier

Qualitatively, we can see that this ROC curve appears to occupy the space in the top-left corner of the diagram, which suggests that it is closer to a perfect classifier than the dashed line representing a useless classifier.

To compare this model's performance to other models making predictions on the same dataset, we can add additional ROC curves to the same plot. Suppose that we had also trained a k-NN model on the SMS data using the knn() function described in *Chapter 3, Lazy Learning – Classification Using Nearest Neighbors*. Using this model, the predicted probabilities of spam were computed for each record in the test set and saved to a CSV file, which we can load here. After loading the file, we'll apply the roc() function as before to compute the ROC curve, then use the plot() function with the parameter add = TRUE to add the curve to the previous plot:

```
> sms_results_knn <- read.csv("sms_results_knn.csv")
> sms_roc_knn <- roc(sms_results$actual_type,
                     sms_results_knn$p_spam)
> plot(sms_roc_knn, col = "red", lwd = 2, add = TRUE)
```

The resulting visualization has a second curve depicting the performance of the k-NN model making predictions on the same test set as the Naive Bayes model. The curve for k-NN is consistently lower, suggesting that it is a consistently worse model than the Naive Bayes approach:

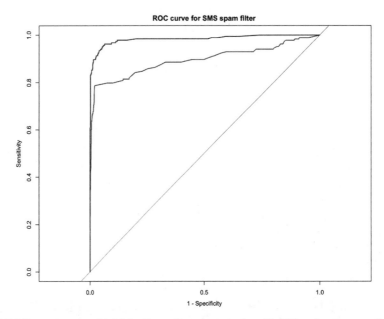

Figure 10.7: ROC curves comparing Naive Bayes (topmost curve) and k-NN performance on the SMS test set

To confirm this quantitatively, we can use the pROC package to calculate the AUC. To do so, we simply apply the package's auc() function to the sms_roc object for each model, as shown in the following code:

```
> auc(sms_roc)
Area under the curve: 0.9836
> auc(sms_roc_knn)
Area under the curve: 0.8942
```

The AUC for the Naive Bayes SMS classifier is 0.98, which is extremely high and substantially better than the k-NN classifier's AUC of 0.89. But how do we know whether the model is just as likely to perform well on another dataset, or whether the difference is greater than expected by chance alone? In order to answer such questions, we need to better understand how far we can extrapolate a model's predictions beyond the test data.

This was mentioned before, but is worth repeating: the AUC value alone is often insufficient to identify a "best" model. In this example, AUC does identify the better model because the ROC curves do not intersect. In other cases, the "best" model will depend on how the model will be used. When the ROC curves do intersect, it is possible to combine them into even stronger models using techniques covered in *Chapter 11, Improving Model Performance*.

Estimating future performance

Some R machine learning packages present confusion matrices and performance measures during the model-building process. The purpose of these statistics is to provide insight about the model's **resubstitution error**, which occurs when the training data is incorrectly predicted in spite of the model being built directly from this data. This information can be used as a rough diagnostic to identify obviously poor performers.

The resubstitution error is not a very useful marker of future performance, however. For example, a model that used rote memorization to perfectly classify every training instance with zero resubstitution error would be unable to generalize its predictions to data it has never seen before. For this reason, the error rate on the training data can be extremely optimistic about a model's future performance.

Instead of relying on resubstitution error, a better practice is to evaluate a model's performance on data it has not yet seen. We used such a method in previous chapters when we split the available data into a set for training and a set for testing. In some cases, however, it is not always ideal to create training and test datasets. For instance, in a situation where you have only a small pool of data, you might not want to reduce the sample any further.

Fortunately, there are other ways to estimate a model's performance on unseen data. The `caret` package we used to calculate performance measures also offers a number of functions to estimate future performance. If you are following along with the R code examples and haven't already installed the `caret` package, please do so. You will also need to load the package to the R session using the `library(caret)` command.

The holdout method

The procedure of partitioning data into training and test datasets that we used in previous chapters is known as the **holdout method**. As shown in the following diagram, the **training dataset** is used to generate the model, which is then applied to the **test dataset** to generate predictions for evaluation. Typically, about one-third of the data is held out for testing and two-thirds is used for training, but this proportion can vary depending on the amount of available data. To ensure that the training and test datasets do not have systematic differences, their examples are randomly divided into the two groups.

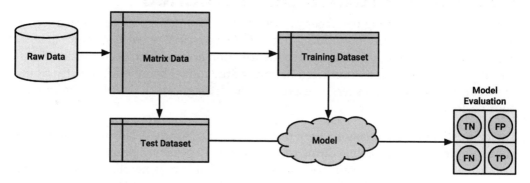

Figure 10.8: The simplest holdout method divides the data into training and test sets

For the holdout method to result in a truly accurate estimate of future performance, at no time should performance on the test dataset be allowed to influence the model. It is easy to unknowingly violate this rule by choosing a "best" model based upon the results of repeated testing. For example, suppose we built several models on the training data and selected the one with the highest accuracy on the test data. In this case, because we have cherry-picked the best result, the test performance is not an unbiased measure of future performance on unseen data.

A keen reader will note that holdout test data was used in previous chapters to both evaluate models and improve model performance. This was done for illustrative purposes but would indeed violate the rule stated previously. Consequently, the model performance statistics shown were not valid estimates of future performance on unseen data.

To avoid this problem, it is better to divide the original data so that in addition to the training and test datasets, a **validation dataset** is available. The validation dataset can be used for iterating and refining the model or models chosen, leaving the test dataset to be used only once as a final step to report an estimated error rate for future predictions. A typical split between training, test, and validation would be 50 percent, 25 percent, and 25 percent respectively.

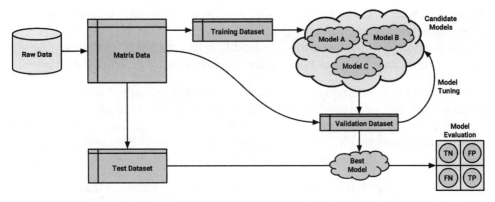

Figure 10.9: A validation dataset can be held out from training to select from multiple candidate models

A simple method for creating holdout samples uses random number generators to assign records to partitions. This technique was first used in *Chapter 5, Divide and Conquer – Classification Using Decision Trees and Rules,* to create training and test datasets.

 If you'd like to follow along with the following examples, download the `credit.csv` dataset from Packt Publishing's website and load it to a data frame using the `credit <- read.csv("credit.csv")` command.

Suppose we have a data frame named `credit` with 1,000 rows of data. We can divide this into three partitions as follows. First, we create a vector of randomly ordered row IDs from one to 1,000 using the `runif()` function, which by default generates a specified number of random values between zero and one. The `runif()` function gets its name from the random uniform distribution, which was discussed in *Chapter 2, Managing and Understanding Data.*

The `order()` function then returns a vector indicating the rank order of the 1,000 random numbers. For instance, `order(c(0.5, 0.25, 0.75, 0.1))` returns the sequence `4 2 1 3` because the smallest number (0.1) appears fourth, the second smallest (0.25) appears second, and so on.

```
> random_ids <- order(runif(1000))
```

Next, the random IDs are used to divide the `credit` data frame into 500, 250, and 250 records comprising the training, validation, and test datasets:

```
> credit_train <- credit[random_ids[1:500], ]
> credit_validate <- credit[random_ids[501:750], ]
> credit_test <- credit[random_ids[751:1000], ]
```

One problem with holdout sampling is that each partition may have a larger or smaller proportion of some classes. In cases where one, or more, class is a very small proportion of the dataset, this can lead it to be omitted from the training dataset—a significant problem because the model cannot then learn this class.

In order to reduce the chance of this occurring, a technique called **stratified random sampling** can be used. Although a random sample should generally contain roughly the same proportion of each class value as the full dataset, stratified random sampling guarantees that the random partitions have nearly the same proportion of each class as the full dataset, even when some classes are small.

The caret package provides a createDataPartition() function that creates partitions based on stratified holdout sampling. Code for creating a stratified sample of training and test data for the credit dataset is shown in the following commands. To use the function, a vector of class values must be specified (here, default refers to whether a loan went into default), in addition to a parameter, p, which specifies the proportion of instances to be included in the partition. The list = FALSE parameter prevents the result from being stored as a list object:

```
> in_train <- createDataPartition(credit$default, p = 0.75,
    list = FALSE)
> credit_train <- credit[in_train, ]
> credit_test <- credit[-in_train, ]
```

The in_train vector indicates the row numbers included in the training sample. We can use these row numbers to select examples for the credit_train data frame. Similarly, by using a negative symbol, we can use the rows not found in the in_train vector for the credit_test dataset.

Although it distributes the classes evenly, stratified sampling does not guarantee other types of representativeness. Some samples may have too many or too few difficult cases, easy-to-predict cases, or outliers. This is especially true for smaller datasets, which may not have a large enough portion of such cases to divide among training and test sets.

In addition to potentially biased samples, another problem with the holdout method is that substantial portions of data must be reserved for testing and validating the model. Since these data cannot be used to train the model until its performance has been measured, the performance estimates are likely to be overly conservative.

Since models trained on larger datasets generally perform better, a common practice is to retrain the model on the full set of data (that is, training plus test and validation) after a final model has been selected and evaluated.

A technique called **repeated holdout** is sometimes used to mitigate the problems of randomly composed training datasets. The repeated holdout method is a special case of the holdout method that uses the average result from several random holdout samples to evaluate a model's performance. As multiple holdout samples are used, it is less likely that the model is trained or tested on non-representative data. We'll expand on this idea in the next section.

Cross-validation

The repeated holdout is the basis of a technique known as **k-fold cross-validation (k-fold CV)**, which has become the industry standard for estimating model performance. Rather than taking repeated random samples that could potentially use the same record more than once, k-fold CV randomly divides the data into *k* completely separate random partitions called **folds**.

Although *k* can be set to any number, by far the most common convention is to use 10-fold CV. Why 10 folds? The reason is that empirical evidence suggests that there is little added benefit to using a greater number. For each of the 10 folds (each comprising 10 percent of the total data), a machine learning model is built on the remaining 90 percent of data. The fold's 10 percent sample is then used for model evaluation. After the process of training and evaluating the model has occurred 10 times (with 10 different training/testing combinations), the average performance across all folds is reported.

An extreme case of k-fold CV is the **leave-one-out method**, which performs k-fold CV using a fold for each one of the data's examples. This ensures that the greatest amount of data is used for training the model. Although this may seem useful, it is so computationally expensive that it is rarely used in practice.

Datasets for cross-validation can be created using the `createFolds()` function in the `caret` package. Similar to stratified random holdout sampling, this function will attempt to maintain the same class balance in each of the folds as in the original dataset. The following is the command to create 10 folds:

```
> folds <- createFolds(credit$default, k = 10)
```

The result of the `createFolds()` function is a list of vectors storing the row numbers for each of the `k = 10` requested folds. We can peek at the contents using `str()`:

```
> str(folds)
List of 10
 $ Fold01: int [1:100] 1 5 12 13 19 21 25 32 36 38 ...
 $ Fold02: int [1:100] 16 49 78 81 84 93 105 108 128 134 ...
 $ Fold03: int [1:100] 15 48 60 67 76 91 102 109 117 123 ...
 $ Fold04: int [1:100] 24 28 59 64 75 85 95 97 99 104 ...
 $ Fold05: int [1:100] 9 10 23 27 29 34 37 39 53 61 ...
 $ Fold06: int [1:100] 4 8 41 55 58 103 118 121 144 146 ...
 $ Fold07: int [1:100] 2 3 7 11 14 33 40 45 51 57 ...
 $ Fold08: int [1:100] 17 30 35 52 70 107 113 129 133 137 ...
 $ Fold09: int [1:100] 6 20 26 31 42 44 46 63 79 101 ...
 $ Fold10: int [1:100] 18 22 43 50 68 77 80 88 106 111 ...
```

Here, we see that the first fold is named `Fold01` and stores 100 integers indicating the 100 rows in the `credit` data frame for the first fold. To create training and test datasets to build and evaluate a model, an additional step is needed. The following commands show how to create data for the first fold. We'll assign the selected 10 percent to the test dataset and use the negative symbol to assign the remaining 90 percent to the training dataset:

```
> credit01_test <- credit[folds$Fold01, ]
> credit01_train <- credit[-folds$Fold01, ]
```

To perform the full 10-fold CV, this step would need to be repeated a total of 10 times, first building a model and then calculating the model's performance each time. At the end, the performance measures would be averaged to obtain the overall performance. Thankfully, we can automate this task by applying several of the techniques we've learned earlier.

To demonstrate the process, we'll estimate the kappa statistic for a C5.0 decision tree model of the credit data using 10-fold CV. First, we need to load some R packages: `caret` (to create the folds), `C50` (to build the decision tree), and `irr` (to calculate kappa). The latter two packages were chosen for illustrative purposes; if you desire, you can use a different model or a different performance measure with the same series of steps.

```
> library(caret)
> library(C50)
> library(irr)
```

Next, we'll create a list of 10 folds as we have done previously. The `set.seed()` function is used here to ensure that the results are consistent if the same code is run again:

```
> RNGversion("3.5.2")
> set.seed(123)
> folds <- createFolds(credit$default, k = 10)
```

Finally, we will apply a series of identical steps to the list of folds using the `lapply()` function. As shown in the following code, because there is no existing function that does exactly what we need, we must define our own function to pass to `lapply()`. Our custom function divides the `credit` data frame into training and test data, builds a decision tree using the `C5.0()` function on the training data, generates a set of predictions from the test data, and compares the predicted and actual values using the `kappa2()` function:

```
> cv_results <- lapply(folds, function(x) {
    credit_train <- credit[-x, ]
    credit_test <- credit[x, ]
    credit_model <- C5.0(default ~ ., data = credit_train)
    credit_pred <- predict(credit_model, credit_test)
    credit_actual <- credit_test$default
    kappa <- kappa2(data.frame(credit_actual, credit_pred))$value
    return(kappa)
  })
```

The resulting kappa statistics are compiled into a list stored in the `cv_results` object, which we can examine using `str()`:

```
> str(cv_results)
List of 10
 $ Fold01: num 0.343
 $ Fold02: num 0.255
 $ Fold03: num 0.109
 $ Fold04: num 0.107
 $ Fold05: num 0.338
 $ Fold06: num 0.474
 $ Fold07: num 0.245
 $ Fold08: num 0.0365
 $ Fold09: num 0.425
 $ Fold10: num 0.505
```

There's just one more step remaining in the 10-fold-CV process: we must calculate the average of these 10 values. Although you will be tempted to type `mean(cv_results)`, because `cv_results` is not a numeric vector, the result would be an error. Instead, use the `unlist()` function, which eliminates the list structure and reduces `cv_results` to a numeric vector. From there, we can calculate the mean kappa as expected:

```
> mean(unlist(cv_results))
[1] 0.283796
```

This kappa statistic is fairly low, corresponding to "fair" on the interpretation scale, which suggests that the credit scoring model performs only marginally better than random chance. In the next chapter, we'll examine automated methods based on 10-fold CV that can assist us with improving the performance of this model.

> Perhaps the current gold-standard method for reliably estimating model performance is **repeated k-fold CV**. As you might guess from the name, this involves repeatedly applying k-fold CV and averaging the results. A common strategy is to perform 10-fold CV 10 times. Although computationally intensive, this provides a very robust estimate.

Bootstrap sampling

A slightly less popular but still fairly widely used alternative to k-fold CV is known as **bootstrap sampling**, the **bootstrap**, or **bootstrapping** for short. Generally speaking, these refer to statistical methods that use random samples of data to estimate properties of a larger set. When this principle is applied to machine learning model performance, it implies the creation of several randomly selected training and test datasets, which are then used to estimate performance statistics. The results from the various random datasets are then averaged to obtain a final estimate of future performance.

So, what makes this procedure different from k-fold CV? Whereas cross-validation divides the data into separate partitions in which each example can appear only once, the bootstrap allows examples to be selected multiple times through a process of **sampling with replacement**. This means that from the original dataset of n examples, the bootstrap procedure will create one or more new training datasets that also contain n examples, some of which are repeated. The corresponding test datasets are then constructed from the set of examples that were not selected for the respective training datasets.

Using sampling with replacement, as described previously, the probability that any given instance is included in the training dataset is 63.2 percent. Consequently, the probability of any instance being in the test dataset is 36.8 percent. In other words, the training data represents only 63.2 percent of available examples, some of which are repeated. In contrast with 10-fold CV, which uses 90 percent of examples for training, the bootstrap sample is less representative of the full dataset.

Because a model trained on only 63.2 percent of the training data is likely to perform worse than a model trained on a larger training set, the bootstrap's performance estimates may be substantially lower than what will be obtained when the model is later trained on the full dataset. A special case of bootstrapping, known as the **0.632 bootstrap**, accounts for this by calculating the final performance measure as a function of performance on both the training data (which is overly optimistic) and the test data (which is overly pessimistic). The final error rate is then estimated as:

$$\text{error} = 0.632 \times \text{error}_{\text{test}} + 0.368 \times \text{error}_{\text{train}}$$

One advantage of bootstrap sampling over cross-validation is that it tends to work better with very small datasets. Additionally, bootstrap sampling has applications beyond performance measurement. In particular, in the following chapter we'll learn how the principles of bootstrap sampling can be used to improve model performance.

Summary

This chapter presented a number of the most common measures and techniques for evaluating the performance of machine learning classification models. Although accuracy provides a simple method for examining how often a model is correct, this can be misleading in the case of rare events because the real-life cost of such events may be inversely proportional to how frequently they appear in the data.

A number of measures based on confusion matrices better capture the balance among the costs of various types of errors. Closely examining the tradeoffs between sensitivity and specificity, or precision and recall, can be a useful tool for thinking about the implications of errors in the real world. Visualizations such as the ROC curve are also helpful to this end.

It is also worth mentioning that sometimes the best measure of a model's performance is to consider how well it meets, or doesn't meet, other objectives. For instance, you may need to explain a model's logic in simple language, which would eliminate some models from consideration. Additionally, even if it performs very well, a model that is too slow or difficult to scale to a production environment is completely useless.

An obvious extension of measuring performance is to identify automated ways to find the "best" models for a particular task. In the next chapter, we will build upon our work so far to investigate ways to make smarter models by systematically iterating, refining, and combining learning algorithms.

11

Improving Model Performance

When a sports team falls short of meeting its goal—whether the goal is to obtain an Olympic gold medal, a league championship, or a world record time—it must search for possible improvements. Imagine that you're the team's coach. How would you spend your practice sessions? Perhaps you'd direct the athletes to train harder or train differently in order to maximize every bit of their potential. Or, you might emphasize better teamwork, utilizing the athletes' strengths and weaknesses more smartly.

Now imagine that you're training a world champion machine learning algorithm. Perhaps you hope to compete in data mining competitions such as those posted on Kaggle (http://www.kaggle.com/). Maybe you simply need to improve business results. Where do you begin? Although the context differs, the strategies one uses to improve a sports team performance can also be used to improve the performance of statistical learners.

As the coach, it is your job to find the combination of training techniques and teamwork skills that allow you to meet your performance goals. This chapter builds upon the material covered throughout this book to introduce a set of techniques for improving the predictive performance of machine learners. You will learn:

- How to automate model performance tuning by systematically searching for the optimal set of training conditions

- Methods for combining models into groups that use teamwork to tackle tough learning tasks

- How to apply a variant of decision trees that has quickly become popular due to its impressive performance

None of these methods will be successful for every problem. Yet looking at the winning entries to machine learning competitions, you'll likely find at least one of them has been employed. To be competitive, you too will need to add these skills to your repertoire.

Tuning stock models for better performance

Some learning problems are well suited to the stock models presented in previous chapters. In such cases, it may not be necessary to spend much time iterating and refining the model; it may perform well enough as it is. On the other hand, some problems are inherently more difficult. The underlying concepts to be learned may be extremely complex, requiring an understanding of many subtle relationships, or the problem may be affected by random variation, making it difficult to define the signal within the noise.

Developing models that perform extremely well on difficult problems is every bit an art as it is a science. Sometimes a bit of intuition is helpful when trying to identify areas where performance can be improved. In other cases, finding improvements will require a brute-force, trial-and-error approach. Of course, the process of searching numerous possible improvements can be aided by the use of automated programs.

In *Chapter 5, Divide and Conquer – Classification Using Decision Trees and Rules*, we attempted a difficult problem: identifying loans that were likely to enter into default. Although we were able to use performance tuning methods to obtain a respectable classification accuracy of about 82 percent, upon a more careful examination in *Chapter 10, Evaluating Model Performance*, we realized that the high accuracy was a bit misleading. In spite of the reasonable accuracy, the kappa statistic was only about 0.28, which suggested that the model was actually performing somewhat poorly. In this section, we'll revisit the credit scoring model to see whether we can improve the results.

 To follow along with the examples, download the credit.csv file from the Packt Publishing website and save it to your R working directory. Load the file into R using the following command: credit <- read. csv("credit.csv").

You will recall that we first used a stock C5.0 decision tree to build the classifier for the credit data. We then attempted to improve its performance by adjusting the trials parameter to increase the number of boosting iterations. By increasing the number of iterations from the default of one up to the value of 10, we were able to increase the model's accuracy. This process of adjusting the model options to identify the best fit is called **parameter tuning**.

Parameter tuning is not limited to decision trees. For instance, we tuned k-NN models when we searched for the best value of *k*. We also tuned neural networks and support vector machines as we adjusted the number of nodes, the number of hidden layers, or chose different kernel functions. Most machine learning algorithms allow the adjustment of at least one parameter, and the most sophisticated models offer a large number of ways to tweak the model fit. Although this allows the model to be tailored closely to the learning task, the complexity of all the possible options can be daunting. A more systematic approach is warranted.

Using caret for automated parameter tuning

Rather than choosing arbitrary values for each of the model's parameters—a task that is not only tedious but also somewhat unscientific—it is better to conduct a search through many possible parameter values to find the best combination.

The `caret` package, which we used extensively in *Chapter 10, Evaluating Model Performance*, provides tools to assist with automated parameter tuning. The core functionality is provided by a `train()` function that serves as a standardized interface for over 200 different machine learning models for both classification and regression tasks. By using this function, it is possible to automate the search for optimal models using a choice of evaluation methods and metrics.

 Do not feel overwhelmed by the large number of models—we've already covered many of them in earlier chapters. Others are simple variants or extensions of the base concepts. Given what you've learned so far, you should be confident that you have the ability to understand all of the available methods.

Automated parameter tuning requires you to consider three questions:

- What type of machine learning model (and specific implementation) should be trained on the data?

- Which model parameters can be adjusted, and how extensively should they be tuned to find the optimal settings?

- What criteria should be used to evaluate the models to find the best candidate?

Answering the first question involves finding a match between the machine learning task and one of the 200+ models available to the `caret` package. Obviously, this requires an understanding of the breadth and depth of machine learning models. It can also help to work through a process of elimination.

Nearly half of the models can be eliminated depending on whether the task is classification or numeric prediction; others can be excluded based on the format of the data or the need to avoid black box models, and so on. In any case, there's also no reason you can't try several approaches and compare the best results of each.

Addressing the second question is a matter largely dictated by the choice of model, since each algorithm utilizes a unique set of parameters. The available tuning parameters for the predictive models covered in this book are listed in the following table. Keep in mind that although some models have additional options not shown, only those listed in the table are supported by `caret` for automatic tuning.

Model	Learning Task	Method Name	Parameters
k-Nearest Neighbors	Classification	`knn`	`k`
Naive Bayes	Classification	`nb`	`fL, usekernel`
Decision Trees	Classification	`C5.0`	`model, trials, winnow`
OneR Rule Learner	Classification	`OneR`	None
RIPPER Rule Learner	Classification	`JRip`	`NumOpt`
Linear Regression	Regression	`lm`	None
Regression Trees	Regression	`rpart`	`cp`
Model Trees	Regression	`M5`	`pruned, smoothed, rules`
Neural Networks	Dual Use	`nnet`	`size, decay`
Support Vector Machines (Linear Kernel)	Dual Use	`svmLinear`	`C`
Support Vector Machines (Radial Basis Kernel)	Dual Use	`svmRadial`	`C, sigma`
Random Forests	Dual Use	`rf`	`mtry`

 For a complete list of the models and corresponding tuning parameters covered by `caret`, refer to the table provided by package author Max Kuhn at `http://topepo.github.io/caret/modelList.html`

If you ever forget the tuning parameters for a particular model, the `modelLookup()` function can be used to find them. Simply supply the method name as illustrated for the C5.0 model:

```
> modelLookup("C5.0")
  model parameter                   label  forReg forClass probModel
1  C5.0    trials # Boosting Iterations   FALSE     TRUE      TRUE
2  C5.0     model            Model Type   FALSE     TRUE      TRUE
3  C5.0    winnow                Winnow   FALSE     TRUE      TRUE
```

The goal of automatic tuning is to search a set of candidate models comprising a matrix, or **grid**, of parameter combinations. Because it is impractical to search every conceivable combination, only a subset of possibilities is used to construct the grid. By default, `caret` searches at most three values for each of the model's p parameters, which means that at most 3^p candidate models will be tested. For example, by default, the automatic tuning of k-NN will compare $3^1 = 3$ candidate models with `k=5`, `k=7`, and `k=9`. Similarly, tuning a decision tree will result in a comparison of up to 27 different candidate models, comprising the grid of $3^3 = 27$ combinations of `model`, `trials`, and `winnow` settings. In practice, however, only 12 models are actually tested. This is because the `model` and `winnow` parameters can only take two values (`tree` versus `rules` and `TRUE` versus `FALSE`, respectively), which makes the grid size $3*2*2 = 12$.

> Since the default search grid may not be ideal for your learning problem, `caret` allows you to provide a custom search grid defined by a simple command that we will cover later.

The third and final step in automatic model tuning involves identifying the best model among the candidates. This uses the methods discussed in *Chapter 10, Evaluating Model Performance*, including the choice of resampling strategy for creating training and test datasets, and the use of model performance statistics to measure the predictive accuracy.

All of the resampling strategies and many of the performance statistics we've learned are supported by `caret`. These include statistics such as accuracy and kappa (for classifiers), and R-squared or RMSE (for numeric models). Cost-sensitive measures like sensitivity, specificity, and AUC can also be used if desired.

By default, `caret` will select the candidate model with the best value of the desired performance measure. Because this practice sometimes results in the selection of models that achieve minor performance improvements via large increases in model complexity, alternative model selection functions are provided.

Given the wide variety of options, it is helpful that many of the defaults are reasonable. For instance, `caret` will use prediction accuracy on a bootstrap sample to choose the best performer for classification models. Beginning with these default values, we can then tweak the `train()` function to design a wide variety of experiments.

Creating a simple tuned model

To illustrate the process of tuning a model, let's begin by observing what happens when we attempt to tune the credit scoring model using the `caret` package's default settings. From there, we will adjust the options to our liking.

The simplest way to tune a learner requires only that you specify a model type via the `method` parameter. Since we used C5.0 decision trees previously with the credit model, we'll continue our work by optimizing this learner. The basic `train()` command for tuning a C5.0 decision tree using the default settings is as follows:

```
> library(caret)
> RNGversion("3.5.2")
> set.seed(300)
> m <- train(default ~ ., data = credit, method = "C5.0")
```

First, the `set.seed()` function is used to initialize R's random number generator to a set starting position. You may recall that we used this function in several prior chapters. By setting the `seed` parameter (in this case to the arbitrary number `300`), the random numbers will follow a predefined sequence. This allows simulations that use random sampling to be repeated with identical results—a very helpful feature if you are sharing code or attempting to replicate a prior result.

Next, we define a tree as `default ~ .` using the R formula interface. This models loan default status (`yes` or `no`) using all of the other features in the `credit` data frame. The parameter `method = "C5.0"` tells `caret` to use the C5.0 decision tree algorithm.

After you've entered the preceding command, there may be a significant delay (depending upon your computer's capabilities) as the tuning process occurs. Even though this is a fairly small dataset, a substantial amount of calculation must occur. R must repeatedly generate random samples of data, build decision trees, compute performance statistics, and evaluate the result.

The result of the experiment is saved in an object named `m`. If you would like to examine the object's contents, the command `str(m)` will list all the associated data, but this can be substantial. Instead, simply type the name of the object for a condensed summary of the results. For instance, typing `m` yields the following output (note that labels have been added for clarity):

(1)
```
1000 samples
  16 predictor
   2 classes: 'no', 'yes'
```

(2)
```
No pre-processing
Resampling: Bootstrapped (25 reps)
Summary of sample sizes: 1000, 1000, 1000, 1000, 1000, 1000, ...
Resampling results across tuning parameters:
```

(3)
```
model  winnow  trials  Accuracy    Kappa
rules  FALSE    1      0.6960037   0.2750983
rules  FALSE   10      0.7147884   0.3181988
rules  FALSE   20      0.7233793   0.3342634
rules  TRUE     1      0.6849914   0.2513442
rules  TRUE    10      0.7126357   0.3156326
rules  TRUE    20      0.7225179   0.3342797
tree   FALSE    1      0.6888248   0.2487963
tree   FALSE   10      0.7310421   0.3148572
tree   FALSE   20      0.7362375   0.3271043
tree   TRUE     1      0.6814831   0.2317101
tree   TRUE    10      0.7285510   0.3093354
tree   TRUE    20      0.7324992   0.3200752
```

(4)
```
Accuracy was used to select the optimal model using the largest value.
The final values used for the model were trials = 20, model = tree
and winnow = FALSE.
```

The labels highlight four main components in the output:

1. **A brief description of the input dataset**: If you are familiar with your data and have applied the `train()` function correctly, this information should not be surprising.

2. **A report of the preprocessing and resampling methods applied**: Here we see that 25 bootstrap samples, each including 1,000 examples, were used to train the models.

3. **A list of the candidate models evaluated**: In this section, we can confirm that 12 different models were tested based on the combinations of three C5.0 tuning parameters: `model`, `trials`, and `winnow`. The average accuracy and kappa statistics for each candidate model are also shown.

4. **The choice of best model**: As the footnote describes, the model with the largest accuracy was selected. This was the C5.0 model that used a decision tree with 20 trials and the setting `winnow = FALSE`.

After identifying the best model, the `train()` function uses its tuning parameters to build a model on the full input dataset, which is stored in the `m` list object as `m$finalModel`. In most cases, you will not need to work directly with the `finalModel` sub-object. Instead, simply use the `predict()` function with the `m` object as follows:

```
> p <- predict(m, credit)
```

The resulting vector of predictions works as expected, allowing us to create a confusion matrix that compares the predicted and actual values:

```
> table(p, credit)
```

```
p      no yes
  no  700   2
  yes   0 298
```

Of the 1,000 examples used for training the final model, only two were misclassified. However, it is very important to note that since the model was built on both the training and test data, this accuracy is optimistic and thus should not be viewed as indicative of performance on unseen data. The bootstrap estimate of 73 percent (shown in the `train()` model summary output) is a more realistic estimate of future performance.

In addition to the automatic parameter tuning, using the `caret` package's `train()` and `predict()` functions also offers a couple of benefits beyond the functions found in the stock packages.

First, any data preparation steps applied by the `train()` function will be similarly applied to the data used for generating predictions. This includes transformations like centering and scaling, as well as imputation of missing values. Allowing `caret` to handle the data preparation will ensure that the steps that contributed to the best model's performance will remain in place when the model is deployed.

Second, the `predict()` function provides a standardized interface for obtaining predicted class values and predicted class probabilities, even for model types that ordinarily would require additional steps to obtain this information. The predicted classes are provided by default:

```
> head(predict(m, credit))
[1] no  yes no  no  yes no
Levels: no yes
```

To obtain the estimated probabilities for each class, use the `type = "prob"` parameter:

```
> head(predict(m, credit, type = "prob"))
          no         yes
1 0.9606970 0.03930299
2 0.1388444 0.86115561
3 1.0000000 0.00000000
4 0.7720279 0.22797208
5 0.2948062 0.70519385
6 0.8583715 0.14162851
```

Even in cases where the underlying model refers to the prediction probabilities using a different string (for example, `"raw"` for a `naiveBayes` model), the `predict()` function will translate `type = "prob"` to the appropriate parameter setting automatically.

Customizing the tuning process

The decision tree we created previously demonstrates the `caret` package's ability to produce an optimized model with minimal intervention. The default settings allow optimized models to be created easily. However, it is also possible to change the default settings to something more specific to a learning task, which may assist with unlocking the upper echelon of performance.

Each step in the model selection process can be customized. To illustrate this flexibility, let's modify our work on the credit decision tree to mirror the process we used in *Chapter 10, Evaluating Model Performance*. If you recall, we had estimated the kappa statistic using 10-fold CV. We'll do the same here, using kappa to optimize the boosting parameter of the decision tree. Note that decision tree boosting was previously covered in *Chapter 5, Divide and Conquer – Classification Using Decision Trees and Rules*, and will also be covered in greater detail later in this chapter.

The `trainControl()` function is used to create a set of configuration options known as a **control object**. This object guides the `train()` function and allows for the selection of model evaluation criteria, such as the resampling strategy and the measure used for choosing the best model. Although this function can be used to modify nearly every aspect of a tuning experiment, we'll focus on two important parameters: `method` and `selectionFunction`.

If you're eager for more details, you can use the `?trainControl` command for a list of all the parameters.

For the `trainControl()` function, the `method` parameter is used to set the resampling method, such as holdout sampling or k-fold CV. The following table lists the possible `method` types, as well as any additional parameters for adjusting the sample size and number of iterations. Although the default options for these resampling methods follow popular convention, you may choose to adjust these depending upon the size of your dataset and the complexity of your model.

Resampling Method	Method Name	Additional Options and Default Values
Holdout sampling	LGOCV	`p = 0.75` (training data proportion)
k-fold CV	cv	`number = 10` (number of folds)
Repeated k-fold CV	repeatedcv	`number = 10` (number of folds)
		`repeats = 10` (number of iterations)
Bootstrap sampling	boot	`number = 25` (resampling iterations)
0.632 bootstrap	boot632	`number = 25` (resampling iterations)
Leave-one-out CV	LOOCV	None

The `selectionFunction` parameter is used to specify the function that will choose the optimal model among the various candidates. Three such functions are included. The `best` function simply chooses the candidate with the best value on the specified performance measure. This is used by default. The other two functions are used to choose the most parsimonious, or simplest, model that is within a certain threshold of the best model's performance. The `oneSE` function chooses the simplest candidate within one standard error of the best performance, and `tolerance` uses the simplest candidate within a user-specified percentage.

Some subjectivity is involved with the `caret` package's ranking of models by simplicity. For information on how models are ranked, see the help page for the selection functions by typing `?best` at the R command prompt.

To create a control object named `ctrl` that uses 10-fold CV and the `oneSE` selection function, use the following command (note that `number = 10` is included only for clarity; since this is the default value for `method = "cv"`, it could have been omitted):

```
> ctrl <- trainControl(method = "cv", number = 10,
                       selectionFunction = "oneSE")
```

We'll use the result of this function shortly.

In the meantime, the next step in defining our experiment is to create the grid of parameters to optimize. The grid must include a column named for each tuning parameter in the desired model. It must also include a row for each desired combination of parameter values. Since we are using a C5.0 decision tree, this means we'll need columns named `model`, `trials`, and `winnow`. For other machine learning models, refer to the table presented earlier in this chapter or use the `modelLookup()` function to lookup the parameters as described previously.

Rather than filling this data frame cell by cell—a tedious task if there are many possible combinations of parameter values—we can use the `expand.grid()` function, which creates data frames from the combinations of all values supplied. For example, suppose we would like to hold constant `model = "tree"` and `winnow = FALSE` while searching eight different values of trials. This can be created as:

```
> grid <- expand.grid(model = "tree",
                       trials = c(1, 5, 10, 15, 20, 25, 30, 35),
                       winnow = FALSE)
```

The resulting `grid` data frame contains *1*8*1 = 8* rows:

```
> grid
  model trials winnow
1  tree      1  FALSE
2  tree      5  FALSE
3  tree     10  FALSE
4  tree     15  FALSE
5  tree     20  FALSE
6  tree     25  FALSE
7  tree     30  FALSE
8  tree     35  FALSE
```

The `train()` function will build a candidate model for evaluation using each row's combination of model parameters.

Given this search grid and the control list created previously, we are ready to run a thoroughly customized `train()` experiment. As before, we'll set the random seed to the arbitrary number `300` in order to ensure repeatable results. But this time, we'll pass our control object and tuning grid while adding a parameter `metric = "Kappa"`, indicating the statistic to be used by the model evaluation function—in this case, `"oneSE"`. The full command is as follows:

```
> RNGversion("3.5.2")
> set.seed(300)
```

```
> m <- train(default ~ ., data = credit, method = "C5.0",
             metric = "Kappa",
             trControl = ctrl,
             tuneGrid = grid)
```

This results in an object that we can view by typing its name:

```
> m
```

```
1000 samples
  16 predictor
   2 classes: 'no', 'yes'

No pre-processing
Resampling: Cross-Validated (10 fold)
Summary of sample sizes: 900, 900, 900, 900, 900, 900, ...
Resampling results across tuning parameters:

  trials  Accuracy  Kappa
  1       0.735     0.3243679
  5       0.722     0.2941429
  10      0.725     0.2954364
  15      0.731     0.3141866
  20      0.737     0.3245897
  25      0.726     0.2972530
  30      0.735     0.3233492
  35      0.736     0.3193931

Tuning parameter 'model' was held constant at a value of tree
Tuning parameter 'winnow' was held constant at  a value of FALSE
Kappa was used to select the optimal model using  the one SE rule.
The final values used for the model were trials = 1, model = tree
and winnow = FALSE.
```

Although much of the output is similar to the automatically tuned model, there are a few differences of note. Because 10-fold CV was used, the sample size to build each candidate model was reduced to 900 rather than the 1,000 used in the bootstrap. As we requested, eight candidate models were tested. Additionally, because model and winnow were held constant, their values are no longer shown in the results; instead, they are listed as a footnote.

The best model here differs quite significantly from the prior experiment. Before, the best model used trials = 20, whereas here, it used trials = 1. This change is due to the fact that we used the oneSE rule rather than the best rule to select the optimal model. Even though the 35-trial model offers the best raw performance according to kappa, the single-trial model offers nearly the same performance with a much simpler algorithm.

> Due to the large number of configuration parameters, `caret` can seem
> overwhelming at first. Don't let this deter you—there is no easier way
> to test the performance of models using 10-fold CV. Instead, think of
> the experiment as defined by two parts: a `trainControl()` object that
> dictates the testing criteria, and a tuning grid that determines what model
> parameters to evaluate. Supply these to the `train()` function and with
> a bit of computing time, your experiment will be complete!

Improving model performance with meta-learning

As an alternative to increasing the performance of a single model, it is possible
to combine several models to form a powerful team. Just as the best sports teams
have players with complementary rather than overlapping skillsets, some of the best
machine learning algorithms utilize teams of complementary models. Since a model
brings a unique bias to a learning task, it may readily learn one subset of examples,
but have trouble with another. Therefore, by intelligently using the talents of several
diverse team members, it is possible to create a strong team of multiple weak learners.

This technique of combining and managing the predictions of multiple models
falls into a wider set of **meta-learning** methods, which are techniques that involve
learning how to learn. This includes anything from simple algorithms that
gradually improve performance by iterating over design decisions—for instance,
the automated parameter tuning used earlier in this chapter—to highly complex
algorithms that use concepts borrowed from evolutionary biology and genetics
for self-modifying and adapting to learning tasks.

For the remainder of this chapter, we'll focus on meta-learning only as it pertains
to modeling a relationship between the predictions of several models and the desired
outcome. The teamwork-based techniques covered here are quite powerful and are
used quite often to build more effective classifiers.

Understanding ensembles

Suppose you were a contestant on a television trivia show that allowed you to choose
a panel of five friends to assist you with answering the final question for the million-
dollar prize. Most people would try to stack the panel with a diverse set of subject
matter experts. A panel containing professors of literature, science, history, and art,
along with a current pop-culture expert would be safely well rounded. Given their
breadth of knowledge, it would be unlikely that a question would stump the group.

The meta-learning approach that utilizes a similar principle of creating a varied team of experts is known as an **ensemble**. All ensemble methods are based on the idea that by combining multiple weaker learners, a stronger learner is created. The various ensemble methods can be distinguished, in large part, by the answers to two questions:

- How are the weak learning models chosen and/or constructed?
- How are the weak learners' predictions combined to make a single final prediction?

When answering these questions, it can be helpful to imagine the ensemble in terms of the following process diagram; nearly all ensemble approaches follow this pattern:

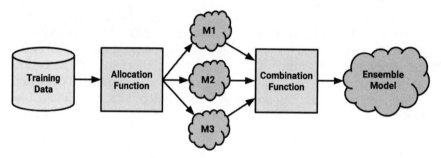

Figure 11.1: Ensembles combine multiple weaker models into a single stronger model

First, input training data is used to build a number of models. The **allocation function** dictates how much of the training data each model receives. Do they each receive the full training dataset or merely a sample? Do they each receive every feature or a subset?

Although the ideal ensemble includes a diverse set of models, the allocation function can increase diversity by artificially varying the input data to bias the resulting learners, even if they are the same type. For instance, in an ensemble of decision trees, the allocation function might use bootstrap sampling to construct unique training datasets for each tree, or it may pass each one a different subset of features. On the other hand, if the ensemble already includes a diverse set of algorithms — such as a neural network, a decision tree, and a k-NN classifier — then the allocation function might pass the data on to each algorithm relatively unchanged.

After the ensemble's models are constructed, they can be used to generate a set of predictions, which must be managed in some way. The **combination function** governs how disagreements among the predictions are reconciled. For example, the ensemble might use a majority vote to determine the final prediction, or it could use a more complex strategy such as weighting each model's votes based on its prior performance.

Some ensembles even utilize another model to learn a combination function from various combinations of predictions. For example, suppose that when M_1 and M_2 both vote yes, the actual class value is usually no. In this case, the ensemble could learn to ignore the vote of M_1 and M_2 when they agree. This process of using the predictions of several models to train a final arbiter model is known as **stacking**.

Figure 11.2: Stacking is a sophisticated ensemble that uses a learning algorithm to combine predictions

One of the benefits of using ensembles is that they may allow you to spend less time in pursuit of a single best model. Instead, you can train a number of reasonably strong candidates and combine them. Yet convenience isn't the only reason why ensemble-based methods continue to rack up wins in machine learning competitions; ensembles also offer a number of performance advantages over single models:

- **Better generalizability to future problems**: As the opinions of several learners are incorporated into a single final prediction, no single bias is able to dominate. This reduces the chance of overfitting to a learning task.

- **Improved performance on massive or miniscule datasets**: Many models run into memory or complexity limits when an extremely large set of features or examples are used, making it more efficient to train several small models than a single full model. Conversely, ensembles also do well on the smallest datasets because resampling methods like bootstrapping are inherently part of many ensemble designs. Perhaps most importantly, it is often possible to train an ensemble in parallel using distributed computing methods.

- **The ability to synthesize data from distinct domains**: Since there is no one-size-fits-all learning algorithm, the ensemble's ability to incorporate evidence from multiple types of learners is increasingly important as complex phenomena rely on data drawn from diverse domains.

- **A more nuanced understanding of difficult learning tasks**: Real-world phenomena are often extremely complex, with many interacting intricacies. Models that divide the task into smaller portions are likely to more accurately capture subtle patterns that a single global model might miss.

None of these benefits would be very helpful if you weren't able to easily apply ensemble methods in R, and there are many packages available to do just that. Let's take a look at several of the most popular ensemble methods and how they can be used to improve the performance of the credit model we've been working on.

Bagging

One of the first ensemble methods to gain widespread acceptance used a technique called **bootstrap aggregating**, or **bagging** for short. As described by Leo Breiman in 1994, bagging generates a number of training datasets by bootstrap sampling the original training data. These datasets are then used to generate a set of models using a single learning algorithm. The models' predictions are combined using voting (for classification) or averaging (for numeric prediction).

 For additional information on bagging, refer to *Bagging predictors,
Breiman, L, Machine Learning, 1996, Vol. 24, pp. 123-140.*

Although bagging is a relatively simple ensemble, it can perform quite well as long as it is used with relatively **unstable** learners, that is, those generating models that tend to change substantially when the input data changes only slightly. Unstable models are essential in order to ensure the ensemble's diversity in spite of only minor variations between the bootstrap training datasets. For this reason, bagging is often used with decision trees, which have the tendency to vary dramatically given minor changes in input data.

The `ipred` package offers a classic implementation of bagged decision trees. To train the model, the `bagging()` function works similarly to many of the models used previously. The `nbagg` parameter is used to control the number of decision trees voting in the ensemble (with a default value of 25). Depending on the difficulty of the learning task and the amount of training data, increasing this number may improve the model's performance, up to a limit. The downside is that this creates additional computational expense, and a large number of trees may take some time to train.

After installing the `ipred` package, we can create the ensemble as follows. We'll stick to the default value of 25 decision trees:

```
> library(ipred)
> RNGversion("3.5.2")
> set.seed(300)
> mybag <- bagging(default ~ ., data = credit, nbagg = 25)
```

The resulting model works as expected with the `predict()` function:

```
> credit_pred <- predict(mybag, credit)
> table(credit_pred, credit$default)


credit_pred  no yes
        no  699   2
        yes   1 298
```

Given the preceding results, the model seems to have fit the training data extremely well. To see how this translates into future performance, we can use the bagged trees with 10-fold CV using the `train()` function in the `caret` package. Note that the method name for the `ipred` bagged trees function is `treebag`:

```
> library(caret)
> RNGversion("3.5.2")
> set.seed(300)
> ctrl <- trainControl(method = "cv", number = 10)
> train(default ~ ., data = credit, method = "treebag",
         trControl = ctrl)


Bagged CART


1000 samples
  16 predictor
   2 classes: 'no', 'yes'


No pre-processing
Resampling: Cross-Validated (10 fold)
Summary of sample sizes: 900, 900, 900, 900, 900, 900, ...
Resampling results:


  Accuracy  Kappa
  0.746     0.3540389
```

The kappa statistic of 0.35 for this model suggests that the bagged tree model performs at least as well as the best C5.0 decision tree we tuned earlier in this chapter, which had a 0.32 kappa statistic. This illustrates the power of ensemble methods; a set of simple learners, working together, can outperform very sophisticated models.

Boosting

Another common ensemble-based method is called **boosting**, because it boosts the performance of weak learners to attain the performance of stronger learners. This method is based largely on the work of Robert Schapire and Yoav Freund, who have published extensively on the topic.

 For additional information on boosting, refer to *Boosting: Foundations and Algorithms, Schapire, RE, Freund, Y, Cambridge, MA: The MIT Press*, 2012.

Similar to bagging, boosting uses ensembles of models trained on resampled data and a vote to determine the final prediction. There are two key distinctions. First, the resampled datasets in boosting are constructed specifically to generate complementary learners. Second, rather than giving each learner an equal vote, boosting gives each learner's vote a weight based on its past performance. Models that perform better have greater influence over the ensemble's final prediction.

Boosting will result in performance that is often better and certainly no worse than the best of the models in the ensemble. Since the models in the ensemble are built to be complementary, it is possible to increase ensemble performance to an arbitrary threshold simply by adding additional classifiers to the group, assuming that each additional classifier performs better than random chance. Given the obvious utility of this finding, boosting is thought to be one of the most significant discoveries in machine learning.

 Although boosting can create a model that meets an arbitrarily low error rate, this may not always be reasonable in practice. For one, the performance gains are incrementally smaller as additional learners are gained, making some thresholds practically infeasible. Additionally, the pursuit of pure accuracy may result in the model being overfitted to the training data and not generalizable to unseen data.

A boosting algorithm called **AdaBoost**, or **adaptive boosting**, was proposed by Freund and Schapire in 1997. The algorithm is based on the idea of generating weak leaners that iteratively learn a larger portion of the difficult-to-classify examples in the training data by paying more attention (that is, giving more weight) to often misclassified examples.

Beginning from an unweighted dataset, the first classifier attempts to model the outcome. Examples that the classifier predicted correctly will be less likely to appear in the training dataset for the following classifier, and conversely, the difficult-to-classify examples will appear more frequently.

As additional rounds of weak learners are added, they are trained on data with successively more difficult examples. The process continues until the desired overall error rate is reached or performance no longer improves. At that point, each classifier's vote is weighted according to its accuracy on the training data on which it was built.

Though boosting principles can be applied to nearly any type of model, the principles are most commonly used with decision trees. We already used boosting in this way in *Chapter 5, Divide and Conquer – Classification Using Decision Trees and Rules,* as a method to improve the performance of a C5.0 decision tree.

The **AdaBoost.M1** algorithm provides another tree-based implementation of AdaBoost for classification. The AdaBoost.M1 algorithm can be found in the adabag package.

 For more information about the adabag package, refer to *adabag: An R Package for Classification with Boosting and Bagging, Alfaro, E, Gamez, M, Garcia, N, Journal of Statistical Software, 2013, Vol. 54, pp. 1-35.*

Let's create an AdaBoost.M1 classifier for the credit data. The general syntax for this algorithm is similar to other modeling techniques:

```
> RNGversion("3.5.2")
> set.seed(300)
> m_adaboost <- boosting(default ~ ., data = credit)
```

As usual, the predict() function is applied to the resulting object to make predictions:

```
> p_adaboost <- predict(m_adaboost, credit)
```

Departing from convention, rather than returning a vector of predictions, this returns an object with information about the model. The predictions are stored in a sub-object called class:

```
> head(p_adaboost$class)
[1] "no"  "yes" "no"  "no"  "yes" "no"
```

A confusion matrix can be found in the confusion sub-object:

```
> p_adaboost$confusion
                Observed Class
Predicted Class  no yes
            no  700   0
            yes   0 300
```

Did you notice that the AdaBoost model made no mistakes? Before you get your hopes up, remember that the preceding confusion matrix is based on the model's performance on the training data. Since boosting allows the error rate to be reduced to an arbitrarily low level, the learner simply continued until it made no more errors. This likely resulted in overfitting on the training dataset.

For a more accurate assessment of performance on unseen data, we need to use another evaluation method. The `adabag` package provides a simple function to use 10-fold CV:

```
> RNGversion("3.5.2")
> set.seed(300)
> adaboost_cv <- boosting.cv(default ~ ., data = credit)
```

Depending on your computer's capabilities, this may take some time to run, during which it will log each iteration to screen—on my recent MacBook Pro computer, it took about four minutes. After it completes, we can view a more reasonable confusion matrix:

```
> adaboost_cv$confusion
                Observed Class
Predicted Class   no yes
            no   594 151
            yes 106 149
```

We can find the kappa statistic using the `vcd` package as described in *Chapter 10, Evaluating Model Performance*:

```
> library(vcd)
> Kappa(adaboost_cv$confusion)
             value    ASE     z  Pr(>|z|)
Unweighted 0.3607 0.0323 11.17 5.914e-29
Weighted   0.3607 0.0323 11.17 5.914e-29
```

With a kappa of about 0.361, this is our best-performing credit scoring model yet. Let's see how it compares to one last ensemble method.

[The AdaBoost.M1 algorithm can be tuned in `caret` by specifying `method = "AdaBoost.M1"`.]

Random forests

Another ensemble-based method called **random forests** (or **decision tree forests**) focuses only on ensembles of decision trees. This method was championed by Leo Breiman and Adele Cutler, and combines the base principles of bagging with random feature selection to add additional diversity to the decision tree models. After the ensemble of trees (the forest) is generated, the model uses a vote to combine the trees' predictions.

For more detail on how random forests are constructed, refer to *Random Forests, Breiman, L, Machine Learning, 2001, Vol. 45, pp. 5-32.*

Random forests combine versatility and power into a single machine learning approach. Because the ensemble uses only a small, random portion of the full feature set, random forests can handle extremely large datasets, where the so-called "curse of dimensionality" might cause other models to fail. At the same time, its error rates for most learning tasks are on a par with nearly any other method.

Although the term "random forests" is trademarked by Breiman and Cutler, the term is sometimes used colloquially to refer to any type of decision tree ensemble. A pedant would use the more general term "decision tree forests" except when referring to the specific implementation by Breiman and Cutler.

It's worth noting that relative to other ensemble-based methods, random forests are quite competitive and offer key advantages. For instance, random forests tend to be easier to use and less prone to overfitting. The following table lists the general strengths and weaknesses of random forest models:

Strengths	Weaknesses
• An all-purpose model that performs well on most problems	• Unlike a decision tree, the model is not easily interpretable
• Can handle noisy or missing data, as well as categorical or continuous features	
• Selects only the most important features	
• Can be used on data with an extremely large number of features or examples	

Due to their power, versatility, and ease of use, random forests are one of the most popular machine learning methods. Later in this chapter, we'll compare a random forest model head-to-head against the boosted C5.0 tree.

Training random forests

Though there are several packages to create random forests in R, the randomForest package is perhaps the implementation most faithful to the specification by Breiman and Cutler, and is also supported by caret for automated tuning. The syntax for training this model is as follows:

Random forest syntax

using the `randomForest()` function in the `randomForest` package

Building the classifier:

```
m <- randomForest(train, class, ntree = 500, mtry = sqrt(p))
```

* `train` is a data frame containing training data
* `class` is a factor vector with the class for each row in the training data
* `ntree` is an integer specifying the number of trees to grow
* `mtry` is an optional integer specifying the number of features to randomly select at each split (uses `sqrt(p)` by default, where `p` is the number of features in the data)

The function will return a random forest object that can be used to make predictions.

Making predictions:

```
p <- predict(m, test, type = "response")
```

* `m` is a model trained by the `randomForest()` function
* `test` is a data frame containing test data with the same features as the training data used to build the classifier
* `type` is either `"response"`, `"prob"`, or `"votes"` and is used to indicate whether the predictions vector should contain the predicted class, the predicted probabilities, or a matrix of vote counts, respectively.

The function will return predictions according to the value of the `type` parameter.

Example:

```
credit_model <- randomForest(credit_train, loan_default)
credit_prediction <- predict(credit_model, credit_test)
```

By default, the randomForest() function creates an ensemble of 500 trees that consider sqrt(p) random features at each split, where p is the number of features in the training dataset and sqrt() refers to R's square root function. Whether or not these default parameters are appropriate depends on the nature of the learning task and training data. Generally, more complex learning problems and larger datasets (both more features as well as more examples) work better with a larger number of trees, though this needs to be balanced with the computational expense of training more trees.

The goal of using a large number of trees is to train enough that each feature has a chance to appear in several models. This is the basis of the sqrt (p) default value for the mtry parameter; using this value limits the features sufficiently such that substantial random variation occurs from tree to tree. For example, since the credit data has 16 features, each tree would be limited to splitting on four features at any time.

Let's see how the default randomForest () parameters work with the credit data. We'll train the model just as we have done with other learners. Again, the set. seed() function ensures that the result can be replicated:

```
> library(randomForest)
> RNGversion("3.5.2")
> set.seed(300)
> rf <- randomForest(default ~ ., data = credit)
```

To look at a summary of the model's performance, we can simply type the resulting object's name:

```
> rf


Call:
 randomForest(formula = default ~ ., data = credit)
               Type of random forest: classification
                     Number of trees: 500
No. of variables tried at each split: 4


        OOB estimate of  error rate: 23.3%
Confusion matrix:
      no yes class.error
no   638  62  0.08857143
yes 171 129  0.57000000
```

The output shows that the random forest included 500 trees and tried four variables at each split, as expected. At first glance, you might be alarmed at the seemingly poor performance according to the confusion matrix—the error rate of 23.3 percent is far worse than the resubstitution error of any of the other ensemble methods so far. However, this confusion matrix does not show resubstitution error. Instead, it reflects the **out-of-bag error rate** (listed in the output as OOB estimate of error rate), which, unlike resubstitution error, is an unbiased estimate of the test set error. This means that it should be a fairly reasonable estimate of future performance.

The out-of-bag estimate is computed during the construction of the random forest. Essentially, any example not selected for a single tree's bootstrap sample can be used to test the model's performance on unseen data. At the end of the forest construction, for each of the 1,000 examples in the dataset, the trees that did not use the example in training are allowed to make a prediction. These predictions are tallied for each example and a vote is taken to determine the single final prediction for the example. The total error rate of such predictions becomes the out-of-bag error rate.

> In *Chapter 10, Evaluating Model Performance*, it was stated that any given example has a 63.2 percent chance of being included in a bootstrap sample. This implies that an average of 36.8 percent of the 500 trees in the random forest voted for each of the 1,000 examples in the out-of-bag estimate.

To calculate the kappa statistic on the out-of-bag predictions, we can use the function in the vcd package as follows. The code applies the Kappa() function to the first two rows and columns of the confusion object, which stores the confusion matrix of the out-of-bag predictions for the rf random forest model object:

```
> library(vcd)
> Kappa(rf$confusion[1:2,1:2])
           value     ASE      z   Pr(>|z|)
Unweighted 0.381  0.03215  11.85  2.197e-32
Weighted   0.381  0.03215  11.85  2.197e-32
```

With a kappa statistic of 0.381, the random forest is our best-performing model yet. It was higher than the best boosted C5.0 decision tree, which had a kappa of about 0.325, and also higher than the AdaBoost.M1 model, which had a kappa of about 0.361. Given this impressive initial result, we should attempt a more formal evaluation of its performance.

Evaluating random forest performance in a simulated competition

As mentioned previously, the randomForest() function is supported by caret, which allows us to optimize the model while at the same time calculating performance measures beyond the out-of-bag error rate. To make things interesting, let's compare an auto-tuned random forest to the best auto-tuned boosted C5.0 model we've developed. We'll treat this experiment as if we were hoping to identify a candidate model for submission to a machine learning competition.

We must first load `caret` and set our training control options. For the most accurate comparison of model performance, we'll use repeated 10-fold CV, or 10-fold CV repeated 10 times. This means that the models will take a much longer time to build and will be more computationally intensive to evaluate, but since this is our final comparison, we should be *very* sure that we're making the right choice; the winner of this showdown, selected via the "best" performance metric, will be our only entry into the machine learning competition.

Additionally, we'll add a few new options to the `trainControl()` function. First, we'll set the `savePredictions` and `classProbs` parameters to `TRUE`, which saves the holdout sample predictions and predicted probabilities for plotting the ROC curve later. Then, we'll also set the `summaryFunction` to `twoClassSummary`, which is a `caret` function that computes performance metrics like AUC. The full control object is defined as follows:

```
> library(caret)
> ctrl <- trainControl(method = "repeatedcv",
                       number = 10, repeats = 10,
                       selectionFunction = "best",
                       savePredictions = TRUE,
                       classProbs = TRUE,
                       summaryFunction = twoClassSummary)
```

Next, we'll set up the tuning grid for the random forest. The only tuning parameter for this model is `mtry`, which defines how many features are randomly selected at each split. By default, we know that the random forest will use `sqrt(16)`, or four features per tree. To be thorough, we'll also test values half of that, and twice that, as well as the full set of 16 features. Thus, we need to create a grid with values of 2, 4, 8, and 16 as follows:

```
> grid_rf <- expand.grid(mtry = c(2, 4, 8, 16))
```

 A random forest that considers the full set of features at each split is essentially the same as a bagged decision tree model.

We'll supply the resulting grid to the `train()` function with the `ctrl` object, and use the `"ROC"` metric to select the best model. This metric refers to the area under the ROC curve. The complete experiment can be run as follows:

```
> RNGversion("3.5.2"); set.seed(300)
> m_rf <- train(default ~ ., data = credit, method = "rf",
                metric = "ROC", trControl = ctrl,
                tuneGrid = grid_rf)
```

The preceding command may take some time to complete, as it has quite a bit of work to do—on my recent MacBook Pro computer, it took about seven minutes! When the random forests are finished training, we'll compare the best forest to the best boosted decision tree among trees with 10, 25, 50, and 100 iterations using the following caret experiment:

```
> grid_c50 <- expand.grid(model = "tree",
                          trials = c(10, 25, 50, 100),
                          winnow = FALSE)
> RNGversion("3.5.2"); set.seed(300)
> m_c50 <- train(default ~ ., data = credit, method = "C5.0",
               metric = "ROC", trControl = ctrl,
               tuneGrid = grid_c50)
```

When the C5.0 decision tree finally completes, we can compare the two approaches side by side. For the random forest model, the results are:

```
> m_rf
```

Resampling results across tuning parameters:

mtry	ROC	Sens	Spec
2	0.7579643	0.9900000	0.09766667
4	0.7695071	0.9377143	0.30166667
8	0.7739714	0.9064286	0.38633333
16	0.7747905	0.8921429	0.44100000

For the boosted C5.0 model, the results are:

```
> m_c50
```

Resampling results across tuning parameters:

trials	ROC	Sens	Spec
10	0.7399571	0.8555714	0.4346667
25	0.7523238	0.8594286	0.4390000
50	0.7559857	0.8635714	0.4436667
100	0.7566286	0.8630000	0.4450000

Based on these head-to-head results, the random forest with mtry = 16 appears to be our winner, as its best AUC of 0.775 is better than the AUC of 0.757 for the best boosted C5.0 model.

To visualize their performance, we can use the pROC package to plot the ROC curves. We'll supply the roc() function with the observed (obs) values of the loan default, as well as the estimated probability of "yes" for the loan default. Note that these were saved by caret because we requested them via the trainControl() function. We can then plot() the ROC curves as follows:

```
> library(pROC)
> roc_rf <- roc(m_rf$pred$obs, m_rf$pred$yes)
> roc_c50 <- roc(m_c50$pred$obs, m_c50$pred$yes)
> plot(roc_rf, col = "red", legacy.axes = TRUE)
> plot(roc_c50, col = "blue", add = TRUE)
```

As anticipated, the resulting curves show that the random forest with an AUC of 0.775 slightly outperforms the boosted C5.0 model with an AUC of 0.757. The random forest is the outermost curve in the following R plot:

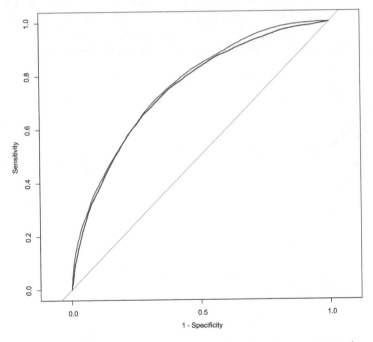

Figure 11.3: ROC curves comparing a random forest (outermost curve) to a boosted C5.0 decision tree on the loan default dataset

Based on our experiment, we would submit a random forest to the competition as our top-performing model. However, until it actually makes predictions on the competition test set, we have no way of knowing for sure whether our random forest will end up winning. Given our performance estimates, it's the safest bet of the models we evaluated, and with a bit of luck, perhaps we'll come away with the prize.

Summary

After reading this chapter, you should now know the base techniques that are used to win data mining and machine learning competitions. Automated tuning methods can assist with squeezing every bit of performance out of a single model. On the other hand, performance gains are also possible by creating groups of machine learning models that work together.

Although this chapter was designed to help you prepare competition-ready models, note that your fellow competitors have access to the same techniques. You won't be able to get away with stagnancy; therefore, continue to add proprietary methods to your bag of tricks. Perhaps you can bring unique subject-matter expertise to the table, or perhaps your strengths include an eye for detail in data preparation. In any case, practice makes perfect, so take advantage of open competitions to test, evaluate, and improve your own machine learning skillset.

In the next chapter—the last in this book—we'll take a bird's-eye look at ways to apply machine learning to some highly specialized and difficult domains using R. You'll gain the knowledge needed to apply machine learning to tasks at the cutting edge of the field, involving extremely large, challenging, or unusual datasets.

12

Specialized Machine Learning Topics

Congratulations on reaching this point in your machine learning journey! If you have not already started work on your own projects, you will do so soon. And in doing so, you may find that the task of turning data into action is more difficult than it first appeared.

As you gathered data, you might have realized that the information was trapped in a proprietary format or spread across pages on the web. Making matters worse, after spending hours reformatting the data, maybe your computer slowed to a crawl after it ran out of memory. Perhaps R even crashed or froze your machine. Hopefully you were undeterred, as these issues can be remedied with a bit more effort.

This chapter covers techniques that may not apply to every project, but will prove useful for working around such specialized issues. You might find the information particularly useful if you tend to work with data that is:

- Stored in unstructured or proprietary formats such as web pages, web APIs, spreadsheets, or databases

- From a specialized domain such as bioinformatics or social network analysis

- Too large to fit in memory or unable to complete analyses in a reasonable time

You're not alone if you suffer from any of these problems. Although there is no panacea — these issues are the bane of many data scientists as well as the reason data skills are in high demand — through the dedicated efforts of the R community, a number of R packages provide a head start toward solving these issues.

This chapter provides a cookbook of solutions. Even if you are an experienced R veteran, you may discover a package that simplifies your workflow. Or, perhaps one day you will author a package that makes work easier for everybody else!

Managing and preparing real-world data

Unlike the examples in this book, real-world data is rarely packaged in a simple CSV form that can be downloaded from a website. Instead, significant effort is needed to prepare data for analysis. Data must be collected, merged, sorted, filtered, or reformatted to meet the requirements of the learning algorithm. This process is known informally as **data munging** or **data wrangling**.

Data preparation has become even more important as the size of typical datasets has grown from megabytes to gigabytes and data is gathered from unrelated and messy sources, many of which are stored in massive databases. Several packages and resources for retrieving and working with proprietary data formats and databases are listed in the following sections.

Making data "tidy" with the tidyverse packages

A new approach has been rapidly taking shape as the dominant paradigm for working with data in R. Championed by Hadley Wickham, the mind behind many of the packages that drove much of R's initial surge of interest, this new wave is now backed by a much larger team at RStudio. The company's RStudio desktop application, which makes R substantially more user-friendly, integrates nicely into this new ecosystem, which is known as the **tidyverse** because it provides a universe of packages devoted to tidy data. The entire set can be installed with the `install.packages("tidyverse")` command.

A growing number of resources are available online to learn more about the tidyverse, starting with its homepage at `https://www.tidyverse.org`. Here, you can learn about the various packages included in the set, a few of which will be described in this chapter. Additionally, the book *R for Data Science* by Hadley Wickham and Garrett Grolemund is available freely online at `https://r4ds.had.co.nz` and illustrates how the tidyverse's "opinionated" approach simplifies data science projects.

 I am often asked the question of how R compares to Python for data science and machine learning. RStudio and the tidyverse are perhaps R's greatest asset and point of distinction. There is arguably no easier way to begin a data science journey. Once you've learned the "tidy" way of doing data analysis, you are likely to wish the tidyverse functionality existed everywhere!

Generalizing tabular data structures with tibble

The tidyverse collection includes the `tibble` package and data structure—named as a pun on the word "table." A **tibble** acts almost exactly like a data frame but includes additional functionality for convenience and simplicity. They can be used almost everywhere a data frame can be used. Detailed information about tibbles can be found in the corresponding *R for Data Science* chapter at https://r4ds.had.co.nz/tibbles.html or by typing the `vignette("tibble")` command in R.

Most of the time, using tibbles will be transparent and seamless. However, in case you need to convert a tibble to a data frame, use the `as.data.frame()` function. To go in the other direction and convert a data frame to a tibble, use the `as_tibble()` function as follows:

```
> library(tibble)
> credit <- read.csv("credit.csv")
> credit_tbl <- as_tibble(credit)
```

Typing the name of this object demonstrates the tibble's cleaner and more informative output than a standard data frame:

```
> credit_tbl
# A tibble: 1,000 x 17
   checking_balance months_loan_dura… credit_history purpose    amount savings_balance employment_dura…
   <fct>                        <int> <fct>          <fct>       <int> <fct>           <fct>
 1 < 0 DM                           6 critical       furnitur…    1169 unknown         > 7 years
 2 1 - 200 DM                      48 good           furnitur…    5951 < 100 DM        1 - 4 years
 3 unknown                         12 critical       education    2096 < 100 DM        4 - 7 years
 4 < 0 DM                          42 good           furnitur…    7882 < 100 DM        4 - 7 years
 5 < 0 DM                          24 poor           car          4870 < 100 DM        1 - 4 years
 6 unknown                         36 good           education    9055 unknown         1 - 4 years
 7 unknown                         24 good           furnitur…    2835 500 - 1000 DM   > 7 years
 8 1 - 200 DM                      36 good           car          6948 < 100 DM        1 - 4 years
 9 unknown                         12 good           furnitur…    3059 > 1000 DM       4 - 7 years
10 1 - 200 DM                      30 critical       car          5234 < 100 DM        unemployed
# … with 990 more rows, and 10 more variables: percent_of_income <int>, years_at_residence <int>,
#   age <int>, other_credit <fct>, housing <fct>, existing_loans_count <int>, job <fct>,
#   dependents <int>, phone <fct>, default <fct>
```

Figure 12.1: Displaying a tibble object results in more informative output than a standard data frame

It is important to note the distinctions between tibbles and data frames, as the tidyverse will automatically create a tibble object for many of its operations. Overall, you are likely to find that tibbles are less annoying than data frames. They generally make smarter assumptions about the data, which means you will spend less time redoing R's work—like recoding strings as factors or vice versa. Indeed, a key distinction between tibbles and data frames is that a tibble never assumes `stringsAsFactors = TRUE`. Additionally, a tibble can also use column names that would be invalid in base R, such as `` `my var` ``, as long as they are surrounded by the backtick (`` ` ``) character.

Tibbles are the base object in the tidyverse and enable the additional benefits of the complementary packages outlined in the sections that follow.

Speeding and simplifying data preparation with dplyr

The dplyr package is the heart of the tidyverse, as it provides the basic functionality that allows data to be transformed and manipulated. It also provides a straightforward way to begin working with larger datasets in R. Though there are other packages that have greater raw speed or are capable of handling even more massive datasets, dplyr is still quite capable, and is a good first step to take if you run into limitations with base R.

Combined with tidyverse tibble objects, dplyr unlocks some impressive functionality:

- With a focus on data frames rather than vectors, new operators are introduced that allow common data transformations to be performed with much less code while remaining highly readable.

- The dplyr package makes reasonable assumptions about data frames that optimize your effort as well as memory use. If possible, it avoids making copies of data by pointing to the original value instead.

- Key portions of the code are written in C++, which according to the authors, yields a 20x to 1,000x performance increase over base R for many operations.

- R data frames are limited by available memory. With dplyr, tibbles can be linked transparently to disk-based databases that can exceed what can be stored in memory.

The dplyr grammar of working with data becomes second nature after the initial learning curve has been passed. There are five key verbs in the dplyr grammar, which perform many of the most common transformations to data tables. Beginning with a tibble, one may choose to:

- filter() rows of data by values of the columns
- select() columns of data by name
- mutate() columns into new columns by transforming the values
- summarize() rows of data by aggregating values into a summary
- arrange() rows of data by sorting the values

These five dplyr verbs are brought together in sequences using a **pipe** operator. Represented by the %>% symbols, the pipe operator literally "pipes" data from one function to another. The use of pipes allows the creation of powerful chains of functions for processing tables of data.

The pipe operator is part of the `magrittr` package by Stefan Milton Bache and Hadley Wickham, installed by default with the tidyverse collection. The name is a play on René Magritte's famous painting of a pipe (you may recall seeing it in *Chapter 1, Introducing Machine Learning*). For more information, visit its tidyverse page at `https://magrittr.tidyverse.org`.

To illustrate the power of `dplyr`, imagine a scenario in which you are asked to examine loan applicants 21 years old or older, and find their average loan duration (in years), grouped by whether or not the applicants defaulted. With this in mind, it is not difficult to follow the following `dplyr` grammar, which reads almost like a pseudocode retelling of that task:

```
> credit %>%
    filter(age >= 21) %>%
    mutate(years_loan_duration =
            months_loan_duration / 12) %>%
    select(default, years_loan_duration) %>%
    group_by(default) %>%
    summarize(mean_duration = mean(years_loan_duration))
# A tibble: 2 x 2
  default mean_duration
  <fct>           <dbl>
1 no               1.61
2 yes              2.09
```

This is just one small example of how sequences of `dplyr` commands can make complex data manipulation tasks simpler. This is on top of the fact that, due to `dplyr`'s more efficient code, the steps often execute more quickly than the equivalent commands in base R! Providing a complete `dplyr` tutorial is beyond the scope of this book, but there are many learning resources available online, including an *R for Data Science* chapter at `https://r4ds.had.co.nz/transform.html`.

Reading and writing to external data files

A frustrating aspect of data analysis is the large amount of work required to pull and combine data from various proprietary formats. Vast troves of data exist in files and databases that simply need to be unlocked for use in R. Thankfully, packages exist for exactly this purpose.

Importing tidy tables with readr

The tidyverse includes the `readr` package as a faster solution for loading tabular data like CSV files into R. This is described in the data import chapter in *R for Data Science* at `https://r4ds.had.co.nz/data-import.html`, but the basic functionality is simple.

The package provides a `read_csv()` function, much like base R's `read.csv()`, that loads data from CSV files. The key difference is that the tidyverse is much speedier — about 10x faster according to the package authors — and is smarter about the format of the columns to be loaded. For example, it has the capability to handle numbers with currency characters, parse date columns, and is better at handling international data.

To create a tibble from a CSV file, simply use the `read_csv()` function as follows:

```
> library(readr)
> credit <- read_csv("credit.csv")
```

This will use the default parsing settings, which will be displayed in the R output. The defaults may be overridden by providing the column specifications via a `col()` function call passed to `read_csv()`.

Importing Microsoft Excel, SAS, SPSS, and Stata files with rio

What used to be a tedious and time-consuming process, requiring knowledge of specific tricks and tools across multiple R packages, has been made trivial by an R package called `rio` (an acronym for R input and output). This package, by Chung-hong Chan, Geoffrey CH Chan, Thomas J. Leeper, and Christopher Gandrud, is described as a "Swiss-Army Knife for Data I/O." It is capable of importing and exporting a large variety of file formats including, but not limited to, tab-separated (`.tsv`) and comma-separated (`.csv`), JSON, Stata (`.dta`), SPSS (`.sav` and `.por`), Microsoft Excel (`.xls` and `.xlsx`), Weka (`.arff`), and SAS (`.sas7bdat` and `.xpt`).

 For the complete list of file types `rio` can import and export, as well as more detailed usage examples, see `http://cran.r-project.org/web/packages/rio/vignettes/rio.html`.

The `rio` package consists of three functions for working with proprietary data formats: `import()`, `export()`, and `convert()`. Each does exactly what you'd expect from their name. Consistent with the package's philosophy of keeping things simple, each function uses the file name extension (such as `.csv` or `.xlsx`) to guess the type of file to import, export, or convert.

For example, to import the credit data CSV file used in previous chapters, simply type:

```
> library(rio)
> credit <- import("credit.csv")
```

This creates the `credit` data frame as expected; as a bonus, not only did we not have to specify the CSV file type, `rio` automatically set `stringsAsFactors = FALSE` as well as other reasonable defaults.

To export the `credit` data frame to Microsoft Excel (`.xlsx`) format, use the `export()` function while specifying the desired filename as follows. For other formats, simply change the file extension to the desired output type.

```
> export(credit, "credit.xlsx")
```

It is also possible to convert the CSV file to another format directly, without an import step, using the `convert()` function. For example, this converts the credit CSV file to Stata (`.dta`) format:

```
> convert("credit.csv", "credit.dta")
```

Though the `rio` package covers many common proprietary data formats, it does not do everything. The next section covers another way to get data into R via database queries.

Querying data in SQL databases

Large datasets are often stored in a **database management system** (**DBMS**) such as Oracle, MySQL, PostgreSQL, Microsoft SQL, or SQLite. These systems allow the datasets to be accessed using the **Structured Query Language** (**SQL**), a programming language designed to pull data from databases.

The tidy approach to managing database connections

RStudio version 1.1 introduced a graphical approach for connecting to databases. The **connections** tab in the top-right portion of the interface lists all of the existing database connections found on your system. The creation of these connections is typically performed by a database administrator, and is specific to the type of database as well as the operating system. For instance, on Microsoft Windows, you may need to install the appropriate database drivers as well as use the ODBC Data Source Administrator application; on MacOS and Unix/Linux you may need to install the drivers and edit an `odbc.ini` file.

Complete documentation about the potential connection types and installation instructions is available at https://db.rstudio.com.

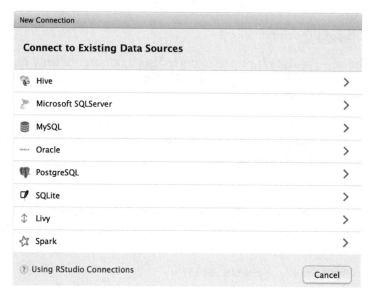

Figure 12.2: The "New Connection" button in RStudio v1.1 or greater opens an interface that will assist you with connecting to any pre-defined data sources

Behind the scenes, the graphical interface uses a variety of R packages to manage the connections to these data sources. At the core of this functionality is the DBI package, which provides a tidyverse-compliant front-end interface to the database. The DBI package also manages the back-end database driver, which must be provided by another R package. Such packages let R connect to Oracle (ROracle), MySQL (RMySQL), PostgreSQL (RPostgreSQL), and SQLite (RSQLite), among many others.

To illustrate this functionality, we'll use the DBI and RSQLite packages to connect to a SQLite database containing the credit dataset used previously. SQLite is a simple database that doesn't require running a server. It simply connects to a database file on a machine, which here is named credit.sqlite3. Before starting, be sure you've installed both of the required packages and saved the database file into your R working directory. After doing this, you can connect to the database using the following command:

```
> con <- dbConnect(RSQLite::SQLite(), "credit.sqlite3")
```

To prove the connection has succeeded, we can list the database tables to confirm the credit table exists as expected:

```
> dbListTables(con)
[1] "credit"
```

From here, we can send SQL query commands to the database and return records as R data frames. For instance, to return the loan applicants with an age of 45 years or greater, we would query the database as follows:

```
> res <- dbSendQuery(con, "SELECT * FROM credit WHERE age >= 45")
```

The entire result set can be fetched as a data frame using the following command:

```
> credit_age45 <- dbFetch(res)
```

To confirm that it worked, we'll examine the summary statistics, which confirm that the ages begin at 45 years:

```
> summary(credit_age45$age)
   Min. 1st Qu.  Median    Mean 3rd Qu.    Max.
  45.00   48.00   52.00   53.98   60.00   75.00
```

When our work is done, it is advisable to clear the query result set and close the database connection to free these resources:

```
> dbClearResult(res)
> dbDisconnect(con)
```

In addition to SQLite and the database-specific R packages, the odbc package allows R to connect to many different types of databases using a single protocol known as the **Open Database Connectivity (ODBC)** standard. The ODBC standard can be used regardless of operating system or DBMS.

If you have previously connected to an ODBC database, you may have referred to it via its **data source name (DSN)**. You can use the DSN to create a database connection with a single line of R code:

```
> con <- dbConnect(odbc:odbc(), "my_data_source_name")
```

If you have a more complicated setup, or want to specify the connection properties manually, you can specify a full connection string as arguments to the DBI package dbConnect() function as follows:

```
> library(DBI)
> con <- dbConnect(odbc::odbc(),
                    database = "my_database",
                    uid = "my_username",
                    pwd = "my_password",
                    host = "my.server.address",
                    port = 1234)
```

With the connection established, queries can be sent to the ODBC database and tables can be returned as data frames using the same functions that were used for the SQLite example previously.

 Due to security and firewall settings, the instructions for configuring an ODBC network connection are highly specific to each situation. If you are having trouble setting up the connection, check with your database administrator. The RStudio team also provides helpful information at `https://db.rstudio.com/best-practices/drivers/`.

Using a database backend with dplyr

Connecting `dplyr` to an external database is no more difficult than using it with a traditional data frame. The `dbplyr` package (database plyr) allows any database supported by the `DBI` package to be used as a backend for `dplyr`. The connection allows tibble objects to be pulled from the database. Generally, one does not need to do more than simply install the `dbplyr` package, and `dplyr` can then take advantage of its functionality.

For example, let's connect to the SQLite `credit.sqlite3` database used previously, then save its `credit` table as a tibble object using the `tbl()` function as follows:

```
> library(DBI)
> con <- dbConnect(RSQLite::SQLite(), "credit.sqlite3")
> credit_tbl <- con %>% tbl("credit")
```

In spite of the fact that `dplyr` has been routed through a database, the `credit_tbl` object here will act exactly like any other tibble and will gain all the other benefits of the `dplyr` package. Note that the steps would be largely similar if the SQLite database was replaced with a database residing across a network on a more traditional SQL server.

For example, to query the database for credit applicants with age at least 45 years, and display the age summary statistics for this group, we can pipe the tibble through the following sequence of functions:

```
> library(dplyr)
> credit_tbl %>%
    filter(age >= 45) %>%
    select(age) %>%
    collect() %>%
    summary()

      age
```

```
Min.    :45.00
1st Qu.:48.00
Median :52.00
Mean    :53.98
3rd Qu.:60.00
Max.    :75.00
```

Note that the dbplyr functions are "lazy," which means that no work is done in the database until it is necessary. Thus, the collect() function forces dplyr to retrieve the results from the server so that the summary statistics may be calculated.

Given a database connection, many dplyr commands will be translated seamlessly into SQL on the backend. This means that the same R code used on smaller data frames can also be used to prepare larger datasets stored in SQL databases—the heavy lifting is done on the remote server, rather than your local laptop or desktop machine. In this way, learning the tidyverse suite of packages ensures your code will apply to any type of project from small to massive.

A traditional approach to SQL connectivity with RODBC

As an alternative to the RStudio and the tidyverse approach, it is also possible to connect to SQL servers using the RODBC package by Brian Ripley. The RODBC functions retrieve data from an ODBC-compliant SQL server and create an R data frame. Although this package is still widely used, it is substantially slower in benchmarking tests than the newer odbc package, and is listed here primarily for reference.

The RODBC package vignette, which you can access in R with the command vignette("RODBC"), provides a wealth of information about connecting to various databases. Be sure to check here if you get stuck.

To open a connection called mydb to the database with the DSN my_dsn, use the odbcConnect() function:

```
> library(RODBC)
> my_db <- odbcConnect("my_dsn")
```

Alternatively, if your ODBC connection requires a username and password, they should be specified when calling the odbcConnect() function:

```
> my_db <- odbcConnect("my_dsn",
    uid = "my_username",
    pwd = "my_password")
```

With an open database connection, we can use the `sqlQuery()` function to create an R data frame from the database rows pulled by SQL queries. This function, like many functions that create data frames, allows us to specify `stringsAsFactors = FALSE` to prevent R from automatically converting character data to factors.

The `sqlQuery()` function uses typical SQL queries as shown in the following command:

```
> my_query <- "select * from my_table where my_value = 1"
> results_df <- sqlQuery(channel = my_db, query = my_query,
    stringsAsFactors = FALSE)
```

The resulting `results_df` object is a data frame containing all of the rows selected using the SQL query stored in `my_query`.

When you are done using the database, the connection can be closed as shown in the following command:

```
> odbcClose(my_db)
```

This will close the `my_db` connection. Although R will automatically close ODBC connections at the end of an R session, it is better practice to do so explicitly.

Working with online data and services

With growing amounts of data available from web-based sources, it is increasingly important for machine learning projects to be able to access and interact with online services. R is able to read data from online sources natively, with some caveats. First, by default, R cannot access secure websites (those using `https://` rather than the `http://` protocol). Secondly, it is important to note that most web pages do not provide data in a form that R can understand. The data will need to be **parsed**, or broken apart and rebuilt into a structured form before it can be useful. We'll discuss the workarounds shortly.

However, if neither of these caveats apply, that is, if the data are already online in a non-secure website and in a tabular form like CSV that R can understand natively, then R's `read.csv()` and `read.table()` functions can access it from the web just as if it were on your local machine. Simply supply the full **Uniform Resource Locator (URL)** for the dataset as follows:

```
> mydata <- read.csv("http://www.mysite.com/mydata.csv")
```

R also provides functionality for downloading other files from the web, even if R cannot use them directly. For a text file, try the `readLines()` function as follows:

```
> mytext <- readLines("http://www.mysite.com/myfile.txt")
```

For other types of files, the `download.file()` function can be used. To download a file to R's current working directory, simply supply the URL and destination filename as follows:

```
> download.file("http://www.mysite.com/myfile.zip", "myfile.zip")
```

Beyond this base functionality, there are numerous packages that extend R's capabilities for working with online data, the most basic of which will be covered in the sections that follow. Because the web is massive and ever changing, these sections are far from a comprehensive set of all the ways R can connect to online data. There are literally hundreds of packages for everything from niche to massive projects.

 For the most complete and up-to-date list of packages, refer to the regularly updated CRAN Web Technologies and Services task view at http://cran.r-project.org/web/views/WebTechnologies.html.

Downloading the complete text of web pages

The `RCurl` package by Duncan Temple Lang provides a more robust way of accessing web pages by providing an R interface to the **curl** (client for URLs) utility, a command-line tool for transferring data over networks. The curl program is a widely used tool that acts much like a programmable web browser; given a set of commands, it can access and download the content of nearly anything available on the web. And unlike R, it can access secure websites, as well as post instructions to online forms. It is an incredibly powerful utility.

 Precisely because it is so powerful, a complete curl tutorial is outside the scope of this chapter. Instead, refer to the online `RCurl` documentation at http://www.omegahat.net/RCurl/.

After installing and loading the `RCurl` package, downloading a page is as simple as typing:

```
> packt_page <- getURL("https://www.packtpub.com/")
```

This will save the full text of the Packt Publishing homepage (including all web markup) into the R character object named `packt_page`. As shown in the following lines, this is not very useful as-is:

```
> str(packt_page, nchar.max = 200)
 chr "<!DOCTYPE html>\n    <html xmlns=\"http://www.w3.org/1999/xhtml\"
lang=\"en\" xml:lang=\"en\">\n    <head>\n          <title>Packt Publishing
| Technology Books, eBooks & Videos</title>"| __truncated__
```

The reason that the first 200 characters of the page look like nonsense is that websites are written using **Hypertext Markup Language** (**HTML**), which combines the page text with special tags that tell web browsers how to display it. The `<title>` and `</title>` tags here surround the page's title, telling the browser that this is the Packt Publishing homepage. Similar tags are used to denote other portions of the page.

Though curl is the cross-platform standard for accessing online content, if you work with web data frequently in R, the `httr` package by Hadley Wickham builds upon the foundation of RCurl to make accessing HTML web data more convenient and R-like. Rather than using RCurl, the `httr` package uses its own `curl` package behind the scenes to retrieve the website data. We can see some of the differences immediately by attempting to download the Packt Publishing homepage using the `httr` package's GET() function:

```
> library(httr)
> packt_page <- GET("https://www.packtpub.com")
> str(packt_page, max.level = 1)
List of 10
 $ url         : chr "https://www.packtpub.com/"
 $ status_code: int 200
 $ headers     :List of 11
  ..- attr(*, "class")= chr [1:2] "insensitive" "list"
 $ all_headers:List of 1
 $ cookies     :'data.frame':    0 obs. of   7 variables:
 $ content     : raw [1:162392] 3c 21 44 4f ...
 $ date        : POSIXct[1:1], format: "2019-02-24 23:41:59"
 $ times       : Named num [1:6] 0 0.00372 0.16185 0.45156...
  ..- attr(*, "names")= chr [1:6] "redirect" "namelookup" "connect"
"pretransfer" ...
 $ request     :List of 7
  ..- attr(*, "class")= chr "request"
 $ handle      :Class 'curl_handle' <externalptr>
 - attr(*, "class")= chr "response"
```

Where the getURL() function in RCurl downloaded only the HTML, the `httr` package's GET() function returns a list with properties of the query in addition to the HTML. To access the page content itself, we need to use the content() function:

```
> str(content(packt_page, type = "text"), nchar.max = 200)
 chr "<!DOCTYPE html>\n<html xmlns=\"http://www.w3.org/1999/xhtml\"
lang=\"en\" xml:lang=\"en\">\n\t<head>\n\t\t<title>Packt Publishing |
Technology Books, eBooks & Videos</title>\n\t\t<script>\n\t\t\tdata"|
truncated__
```

In order to use this data in an R program, it is necessary to process the HTML data to structure it in a format such as a list or data frame. Functions for doing so are discussed in the sections that follow.

 For detailed `httr` documentation and tutorials, visit the project homepage at `https://httr.r-lib.org`. The quick start guide is particularly helpful for learning the base functionality.

Parsing the data within web pages

Because there is a consistent structure to the HTML tags of many web pages, it is possible to write programs that look for desired sections of the page and extracts them for compilation into a dataset. This practice of harvesting data from websites and transforming it into a structured form is known as **web scraping**.

 Though frequently used, web scraping should be considered a last resort for getting data from the web. This is because any changes to the underlying HTML structure may break your code, requiring effort to fix, or worse, introducing unnoticed errors into your data. Additionally, many websites' terms of use agreements explicitly forbid automated data extraction, not to mention the fact that your program's traffic may overload their servers. Always check the site's terms of use before beginning your project; you may even find that the site offers their data freely via a developer agreement. You may also keep a lookout for a file named `robots.txt`, which is a web standard that describes what parts of a site bots are allowed to crawl.

The `rvest` package (a pun on the term "harvest") by Hadley Wickham makes web scraping a largely effortless process, assuming the data you want can be found in a consistent place within the HTML.

Let's start with a simple example using the Packt Publishing homepage. We begin by downloading the page as before, using the `read_html()` function in the `rvest` package. Note that this function, when supplied with a URL, simply calls the `GET()` function in Hadley Wickham's `httr` package:

```
> library(rvest)
> packt_page <- read_html("https://www.packtpub.com")
```

Suppose we'd like to scrape the page title; looking at the previous HTML code, we know that there is only one title per page, wrapped within `<title>` and `</title>` tags. To pull the title, we supply the tag name to the `html_node()` function, then a second `html_text()` function that translates the result to plain text:

```
> html_node(packt_page, "title") %>% html_text()
[1] "Packt Publishing | Technology Books, eBooks & Videos"
```

Notice the use of the `%>%` pipe operator. Just like with the base `dplyr` package, the use of pipes allows the creation of powerful chains of functions for processing HTML data with `rvest`.

Let's try a slightly more interesting example. Suppose we'd like to scrape a list of all packages on the CRAN machine learning task view. We begin as before by downloading the HTML page using the `read_html()` function:

```
> library(rvest)

> cran_ml <- read_html("http://cran.r-project.org/web/views/
MachineLearning.html")
```

If we view the source of the website in a web browser, one section appears to have the data we're interested in. Note that only a subset of the output is shown here:

```
<h3>CRAN packages:</h3>
<ul>
  <li><a href="../packages/ahaz/index.html">ahaz</a></li>
  <li><a href="../packages/arules/index.html">arules</a></li>
  <li><a href="../packages/bigrf/index.html">bigrf</a></li>
  <li><a href="../packages/bigRR/index.html">bigRR</a></li>
  <li><a href="../packages/bmrm/index.html">bmrm</a></li>
  <li><a href="../packages/Boruta/index.html">Boruta</a></li>
  <li><a href="../packages/bst/index.html">bst</a></li>
  <li><a href="../packages/C50/index.html">C50</a></li>
  <li><a href="../packages/caret/index.html">caret</a></li>
```

The `<h3>` tags imply a heading of level 3, while the `<ul>` and `<li>` tags refer to the creation of an unordered list (that is, bulleted as opposed to ordered/numbered) and list items, respectively. The data elements we want are surrounded by `<a>` tags, which are hyperlink anchor tags that link to the CRAN page for each package.

 Because the CRAN page is actively maintained and may be changed at any time, do not be surprised if your results differ from those shown here.

With this knowledge in hand, we can scrape the links much like we did previously. The one exception is that because we expect to find more than one result, we need to use the `html_nodes()` function to return a vector of results rather than `html_node()`, which returns only a single item. The following function call returns `<a>` tags nested within `<li>` tags:

```
> ml_packages <- html_nodes(cran_ml, "li a")
```

Let's peek at the result using the `head()` function:

```
> head(ml_packages, n = 5)
{xml_nodeset (5)}
[1] <a href="../packages/nnet/index.html">nnet</a>
[2] <a href="../packages/RSNNS/index.html">RSNNS</a>
[3] <a href="../packages/rnn/index.html">rnn</a>
[4] <a href="../packages/deepnet/index.html">deepnet</a>
[5] <a href="../packages/RcppDL/index.html">RcppDL</a>
```

The result includes the `<a>` HTML output. To eliminate this, simply pipe the result into the `html_text()` function. The result is a vector containing the names of all packages listed in the CRAN machine learning task view, here piped into the `head()` function to display only its first few values:

```
> ml_packages %>% html_text() %>% head()
[1] "nnet"     "RSNNS"     "rnn"        "deepnet"  "RcppDL"  "h2o"
```

These are simple examples that merely scratch the surface of what is possible with the `rvest` package. Using the pipe functionality, it is possible to look for tags nested within tags, or specific classes of HTML tags. For these types of complex examples, refer to the package's documentation.

> In general, web scraping is always a process of iterate-and-refine as you identify more specific criteria to exclude or include specific cases. The most difficult cases may even require a human eye to achieve 100 percent accuracy.

Parsing XML documents

XML is a plaintext, human-readable, but structured markup language upon which many document formats have been based. It employs a tagging structure in some ways similar to HTML, but is far stricter about formatting. For this reason, it is a popular online format for storing structured datasets.

The XML package by Duncan Temple Lang provides a suite of R functionality based on the popular C-based `libxml2` parser for reading and writing XML documents. It is the grandfather of XML parsing packages in R and is still widely used.

> Information on the XML package, including simple examples to get you started quickly, can be found at the project's website at http://www.omegahat.net/RSXML/.

More recently, the xml2 package by Hadley Wickham has surfaced as an easier and more R-like interface to the libxml2 library. The rvest package, which was covered earlier in this chapter, utilizes xml2 behind-the-scenes to parse HTML; therefore, rvest can obviously also be used to parse XML.

 The xml2 homepage is found at http://xml2.r-lib.org.

Because parsing XML is so closely related to parsing HTML, the exact syntax is not covered here. Please refer to these packages' documentation for examples.

Parsing JSON from web APIs

Online applications communicate to one another using web-accessible functions known as **application programming interfaces (API)**. These interfaces act much like a typical website; they receive a request from a client at a particular URL and return a response. The difference is that where a normal website returns HTML meant for display in a web browser, an API typically returns data in a structured form meant for processing by a machine.

Though it is not uncommon to find XML-based APIs, perhaps the most common API data structure today is **JavaScript Object Notation (JSON)**. Like XML, this is a standard, plaintext format, most often used for data structures and objects on the web. The format has become popular recently due to its roots in browser-based JavaScript applications, but despite the pedigree, its utility is not limited to the web. The ease with which JSON data structures can be understood by humans and parsed by machines makes it an appealing data structure for many types of projects.

JSON is based on a simple {key: value} format. The { } brackets denote a JSON object, and the key and value denote a property of the object and the status of that property. An object can have any number of properties, and the properties themselves may be objects. For example, a JSON object for this book might look something like this:

```
{
    "title": "Machine Learning with R",
    "author": "Brett Lantz",
    "publisher": {
        "name": "Packt Publishing",
        "url": "https://www.packtpub.com"
    },
    "topics": ["R", "machine learning", "data mining"],
    "MSRP": 54.99
}
```

This example illustrates the data types available to JSON: numeric, character, array (surrounded by [and] characters), and object. Not shown are the null and Boolean (true or false) values. The transmission of these types of objects from application to application, and application to web browser, is what powers many of the most popular websites.

 For details about the JSON format, visit http://www.json.org/.

There are a number of packages that can convert to and from JSON data. The jsonlite package by Jeroen Ooms quickly gained prominence because it creates data structures that are more consistent and R-like than other packages, especially when using data from web APIs. For detailed information on how to use this package, visit its GitHub page at https://github.com/jeroen/jsonlite.

After installing the jsonlite package, to convert from an R object to a JSON string, we use the toJSON() function. Notice that in the output the quote characters have been escaped using the \" notation:

```
> library(jsonlite)
> ml_book <- list(book_title = "Machine Learning with R",
                  author = "Brett Lantz")
> toJSON(ml_book)
{"book_title":["Machine Learning with R"],
 "author":["Brett Lantz"]}
```

To convert a JSON string to an R object, use the fromJSON() function. Quotes in the string need to be escaped as shown:

```
> ml_book_json <- "{
  \"title\": \"Machine Learning with R\",
  \"author\": \"Brett Lantz\",
  \"publisher\": {
    \"name\": \"Packt Publishing\",
    \"url\": \"https://www.packtpub.com\"
  },
  \"topics\": [\"R\", \"machine learning\", \"data mining\"],
  \"MSRP\": 54.99
}"

> ml_book_r <- fromJSON(ml_book_json)
```

This results in a list structure in a form much like the JSON:

```
> str(ml_book_r)
List of 5
 $ title    : chr "Machine Learning with R"
 $ author   : chr "Brett Lantz"
 $ publisher:List of 2
  ..$ name: chr "Packt Publishing"
  ..$ url : chr "https://www.packtpub.com"
 $ topics   : chr [1:3] "R" "machine learning" "data mining"
 $ MSRP     : num 55
```

 For more information on the `jsonlite` package, see *The jsonlite Package: A Practical and Consistent Mapping Between JSON Data and R Objects, Ooms, J, 2014.* Available at `http://arxiv.org/abs/1403.2805`.

Public facing APIs allow programs like R to systematically query websites to retrieve results in JSON format using packages like `Rcurl` and `httr`. Nearly all websites that provide interesting data offer APIs for querying, although some charge fees and require an access key for doing so. Popular examples include the APIs for Twitter, Google Maps, and Facebook. Though a full tutorial for using web APIs is worthy of a separate book, the basic process relies on only a couple of steps – it's the details that are tricky.

Suppose we wanted to query the Apple iTunes API to find the albums released by *The Beatles*. We first need to review the iTunes API documentation at `https://affiliate.itunes.apple.com/resources/documentation/itunes-store-web-service-search-api/` to determine the URL and parameters needed to make this query. We then supply this information to the `httr` package `GET()` function, adding a list of query parameters in to send to the search URL:

```
> library(httr)
> music_search <- GET("https://itunes.apple.com/search",
                 query = list(term = "Beatles",
                             media = "music",
                             entity = "album",
                             limit = 10))
```

By typing the name of the resulting object, we can see some details about the request:

```
> music_search
Response [https://itunes.apple.com/search?term=Beatles&media=music&entity
=album&limit=10]
```

```
Date: 2019-02-25 00:33

Status: 200

Content-Type: text/javascript; charset=utf-8

Size: 9.75 kB
```

```
{
 "resultCount":10,
 "results": [
{"wrapperType":"collection", "collectionType":"Album", "artistId":136975,
"collectionId":402060584, "amgArtistId":3644, "artistName":"The B...
```

To access the resulting JSON, we use the `content()` function, which we can then convert to an R object with the `jsonlite` package's `fromJSON()` function:

```
> library(jsonlite)
> music_results <- fromJSON(content(music_search))
```

The `music_results` object is a list containing the data returned from the iTunes API. Although the results are too large to print here, the `str(music_results)` command will display the structure of this object, which shows that the interesting data is stored as sub-objects within the `results` object. For example, the vector of album titles can be found as the `collectionName` sub-object:

```
> music_results$results$collectionName
 [1] "The Beatles Box Set"
 [2] "Abbey Road"
 [3] "The Beatles (White Album)"
 [4] "The Beatles 1967-1970 (The Blue Album)"
 [5] "1 (2015 Version)"
 [6] "Sgt. Pepper's Lonely Hearts Club Band"
 [7] "The Beatles 1962-1966 (The Red Album)"
 [8] "Revolver"
 [9] "Rubber Soul"
[10] "Love"
```

These data elements could then be used in an R program as desired.

Because the Apple iTunes API may be updated in the future, if you find that your results differ from those shown here, please check the Packt Publishing support page for this book for updated code.

Working with domain-specific data

Machine learning has undoubtedly been applied to problems across every discipline. Although the basic techniques are similar across all domains, some are so specialized that communities have formed to develop solutions to the challenges unique to the field. This leads to the discovery of new techniques and new terminology that is relevant only to domain-specific problems.

This section covers a pair of domains that use machine learning techniques extensively, but require specialized knowledge to unlock their full potential. Since entire books have been written on these topics, this will serve as only the briefest of introductions. For more detail, seek out the help provided by the resources cited in each section.

Analyzing bioinformatics data

The field of **bioinformatics** is concerned with the application of computers and data analysis to the biological domain, particularly with regard to better understanding the genome. As genetic data is unique compared to many other types, data analysis in the field of bioinformatics offers a number of unique challenges. For example, because living creatures have a tremendous number of genes and genetic sequencing is still relatively expensive, typical datasets are much wider than they are long; that is, they have more features (genes) than examples (creatures that have been sequenced). This creates problems when attempting to apply conventional visualizations, statistical tests, and machine learning methods to such data. Additionally, the use of proprietary **microarray** "lab-on-a-chip" techniques requires highly specialized knowledge simply to load the genetic data.

 A CRAN task view listing some of R's specialized packages for statistical genetics and bioinformatics is available at http://cran.r-project.org/web/views/Genetics.html.

The **Bioconductor** project of the Fred Hutchinson Cancer Research Center in Seattle, Washington, aims to solve some of these problems by providing a standardized set of methods of analyzing genomic data. Using R as its foundation, Bioconductor adds bioinformatics-specific packages and documentation specific to base R software.

Bioconductor provides workflows for analyzing DNA and protein microarray data from common microarray platforms such as Affymetrix, Illumina, NimbleGen, and Agilent. Additional functionality includes sequence annotation, multiple testing procedures, specialized visualizations, tutorials, documentation, and much more.

For more information about the Bioconductor project, visit the project website at http://www.bioconductor.org.

Analyzing and visualizing network data

Social network data and graph datasets consist of structures that describe connections, or **links** (sometimes also called **edges**), between people or objects known as **nodes**. With N nodes, an $N \times N = N^2$ matrix of potential links can be created. This creates tremendous computational complexity as the number of nodes grows.

The field of **network analysis** is concerned with statistical measures and visualizations that identify meaningful patterns of connections. For example, the following figure shows three clusters of circular nodes all connected via a square node at the center. A network analysis may reveal the importance of the square node, among other key metrics.

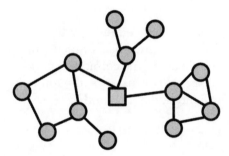

Figure 12.3: A social network showing nodes in three clusters around the central square node

The network package by Carter T. Butts, David Hunter, and Mark S. Handcock offers a specialized data structure for working with networks. This data structure is necessary due to the fact that the matrix needed to store N^2 potential links would quickly run out of memory; the network data structure uses a sparse representation to store only the existent links, saving a great deal of memory if most relationships are non-existent. A closely related package, sna (social network analysis), allows analysis and visualization of the network objects.

 For more information on network and sna, including very detailed tutorials and documentation, refer to the project website hosted by the University of Washington: http://www.statnet.org/. The social network analysis lab at Stanford University also hosts very nice tutorials at https://sna.stanford.edu/rlabs.php.

The igraph package by Gábor Csárdi provides another set of tools for visualizing and analyzing network data. It is capable of handling very large networks and calculating metrics. An additional benefit of igraph is the fact that it has analogous packages for the Python and C programming languages, allowing it to be used virtually anywhere analyses are being performed. As will be demonstrated shortly, it is very easy to use.

 For more information on the igraph package, including demos and tutorials, visit the homepage at http://igraph.org/r/.

Using network data in R requires use of specialized formats, as network data are not typically stored in typical tabular data structures like CSV files and data frames. As mentioned previously, because there are N^2 potential connections between N network nodes, a tabular structure would quickly grow to be unwieldy for all but the smallest N values. Instead, graph data are stored in a form that lists only the connections that are truly present; the absent connections are inferred from the absence of data.

Perhaps the simplest such format is the **edgelist**, which is a text file with one line per network connection. Each node must be assigned a unique identifier, and links between nodes are defined by placing the connected nodes' identifiers together on a single line, separated by a space. For instance, the following edgelist defines three connections between node 0 and nodes 1, 2, and 3.

```
0 1
0 2
0 3
```

To load network data into R, the igraph package provides a read.graph() function that can read edgelist files as well as other, more sophisticated formats like **Graph Modeling Language** (GML). To illustrate this functionality, we'll use a dataset describing friendships among members of a small karate club. To follow along, download the karate.txt file from the Packt Publishing website and save it to your R working directory. Then, after you've installed the igraph package, the karate network can be read into R as follows:

```
> library(igraph)
> karate <- read.graph("karate.txt", "edgelist", directed = FALSE)
```

This will create a sparse matrix object that can be used for graphing and network analysis. Note that the directed = FALSE parameter forces the network to use undirected, bi-directional links between nodes.

Since the karate dataset describes friendships, this means that if Person 1 is friends with Person 2, then Person 2 must be friends with Person 1. On the other hand, if the dataset described fight outcomes, the fact that Person 1 defeated Person 2 would certainly not imply that Person 2 defeated Person 1. In this case, the parameter `directed = TRUE` should be set.

> The karate network dataset used here was compiled by M.E.J. Newman of the University of Michigan. It was first presented in *An Information Flow Model for Conflict and Fission in Small Groups, Zachary, WW, Journal of Anthropological Research, 1977, Vol. 33, pp. 452-473.*

To examine the graph, use the `plot()` function:

```
> plot(karate)
```

This produces the following figure:

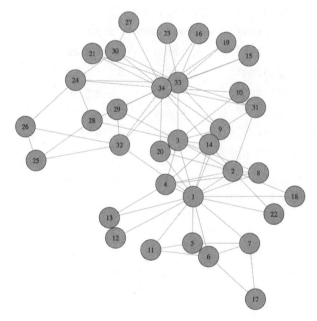

Figure 12.4: A network visualization of the karate dataset. Connections indicate fights between competitors.

Examining the network visualization, it is apparent that there are a few highly connected members of the karate club. Nodes 1, 33, and 34 seem to be more central than the others, which remain at the club periphery.

Using `igraph` to calculate graph metrics, it is possible to demonstrate our hunch analytically. The **degree** of a node measures how many nodes it is linked to.

The `degree()` function confirms our hunch that nodes 1, 33, and 34 are more connected than the others with 16, 12, and 17 connections respectively:

```
> degree(karate)
 [1] 16  9 10  6  3  4  4  4  5  2  3  1  2  5  2  2  2  2
[19]  2  3  2  2  2  5  3  3  2  4  3  4  4  6 12 17
```

Because some connections are more important than others, a variety of network measures have been developed to measure node connectivity with this consideration. A network metric called **betweenness centrality** is intended to capture the number of shortest paths between nodes that pass through each node. Nodes that are truly more central to the entire graph will have a higher betweenness centrality value, because they act as a bridge between other nodes. We obtain a vector of the centrality measures using the `betweenness()` function as follows:

```
> betweenness(karate)
 [1] 231.0714286  28.4785714  75.8507937   6.2880952
 [5]   0.3333333  15.8333333  15.8333333   0.0000000
 [9]  29.5293651   0.4476190   0.3333333   0.0000000
[13]   0.0000000  24.2158730   0.0000000   0.0000000
[17]   0.0000000   0.0000000   0.0000000  17.1468254
[21]   0.0000000   0.0000000   0.0000000   9.3000000
[25]   1.1666667   2.0277778   0.0000000  11.7920635
[29]   0.9476190   1.5428571   7.6095238  73.0095238
[33]  76.6904762 160.5515873
```

As nodes 1 and 34 have much greater betweenness values than the others, they are more central to the karate club's friendship network. These two individuals, with extensive personal friendship networks, may be the "glue" that holds the network together.

 Betweenness centrality is only one of many metrics intended to capture a node's importance, and isn't even the only measure of centrality. Refer to the igraph documentation for definitions of other network properties.

The `sna` and `igraph` packages are capable of computing many such graph metrics, which may then be used as inputs to machine learning functions. For example, suppose we were attempting to build a model predicting who would win an election for club president. The fact that nodes 1 and 34 are well connected suggests that they may have the social capital they would need for such a leadership role. These might be highly valuable predictors of election results.

By combining network analysis with machine learning, services like Facebook, Twitter, and LinkedIn provide vast stores of network data for making predictions about users' future behavior. A high-profile example is the 2012 US Presidential campaign in which Chief Data Scientist Rayid Ghani utilized Facebook data to identify people who might be persuaded to vote for Barack Obama.

Improving the performance of R

Base R has a reputation for being slow and memory inefficient, a reputation that is at least somewhat earned. These faults are largely unnoticed on a modern PC for datasets of many thousands of records, but datasets with a million records or more can exceed the limits of what is currently possible with consumer-grade hardware. The problem is worsened if the dataset contains many features or if complex learning algorithms are being used.

CRAN has a high-performance computing task view that lists packages pushing the boundaries on what is possible in R at http://cran.r-project.org/web/views/HighPerformanceComputing.html.

Packages that extend R past the capabilities of the base package are being developed rapidly. This work comes primarily on two fronts: some packages add the capability to manage extremely large datasets by making data operations faster or by allowing the size of data to exceed the amount of available system memory; others allow R to work faster, perhaps by spreading the work over additional computers or processors, by utilizing specialized computer hardware, or by providing machine learning optimized to big data problems.

Managing very large datasets

Extremely large datasets can cause R to grind to a halt when the system runs out of memory to store the data. Even if the entire dataset can fit into the available memory, additional memory overhead is needed for data processing. Furthermore, very large datasets can take a long amount of time to process for no reason other than the sheer volume of records; even a quick operation can add up when performed many millions of times.

Years ago, many would suggest performing data preparation outside R in another programming language, then using R to perform analyses on a smaller subset of data. However, this is no longer necessary, as several packages have been contributed to R to address these big data problems.

Making data frames faster with data.table

The `data.table` package by Matt Dowle, Tom Short, Steve Lianoglou, and Arun Srinivasan provides an enhanced version of a data frame called a **data table**. The `data.table` objects are typically much faster than data frames for subsetting, joining, and grouping operations. And for the largest datasets — those with many millions of rows — these objects may be substantially faster than even `dplyr` objects. Yet, because a `data.table` object is essentially an improved data frame, the resulting objects can still be used by any R function that accepts a data frame.

> The `data.table` project can be found on GitHub at `https://github.com/Rdatatable/data.table/wiki`.

After installing the `data.table` package, the `fread()` function will read tabular files like CSVs into data table objects. For instance, to load the credit data used previously, type:

```
> library(data.table)
> credit <- fread("credit.csv")
```

The credit data table can then be queried using syntax similar to R's `[row, col]` form but optimized for speed and some additional useful conveniences. In particular, the data table structure allows the `row` portion to select rows using an abbreviated subsetting command, and the `col` portion to use a function that does something with the selected rows. For example, the following command computes the mean requested loan amount for people with a good credit history:

```
> credit[credit_history == "good", mean(amount)]
[1] 3040.958
```

By building larger queries with this simple syntax, very complex operations can be performed on data tables. And since the data structure is optimized for speed, it can be used with large datasets.

One limitation of `data.table` structures is that like data frames, they are limited by the available system memory. The next two sections discuss packages that overcome this shortcoming at the expense of breaking compatibility with many R functions.

> The `dplyr` and `data.table` packages have unique strengths. For an in-depth comparison, see the following Reddit discussion at `https://www.reddit.com/r/rstats/comments/acjr9d/dplyr_performance/` and a similar conversation on Stack Overflow: `https://stackoverflow.com/questions/21435339/data-table-vs-dplyr-can-one-do-something-well-the-other-cant-or-does-poorly`. It is also possible to have the best of both worlds, as `data.table` structures can be loaded into `dplyr` using the `tbl_dt()` function.

Creating disk-based data frames with ff

The ff package by Daniel Adler, Christian Gläser, Oleg Nenadic, Jens Oehlschlägel, and Walter Zucchini provides an alternative to a data frame (ffdf) that allows datasets of over two billion rows to be created, even if this far exceeds the available system memory.

The ffdf structure has a physical component that stores the data on disk in a highly efficient form and a virtual component that acts like a typical R data frame but transparently points to the data stored in the physical component. You can imagine the ffdf object as a map that points to a location of data on a disk.

[The ff project is on the web at http://ff.r-forge.r-project.org/.]

A downside of ffdf data structures is that they cannot be used natively by most R functions. Instead, the data must be processed in small chunks, and the results should be combined later on. The upside of chunking the data is that the task can be divided across several processors simultaneously using the parallel computing methods presented later in this chapter.

After installing the ff package, to read in a large CSV file use the read.csv.ffdf() function as follows:

```
> library(ff)
> credit <- read.csv.ffdf(file = "credit.csv", header = TRUE)
```

Unfortunately, we cannot work directly with the ffdf object, as attempting to treat it like a traditional data frame results in an error message:

```
> mean(credit$amount)
[1] NA
Warning message:
In mean.default(credit$amount) :
  argument is not numeric or logical: returning NA
```

The ffbase package by Edwin de Jonge, Jan Wijffels, and Jan van der Laan addresses this issue somewhat by adding capabilities for basic analyses using ff objects. This makes it possible to use ff objects directly for data exploration. For instance, after installing the ffbase package, the mean function works as expected:

```
> library(ffbase)
> mean(credit$amount)
[1] 3271.258
```

The package also provides other basic functionality such as mathematical operators, query functions, summary statistics, and wrappers for working with optimized machine learning algorithms like `biglm` (described later in this chapter). Though these do not completely eliminate the challenges of working with extremely large datasets, they make the process a bit more seamless.

 For more information about advanced functionality, visit the `ffbase` project site at `http://github.com/edwindj/ffbase`.

Using massive matrices with bigmemory

The `bigmemory` package by Michael J. Kane, John W. Emerson, and Peter Haverty allows extremely large matrices that exceed the amount of available system memory. The matrices can be stored on disk or in shared memory, allowing them to be used by other processes on the same computer or across a network. This facilitates parallel computing methods, such as those covered later in this chapter.

 Additional documentation on the `bigmemory` package can be found at `http://www.bigmemory.org/`.

Because `bigmemory` matrices are intentionally unlike data frames, they cannot be used directly with most of the machine learning methods covered in this book. They also can only be used with numeric data. That said, since they are similar to a typical R matrix, it is easy to create smaller samples or chunks that can be converted to standard R data structures.

The authors also provide `bigalgebra`, `biganalytics`, and `bigtabulate` packages, which allow simple analyses to be performed on the matrices. Of particular note is the `bigkmeans()` function in the `biganalytics` package, which performs k-means clustering as described in *Chapter 9*, *Finding Groups of Data – Clustering with k-means*. Due to the highly specialized nature of these packages, use cases are outside the scope of this chapter.

Learning faster with parallel computing

In the early days of computing, processors executed instructions in **serial**, which meant that they were limited to performing a single task at a time. The next instruction could not be started until the previous instruction was complete. Although it was widely known that many tasks could be completed more efficiently by completing steps simultaneously, the technology simply did not exist yet.

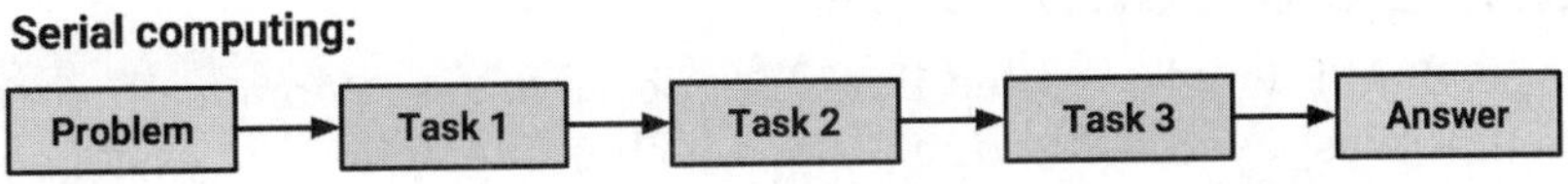

Serial computing:

Problem → Task 1 → Task 2 → Task 3 → Answer

Figure 12.5: In serial computing, tasks cannot begin until prior tasks have completed

This was addressed by the development of **parallel computing** methods, which use a set of two or more processors or computers to solve a larger problem. Many modern computers are designed for parallel computing. Even in the case that they have a single processor, they often have two or more **cores** that are capable of working in parallel. This allows tasks to be accomplished independently from one another.

Parallel computing:

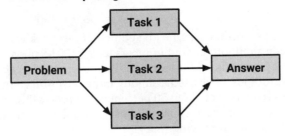

Figure 12.6: In parallel computing, tasks occur simultaneously. Their results must be combined at the end.

Networks of multiple computers called clusters can also be used for parallel computing. A large cluster may include a variety of hardware and be separated over large distances. In this case, the cluster is known as a **grid**. Taken to an extreme, a cluster or grid of hundreds or thousands of computers running commodity hardware could be a very powerful system.

The catch, however, is that not every problem can be parallelized. Certain problems are more conducive to parallel execution than others. One might expect that adding 100 processors would result in 100 times the work being accomplished in the same amount of time (that is, the overall execution time is 1/100), but this is typically not the case. The reason is that it takes effort to manage the workers. Work must be divided into equal, non-overlapping tasks, and each of the workers' results must be combined into one final answer.

So-called **embarrassingly parallel** problems are the ideal. These tasks are easy to reduce into non-overlapping blocks of work, and the results are easy to recombine. An example of an embarrassingly parallel machine learning task would be 10-fold cross-validation; once the 10 samples are divided, each of the 10 blocks of work is independent, meaning that they do not affect the others. As you will soon see, this task can be sped up quite dramatically using parallel computing.

Measuring execution time

Efforts to speed up R will be wasted if it is not possible to systematically measure how much time was saved. Although a stopwatch is one option, an easier solution is to wrap the offending code in a `system.time()` function.

For example, on the author's laptop, the `system.time()` function notes that it takes about 0.080 seconds to generate a million random numbers:

```
> system.time(rnorm(1000000))
   user   system elapsed
  0.079   0.000   0.067
```

The same function can be used for evaluating improvement in performance, obtained with the methods that were just described or any R function.

> For what it's worth, when the first edition of this book was published, generating a million random numbers took 0.130 seconds; the same took about 0.093 seconds for the second edition. Here it takes about 0.067 seconds. Although I've used a slightly more powerful computer each time, this reduction of about 50 percent of the processing time over the course of about six years illustrates just how quickly computer hardware and software are improving.

Working in parallel with multicore and snow

The `parallel` package, included with R version 2.14.0 and later, has lowered the entry barrier to deploying parallel algorithms by providing a standard framework for setting up worker processes that can complete tasks simultaneously. It does this by including components of the `multicore` and `snow` packages, which each take a different approach to multitasking.

If your computer is reasonably recent, you are likely to be able to use parallel processing. To determine the number of cores your machine has, use the `detectCores()` function as follows. Note that your output will differ depending on your hardware specifications:

```
> library(parallel)
> detectCores()
[1] 8
```

The multicore package was developed by Simon Urbanek and allows parallel processing on a single machine that has multiple processors or processor cores. It utilizes the multitasking capabilities of a computer's operating system to **fork** additional R sessions that share the same memory, and is perhaps the simplest way to get started with parallel processing in R. Unfortunately, because Windows does not support forking, this solution will not work everywhere.

An easy way to get started with the multicore functionality is to use the mclapply() function, which is a multi-core version of lapply(). For instance, the following blocks of code illustrate how the task of generating a million random numbers can be divided across 1, 2, 4, and 8 cores. The unlist() function is used to combine the parallel results (a list) into a single vector after each core has completed its chunk of work:

```
> system.time(l1 <- unlist(mclapply(1:10, function(x) {
+   rnorm(1000000)}, mc.cores = 1)))
   user  system elapsed
  0.627   0.015   0.647


> system.time(l2 <- unlist(mclapply(1:10, function(x) {
+   rnorm(1000000)}, mc.cores = 2)))
   user  system elapsed
  0.751   0.211   0.568


> system.time(l4 <- unlist(mclapply(1:10, function(x) {
+   rnorm(1000000) }, mc.cores = 4)))
   user  system elapsed
  0.786   0.270   0.405


> system.time(l8 <- unlist(mclapply(1:10, function(x) {
+   rnorm(1000000) }, mc.cores = 8)))
   user  system elapsed
  1.033   0.315   0.321
```

Notice how as the number of cores increases, the elapsed time decreases, though the benefit tapers off. Though this is a simple example, it can be adapted easily to many other tasks.

The snow package (simple networking of workstations) by Luke Tierney, A. J. Rossini, Na Li, and H. Sevcikova allows parallel computing on multicore or multiprocessor machines as well as on a network of multiple machines. It is slightly more difficult to use, but offers much more power and flexibility. The snow functionality is included in the parallel package, so to set up a cluster on a single machine, use the makeCluster() function with the number of cores to be used:

```
> cl1 <- makeCluster(4)
```

Because snow communicates via network traffic, depending on your operating system, you may receive a message to approve access through your firewall.

To confirm the cluster is operational, we can ask each node to report back its hostname. The clusterCall() function executes a function on each machine in the cluster. In this case, we'll define a function that simply calls the Sys.info() function and returns the nodename parameter:

```
> clusterCall(cl1, function() { Sys.info()["nodename"] } )
[[1]]
                    nodename
"Bretts-Macbook-Pro.local"


[[2]]
                    nodename
"Bretts-Macbook-Pro.local"


[[3]]
                    nodename
"Bretts-Macbook-Pro.local"


[[4]]
                    nodename
"Bretts-Macbook-Pro.local"
```

Unsurprisingly, since all four nodes are running on a single machine, they report back the same hostname. To have the four nodes run a different command, supply them with a unique parameter via the clusterApply() function. Here, we'll supply each node with a different letter. Each node will then perform a simple function on its letter in parallel:

```
> clusterApply(cl1, c('A', 'B', 'C', 'D'),
                function(x) { paste("Cluster", x, "ready!") })
[[1]]
```

```
[1] "Cluster A ready!"

[[2]]
[1] "Cluster B ready!"

[[3]]
[1] "Cluster C ready!"

[[4]]
[1] "Cluster D ready!"
```

When we're done with the cluster, it's important to terminate the processes it spawned. This will free up the resources each node is using:

```
> stopCluster(cl1)
```

Using these simple commands, it is possible to speed up many machine learning tasks. For the largest big data problems, much more complex snow configurations are possible. For instance, you may attempt to configure a **Beowulf cluster** – a network of many consumer-grade machines. In academic and industry research settings with dedicated computing clusters, snow can use the Rmpi package to access these high-performance **message passing interface** (**MPI**) servers. Working with such clusters requires knowledge of network configurations and computing hardware outside the scope of this book.

 For a much more detailed introduction to snow, including some information on how to configure parallel computing on several computers over a network, see the following lecture by Luke Tierney: http://homepage. stat.uiowa.edu/~luke/classes/295-hpc/notes/snow.pdf.

Taking advantage of parallel with foreach and doParallel

The foreach package by Rich Calaway and Steve Weston provides perhaps the easiest way to get started with parallel computing, especially if you are running R on the Windows operating system, as some of the other packages are platform-specific.

The core of the package is a foreach looping construct. If you have worked with other programming languages, this may be familiar. Essentially, it allows looping over a number of items in a set, without explicitly counting the number of items; in other words, *for each* item in the set, *do* something.

If you're thinking that R already provides a set of apply functions to loop over sets of items (for example, `apply()`, `lapply()`, `sapply()`, and so on), you are correct. However, the `foreach` loop has an additional benefit: iterations of the loop can be completed in parallel using a very simple syntax. Let's see how this works.

Recall the command we've been using to generate a million random numbers. To make this more challenging, let's increase the count to a hundred million, which causes the process to take over six seconds:

```
> system.time(11 <- rnorm(100000000))
   user   system elapsed
  5.873    0.204   6.087
```

After the `foreach` package has been installed, the same task can be expressed by a loop that generates four sets of 25,000,000 random numbers. The `.combine` parameter is an optional setting that tells `foreach` which function it should use to combine the final set of results from each loop iteration. In this case, since each iteration generates a set of random numbers, we simply use the `c()` concatenate function to create a single, combined vector:

```
> library(foreach)
> system.time(14 <- foreach(i = 1:4, .combine = 'c')
                %do% rnorm(25000000))
   user   system elapsed
  6.177    0.391   6.578
```

If you noticed that this function didn't result in a speed improvement, good catch! In fact, the process was actually slower. The reason is that by default, the `foreach` package runs each loop iteration in serial, and the function adds a small amount of computational overhead to the process. The sister package `doParallel` provides a parallel backend for `foreach` that utilizes the `parallel` package included with R, described earlier in this chapter. After installing the `doParallel` package, simply register the number of cores and swap the `%do%` command with `%dopar%` as follows:

```
> library(doParallel)
> registerDoParallel(cores = 4)
> system.time(14p <- foreach(i = 1:4, .combine = 'c')
                %dopar% rnorm(25000000))
   user   system elapsed
  7.841    2.288   3.894
```

As shown in the output, this results in the expected performance increase, cutting the execution time by about 40 percent.

To close the `doParallel` cluster, simply type:

```
> stopImplicitCluster()
```

Though the cluster will be closed automatically at the conclusion of the R session, it is better form to do so explicitly.

Training and evaluating models in parallel with caret

The `caret` package by Max Kuhn (covered extensively in *Chapter 10, Evaluating Model Performance* and *Chapter 11, Improving Model Performance*) will transparently utilize a parallel backend if one has been registered with R using the `foreach` package described previously.

Let's take a look at a simple example in which we attempt to train a random forest model on the credit dataset. Without parallelization, the model takes about 79 seconds to train:

```
> library(caret)
> credit <- read.csv("credit.csv")
> system.time(train(default ~ ., data = credit, method = "rf"))
   user  system elapsed
 77.345   1.778  79.205
```

On the other hand, if we use the `doParallel` package to register four cores to be used in parallel, the model takes under 20 seconds to build—less than a quarter of the time—and we didn't need to change even a single line of `caret` code:

```
> library(doParallel)
> registerDoParallel(cores = 8)
> system.time(train(default ~ ., data = credit, method = "rf"))
   user  system elapsed
122.579   3.292  19.034
```

Many of the tasks involved in training and evaluating models, such as creating random samples and repeatedly testing predictions for 10-fold cross-validation are embarrassingly parallel and ripe for performance improvements. With this in mind, it is wise to always register multiple cores before beginning a `caret` project.

Configuration instructions and a case study of the performance improvements for enabling parallel processing in `caret` are available at the project's website: `http://topepo.github.io/caret/parallel.html`.

Parallel cloud computing with MapReduce and Hadoop

The **MapReduce** programming model was developed at Google as a way to process its data on a large cluster of networked computers. MapReduce defined parallel programming as a two-step process:

- A **map** step, in which a problem is divided into smaller tasks that are distributed across the computers in the cluster

- A **reduce** step, in which the results of the small chunks of work are collected and synthesized into a final solution to the original problem

A popular open-source alternative to the proprietary MapReduce framework is **Apache Hadoop**. The Hadoop software comprises of the MapReduce concept plus a distributed filesystem capable of storing large amounts of data across a cluster of computers.

 Packt Publishing has published a large number of books on Hadoop. To search current offerings, visit `https://www.packtpub.com/all/?search=hadoop`.

Several R projects that provide an R interface to Hadoop are in development. The RHadoop project by Revolution Analytics provides an R interface to Hadoop. The project provides a package, `rmr2`, intended to be an easy way for R developers to write MapReduce programs. Another companion package, `plyrmr`, provides functionality similar to the `dplyr` package for processing large datasets. Additional RHadoop packages provide R functions for accessing Hadoop's distributed data stores.

 For more information about the RHadoop project, see `https://github.com/RevolutionAnalytics/RHadoop/wiki`.

Although Hadoop is a mature framework, it requires somewhat specialized programming skills to take advantage of its capabilities and to perform even basic machine learning tasks. Perhaps this explains its apparent lack of popularity with R users. Additionally, although Hadoop is excellent at working with *extremely* large amounts of data, it may not always be the fastest option because it keeps all data on disk rather than utilizing available memory. The next section covers an extension to Hadoop that addresses these speed and usability issues.

Parallel cloud computing with Apache Spark

The **Apache Spark** project is a cluster-computing framework for big data, offering many advantages over Apache Hadoop. Because it takes advantage of the cluster's available memory, it can process data approximately 100x faster than Hadoop. Additionally, it provides high-level libraries for many common data processing, analysis, and modeling tasks. These include the SparkSQL data query language, the MLlib machine learning library, GraphX for graph and network analysis, and the Spark Streaming library for processing real-time data streams.

> Packt Publishing has published a large number of books on Spark. To search current offerings, visit https://www.packtpub.com/all/?search=spark.

Apache Spark is often run remotely on a cloud-hosted cluster of virtual machines, but its benefits can also be seen running on your own hardware. In either case, the sparklyr package connects to the cluster and provides a dplyr interface for analyzing the data using Spark. More detailed instructions for using Spark with R can be found at https://spark.rstudio.com, but the basics of getting up-and-running are fairly straightforward.

To illustrate the fundamentals, let's build a random forest model on the credit dataset to predict loan defaults. We'll begin by installing the sparklyr package. Then, you can instantiate a Spark cluster on your local machine using the following code:

```
> install.packages("sparklyr")
> library(sparklyr)
> spark_install(version = "2.1.0")
> spark_cluster <- spark_connect(master = "local")
```

Next, we'll load the loan dataset from the credit.csv file on our local machine into the Spark instance, then use the Spark data frame partitioning function sdf_partition() to randomly assign 75 and 25 percent of data to training and test sets. The seed parameter is the random seed to ensure the results are identical each time this code is run:

```
> credit_spark <- spark_read_csv(spark_cluster, "credit.csv")

> splits <- sdf_partition(credit_spark,
                          train = 0.75, test = 0.25,
                          seed = 123)
```

Lastly, we'll pipe the training data into the random forest model function, make predictions, and use the classification evaluator to compute the AUC on the test set:

```
> credit_rf <- splits$train %>%
    ml_random_forest(default ~ .)
> pred <- ml_predict(credit_rf, splits$test)
> ml_binary_classification_evaluator(pred,
    metric_name = "areaUnderROC")
[1] 0.7848068
```

With just a few lines of R code, we've built a random forest model using Spark that could expand to model millions of records. If even more computing power is needed, the code can be run in the cloud using a massively parallel Spark cluster simply by pointing the `spark_connect()` function to the correct hostname. The code can also be easily adapted to other modeling approaches using one of the supervised learning functions listed at `https://spark.rstudio.com/mlib/`.

> Perhaps the easiest way to get started using Spark is with Databricks, a cloud platform developed by the creators of Spark that makes it easy to manage and scale clusters via a web-based interface. The free "community edition" provides a small cluster for you to try a number of tutorials or even experiment with your own data. Check it out at `https://databricks.com`.

Deploying optimized learning algorithms

Some of the machine learning algorithms covered in this book are able to work on extremely large datasets with relatively minor modifications. For instance, it would be fairly straightforward to implement Naive Bayes or the Apriori algorithm using one of the data structures for big datasets described in the previous sections. Some types of learners, such as ensembles, lend themselves well to parallelization because the work of each model can be distributed across processors or computers in a cluster. On the other hand, some require larger changes to the data or algorithm, or need to be rethought altogether before they can be used with massive datasets.

The following sections examine packages that provide optimized versions of the learning algorithms we've worked with so far.

Building bigger regression models with biglm

The `biglm` package by Thomas Lumley provides functions for training regression models on datasets that may be too large to fit into memory. It works by an iterative process in which the model is updated little by little using small chunks of data. In spite of the different approach, the results will be nearly identical to what would have been obtained running the conventional `lm()` function on the entire dataset.

For convenience when working with the largest datasets, the `biglm()` function allows the use of a SQL database in place of a data frame. The model can also be trained with chunks obtained from data objects created by the `ff` package described previously.

Growing random forests faster with ranger

The `ranger` package by Marvin N. Wright, Stefan Wager, and Philipp Probst is a faster implementation of the random forest algorithm, particularly for datasets with a large number of features or examples. The function is used much like the earlier random forest:

```
> library(ranger)
> credit <- read.csv("credit.csv")

> m <- ranger(default ~ ., data = credit,
              num.trees = 500,
              mtry = 4)


> p <- predict(m, credit)
```

Note that unlike most of the `predict()` results used previously, the `ranger` predictions are stored as a sub-object in the prediction object:

```
> head(p$predictions)
[1] no  yes no  no  yes no
Levels: no yes
```

Using the `ranger()` function is the easiest way to build bigger and better random forests, without resorting to cluster computing or alternative data structures.

Growing massive random forests with bigrf

The `bigrf` package by Aloysius Lim implements the training of random forests for classification and regression on datasets that are too large to fit into memory using `bigmemory` objects as described earlier in this chapter. For speedier forest growth, the package can be used with the `foreach` and `doParallel` packages described previously to grow trees in parallel.

> For more information, including examples and Windows installation instructions, see the package wiki hosted at https://github.com/aloysius-lim/bigrf.

A faster machine learning computing engine with H2O

The **H2O project** is a big data framework that provides fast in-memory implementations of machine learning algorithms, which can also operate in a cluster computing environment. It includes functions for many of the methods covered in this book, including Naive Bayes, regression, deep neural networks, k-means clustering, ensemble methods, and random forests, among many others.

H2O uses heuristics to find approximate solutions to machine learning problems by iterating repeatedly over smaller chunks of the data. This gives the user the control to determine exactly how much of a massive dataset the learner should use. For some problems, a quick solution may be acceptable, but for others, the complete set may be required, which will require additional training time.

H2O is usually substantially faster and performs better on very massive datasets relative to Apache Spark's machine learning functions (MLlib), which itself is often much faster than base R. However, because Apache Spark is a commonly used cluster computing and big data preparation environment, H2O can be run on Apache Spark using the **Sparkling Water** software. With Sparkling Water, data scientists have the best of both worlds—the benefits of Spark for data preparation, and the benefits of H2O for machine learning.

The `h2o` package provides functionality for accessing an H2O instance from within the R environment. A full tutorial on H2O is outside the scope of this book, and documentation is available at http://docs.h2o.ai, but the basics are straightforward. To get started, install and load the `h2o` package. Then, initialize a local H2O instance using the following code:

```
> library(h2o)
> h2o_instance <- h2o.init()
```

This starts an H2O server on your computer, which can be viewed via **H2O Flow** at `http://localhost:54321`. The H2O Flow web application allows you to administer and send commands to the H2O server, or even build and evaluate models using a simple, browser-based interface:

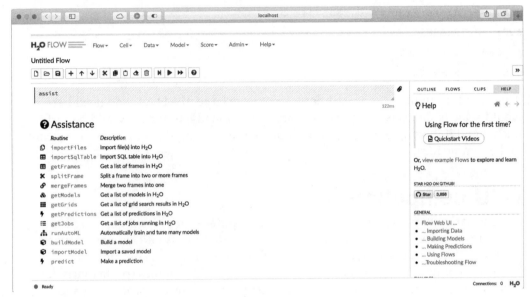

Figure 12.7: H2O Flow is a web application for interacting with the H2O instance

Although you could complete an analysis within this interface, let's go back to R and use H2O on the loan default data that we examined previously. First, we need to upload the `credit.csv` dataset to this instance using the following command:

```
> credit.hex <- h2o.uploadFile("credit.csv")
```

Note that the `.hex` extension is used for referring to an H2O data frame.

Next, we'll apply H2O's random forest implementation to this dataset using the following command:

```
> h2o.randomForest(y = "default",
                   training_frame = credit.hex,
                   ntrees = 500,
                   seed = 123)
```

The output of this command includes information on the out-of-bag estimates of model performance:

```
** Reported on training data. **
** Metrics reported on Out-Of-Bag training samples **
```

```
MSE:    0.1636964
RMSE:   0.4045941
LogLoss:   0.4956524
Mean Per-Class Error:    0.2835714
AUC:   0.7844881
pr_auc:   0.6192192
Gini:   0.5689762
```

Although the credit dataset used here is not very large, the H2O code used here would scale to datasets of almost any size. Additionally, the code would be virtually unchanged if it were to be run in the cloud—simply point the `h2o.init()` function to the remote host.

GPU computing

An alternative to parallel processing uses a computer's graphics processing unit (GPU) to increase the speed of mathematical calculations. A GPU is a specialized processor that is optimized for rapidly displaying images on a computer screen. Because a computer often needs to display complex 3D graphics (particularly for video games), many GPUs use hardware designed for parallel processing and extremely efficient matrix and vector calculations. A side benefit is that they can be used for efficiently solving certain types of mathematical problems. Where a typical laptop or desktop computer processor may have up to 16 cores, a typical GPU may have thousands or even tens of thousands.

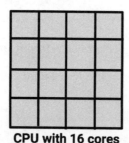

CPU with 16 cores   **GPU with 1000+ cores**

Figure 12.8: A graphics processing unit (GPU) has many times more cores than the typical central processing unit (CPU)

The downside of GPU computing is that it requires specific hardware that is not included with many computers. In most cases, a GPU from the manufacturer NVIDIA is required, as it provides a proprietary framework called **Complete Unified Device Architecture (CUDA)** that makes the GPU programmable using common languages such as C++.

For more information on NVIDIA's role in GPU computing, go
to https://www.nvidia.com/en-us/about-nvidia/ai-
computing/.

The gputools package by Josh Buckner, Mark Seligman, and Justin Wilson
implements several R functions, such as matrix operations, clustering, and
regression modeling using the NVIDIA CUDA toolkit. The package requires
a CUDA 1.3 or higher GPU and the installation of the NVIDIA CUDA toolkit.

Flexible numeric computing and machine learning with TensorFlow

One of the most significant recent innovations in machine learning software is
TensorFlow (https://www.tensorflow.org), an open-source mathematical
library developed at Google for advanced machine learning. TensorFlow provides
a computing interface using directed graphs that "flow" data arrays called **tensors**
through a number of mathematical operations. In this way, a very complex "black
box" method like a deep neural network can be represented as a simpler abstraction.
Additionally, because the graph stores the set of computations as a set of dependent
steps, TensorFlow is able to distribute the work across available CPU or GPU cores
and take advantage of massively parallel computing environments.

Packt Publishing has published a large number of books on TensorFlow.
To search current offerings, visit https://www.packtpub.com/
all/?search=tensorflow.

R interfaces to TensorFlow have been developed by the team at RStudio. The
tensorflow package provides access to the core API, while the tfestimators
package provides access to higher-level machine learning functionality. Note that
TensorFlow's directed graph approach can be used to implement many different
machine learning models, including many of the ones discussed in this book.
However, to do so requires a thorough understanding of the matrix mathematics
that define each model, and thus is well outside the scope of this text. For more
information about these packages and RStudio's ability to interface with TensorFlow,
visit https://tensorflow.rstudio.com.

Due to TensorFlow's unique approach to estimating machine learning models, you may find vast differences in the terminology used by its practitioners, even for simpler methods like linear regression. You may hear phrases like "cost function," "gradient descent," and "optimization." This terminology reflects the fact that machine learning with TensorFlow is in many ways analogous to building a neural network that finds the best approximation of the desired model.

An interface for deep learning with Keras

Because the computing framework provided by TensorFlow is ideal for constructing deep neural networks, the **Keras** library (https://keras.io) was developed to provide a simpler, high-level interface to this widely-used functionality. Keras was developed in Python, and can use TensorFlow or similar frameworks as the backend computing engine. Using Keras, it is possible to do deep learning in just a few lines of code — even for challenging applications such as image classification.

Packt Publishing offers a number of books and videos to learn Keras. To search current offerings, visit https://www.packtpub.com/all/?search=keras.

The keras package, developed by RStudio CEO and founder J.J. Allaire, provides an R interface to Keras. Although there is very little code required to use the keras package, developing useful deep learning models requires extensive knowledge of neural networks as well as familiarity with TensorFlow and the Keras API. Additionally, to build anything but the simplest neural networks, it is essential to use a GPU — the code will simply never finish running without the massively-parallel processing the GPU provides. For these reasons, a tutorial is outside the scope of this book. Instead, refer to the RStudio documentation at https://keras.rstudio.com, or the book *Deep Learning with R* (2018), which was co-authored by Francois Chollet and J.J. Allaire — the creators of Keras and the keras package, respectively. Given their credentials, there is no better place to begin learning about this tool.

At the time of publication, a typical GPU used for deep learning is priced at several hundred US dollars for entry-level models and around $1,000-$3,000 for moderately-priced units with greater performance. High-end units may cost many thousands of dollars. Rather than spend this much up front, many people rent server time by the hour on cloud providers like Amazon AWS and Microsoft Azure, where it costs approximately $1 per hour for a minimal GPU instance—just don't forget to shut it down when your work completes, as it can get expensive quite quickly! The RStudio team also provides information about their preferred host at `https://tensorflow.rstudio.com/tools/cloud_desktop_gpu.html`.

Summary

It is certainly an exciting time to be studying machine learning. Ongoing work on the relatively uncharted frontiers of parallel and distributed computing offers great potential for tapping the knowledge found in the deluge of big data. And the burgeoning data science community is facilitated by the free and open-source R programming language, which provides a very low barrier for entry—you simply need to be willing to learn.

The topics you have learned, both in this chapter as well as previous chapters, provide the foundation for understanding more advanced machine learning methods. It is now your responsibility to keep learning and adding tools to your arsenal. Along the way, be sure to keep in mind the no free lunch theorem—no learning algorithm rules them all, and they all have varying strengths and weaknesses. For this reason, there will always be a human element to machine learning, adding subject-specific knowledge and the ability to match the appropriate algorithm to the task at hand.

In the coming years, it will be interesting to see how the human side changes as the line between machine learning and human learning is blurred. Services such as Amazon's Mechanical Turk provide crowd-sourced intelligence, offering a cluster of human minds ready to perform simple tasks at a moment's notice. Perhaps one day, just as we have used computers to perform tasks that human beings cannot do easily, computers will employ human beings to do the reverse. What interesting food for thought!

Other Books You May Enjoy

If you enjoyed this book, you may be interested in these other books by Packt:

Mastering Machine Learning with R - Third Edition

Cory Lesmeister

ISBN: 978-1-78961-800-6

- Prepare data for machine learning methods with ease
- Learn to write production-ready code and package it for use
- Produce simple and effective data visualizations for improved insights
- Master advanced methods such as Boosted Trees and deep neural networks
- Use natural language processing to extract insights for text
- Implement tree-based classifiers including Random Forest and Boosted Tree

Python Machine Learning - Second Edition

Sebastian Raschka, Vahid Mirjalili

ISBN: 978-1-78712-593-3

- Understand the key frameworks in data science, machine learning, and deep learning
- Harness the power of the latest Python open source libraries in machine learning
- Master machine learning techniques using challenging real-world data
- Master deep neural network implementation using the TensorFlow library
- Ask new questions of your data through machine learning models and neural networks
- Learn the mechanics of classification algorithms to implement the best tool for the job
- Predict continuous target outcomes using regression analysis
- Uncover hidden patterns and structures in data with clustering
- Delve deeper into textual and social media data using sentiment analysis

Architects of Intelligence

Martin Ford

ISBN: 978-1-78995-453-1

- The state of modern AI
- How AI will evolve and the breakthroughs we can expect
- Insights into the minds of AI founders and leaders
- How and when we will achieve human-level AI
- The impact and risks associated with AI and its impact on society and the economy

Leave a review - let other readers know what you think

Please share your thoughts on this book with others by leaving a review on the site that you bought it from. If you purchased the book from Amazon, please leave us an honest review on this book's Amazon page. This is vital so that other potential readers can see and use your unbiased opinion to make purchasing decisions, we can understand what our customers think about our products, and our authors can see your feedback on the title that they have worked with Packt to create. It will only take a few minutes of your time, but is valuable to other potential customers, our authors, and Packt. Thank you!

Index

Symbols

0.632 bootstrap 345
1R algorithm
  about 151
  strengths 152
  weaknesses 152

A

activation function 220-223
AdaBoost 364
AdaBoost.M1 algorithm 365
adaptive boosting 144, 364
adversarial learning 23
algorithms
  input data, matching to 23, 24
allocation function 360
Amazon Web Services (AWS) 241
ANNs, used for modeling concrete strength
  about 229
  data collection 230
  data exploration 230, 231
  data preparation 230, 231
  model performance, evaluating 235, 236
  model performance, improving 236-240
  model, training on data 232-235
antecedent 148
Apache Hadoop 412
Apache Spark
  parallel cloud computing 413, 414
application programming interfaces
    (API) 392
Apriori 263
Apriori algorithm
  for association rule learning 263-265

strengths 264
weaknesses 264
Apriori principle
  set of rules, building 266, 267
Apriori property 264
area under the ROC curve (AUC) 333
arrays 41
artificial neural network (ANN) 218, 219
artificial neurons 220
association rules
  about 262
  applications 263
  left-hand side (LHS) 262
  right-hand side (RHS) 262
  rule interest, measuring 265, 266
automated parameter tuning
  caret, using for 349-351
axis-parallel splits 130
axon 219

B

backpropagation
  about 227
  neural networks, training 227, 228
bagging 362, 363
bag-of-words 105
Bayesian classifiers
  uses 90
Bayesian methods
  about 90
  concepts 90, 91
Bayes' theorem
  conditional probability, computing 94-96
Beowulf cluster 409
betweenness centrality 400

class imbalance problem 314
clustering
  about 22, 288
  as machine learning task 288-290
clusters 288
Cohen's kappa coefficient 324
combination function 360
comma-separated values (CSV) 43
complement 91
complete text of web pages
  downloading 387-389
Complete Unified Device Architecture
    (CUDA) 418
Comprehensive R Archive Network
    (CRAN)
  reference 24
conditional probability
  about 94
  computing, with Bayes' theorem 94-96
confusion matrix
  about 147, 148, 314-320
  used, for measuring performance 320-322
consequent 148
contingency table 62
control object 355
convex hull 244
corpus 106
correlation 61, 175, 176
correlation ellipse 188
correlation matrix 185
cost matrix 146, 147
covariance function 174
covering algorithms 151
CRAN task view, for clustering
  reference 290
CRAN Web Technologies and Services
    task view
  reference 387
crosstab 62
cross-validation 341-344
CSV files
  data, importing from 42, 43
  data, saving from 42, 43
Cubist algorithm 212
cut points 102

D

data
  exploring 44
  importing, from CSV files 42, 43
  managing, with R 41
  parsing, within web pages 389-391
  querying, in SQL databases 381
  saving, from CSV files 42, 43
  structure 44, 45
database backend
  using, with dplyr 384, 385
database connections
  managing 381-383
database management system (DBMS) 381
data frames 36-39
data mining 3
data munging 376
data preparation
  simplifying, with dplyr 378
  speeding up, with dplyr 378
data source name (DSN) 383
data structures, R
  about 30
  arrays 41
  data frames 36-39
  factors 32, 33
  lists 34-36
  loading 41
  matrices 39, 40
  removing 42
  saving 41
  vectors 30, 31
data table
  reference 402
  used, for making data frames faster 402
data wrangling 376
deciles 49
decision nodes 126
decision tree algorithms
  benefits 127
decision tree forests 367
decision trees
  about 126
  accuracy, boosting of 144-146

model performance, evaluating 192, 193
model performance, improving 193
model specification 194
model, training on data 189-191
non-linear relationships, adding 194
numeric variable, converting to binary
 indicator 194, 195
predictions, making with regression
 model 197-200
relationships, exploring among
 features 185
relationships, visualizing among
 features 186-189
transformation 194, 195
message passing interface (MPI) 409
meta-learners 22
meta-learning
model performance, improving with 359
microarray 396
Microsoft Azure 241
Microsoft Excel files
importing, with rio 380
min-max normalization 73
mobile phone filtering, with Naive
 Bayes algorithm
about 103
data collection 103, 104
data exploration 105, 106
data preparation 105, 106
indicator features, creating for frequent
 words 118, 119
model performance, evaluating 121, 122
model performance, improving 122, 123
model, training on data 120, 121
test dataset, creating 114
text data, cleaning 106-112
text data, standardizing 106-112
text data, visualizing 115-117
text documents, splitting into
 words 112-114
training dataset, creating 114
model performance
improving, with meta-learning 359
model trees 200
multicore package 407
multilayer network 224
multilayer perceptron (MLP) 226

multimodal 58
multinomial logistic regression 170
multiple linear regression
about 169, 177-181
strengths 177
weaknesses 177
multiple regression 169
multivariate relationships 60
mutually exclusive event 91

N

Naive Bayes
about 90
numeric features, using with 102, 103
using, in classification 98-100
Naive Bayes algorithm
about 96, 97
strengths 96
weaknesses 96
nearest neighbor classification
about 66
k-NN algorithm 66-69
negative class predictions 319
network analysis 397
network data
analyzing 397-400
visualizing 397-400
network topology
about 220, 223
direction of information travel 225
number of layers 224
number of node, in each layer 226
neural networks
characteristics 220
training, with backpropagation 227, 228
neurons 218
nodes 218, 397
No Free Lunch theorem
reference 16
nominal data 20
non-linear spaces
kernels, using for 247
non-parametric learning methods 76
normal distribution 55
numeric data 20
numeric features

R

data, managing 41
data structures 30
radial basis function (RBF) 223
random-access memory (RAM) 12
random forest models
strengths 367
weaknesses 367
random forest performance
evaluating, in simulated
competition 370-373
random forests
about 367
training 368-370
random sample 138
range 49
ranger
random forests faster, growing 415
RCurl package
reference 387
readr package
tidy tables, importing with 380
real-world data
managing 376
preparing 376
recall 329, 330
receiver operating characteristic (ROC)
curve 332
rectifier 237
rectifier linear unit (ReLU) 237
recurrent network 225
recursive partitioning 127
regression
about 168
adding, to trees 201, 203
multiple linear regression 177-181
simple linear regression 170-172
regression analysis 169
regression trees
about 200
strengths 201
weaknesses 201
reinforcement learning 23
relationships
exploring, between variables 60
visualizing 60

repeated holdout 341
repeated k-fold CV 344
residuals 173
resubstitution error 337
RHadoop project
reference 412
rio package
Microsoft Excel files, importing 380
reference 380
SAS files, importing 380
SPSS files, importing 380
Stata files, importing 380, 381
RIPPER algorithm
about 153
strengths 153
weaknesses 153
ROC curves
performance tradeoffs, visualizing
with 331-336
root node 126
rote learning 76
R packages
installing 25, 26
loading 26
unloading 26
R performance, improving
about 401
GPU computing 418
large datasets, managing 401
optimized learning algorithms,
deploying 414
parallel computing, using 404, 405
RStudio
installing 27
reference 27
rule learner 156
rules
greedy approach 156
RWeka 25

S

sample SAM ham 104
sample SMS spam 104
SAS files
importing, with rio 380
scatterplot matrix 186